Algorithms for Nonlinear Programming and Multiple-Objective Decisions

Wiley-Interscience Series in Systems and Optimization

Advisory Editors

Sheldon Ross

Department of Industrial Engineering and Operations Research, University of California, Berkeley, CA 94720, USA

Richard Weber

Cambridge University, Engineering Department, Management Studies Group, Mill Lane, Cambridge CB2 1RX, UK

GITTINS — Multi-armed Bandit Allocation Indices

KALL/WALLACE — Stochastic Programming

KAMP/HASLER — Recursive Neural Networks for Associative Memory

KIBZUN/KAN — Stochastic Programming Problems with Probability and Quantile Functions

RUSTEM — Algorithms for Nonlinear Programming and Multiple-Objective Decisions

VAN DIJK — Queueing Networks and Product Forms: A Systems Approach

WHITTLE — Optimal Control: Basics and Beyond

WHITTLE — Risk-sensitive Optimal Control

Algorithms for Nonlinear Programming and Multiple-Objective Decisions

Berç Rustem

Imperial College of Science, Technology and Medicine, London, UK

JOHN WILEY & SONS
Chichester · New York · Weinheim · Brisbane · Singapore · Toronto

Baffins Lane, Chichester,
West Sussex PO19 1UD, England

National 01243 779777
International (+44) 1243 779777

e-mail (for orders and customer service enquiries): cs-books@wiley.co.uk.
Visit our Home Page on http://www.wiley.co.uk
or http://www.wiley.com

Other Wiley Editorial Offices

John Wiley & Sons, Inc., 605 Third Avenue,
New York, NY 10158–0012, USA

WILEY-VCH Verlag GmbH, Papelallee 3,
D-69469 Weinheim, Germany

Jacaranda Wiley Ltd, 33 Park Road, Milton,
Queensland 4064, Australia

John Wiley & Sons (Asia) Pte Ltd, 2 Clementi Loop #02-01,
Jin Xing Distripark, Singapore 129809

John Wiley & Sons (Canada) Ltd, 22 Worcester Road,
Rexdale, Ontario, M9W 1L1 Canada

Nonlinear programming

British Library Cataloguing in Publication Data

A catalogue record for this book is available from the British Library

ISBN 0 471 97850 7

Typeset in 10/12pt Times by Pure Tech India Ltd, Pondicherry
Printed and bound in Great Britain by Biddles, Guildford and King's Lynn
This book is printed on acid-free paper responsibly manufactured from sustainable forestation, for which at least two trees are planted for each one used for paper production.

Contents

Preface

This book is a study of algorithms for decision making with multiple objectives. It is addressed to researchers in computational methods for decision making and optimal design: computer scientists interested in quantitative decision support, in particular numerical optimization; engineers; mathematicians; and those working in management science, operations research, economics and finance. Although most of the mathematics required is reviewed in a series of comments and notes, located at the ends of relevant chapters, the reader is expected to have developed some insight in decision making and associated concepts.

The starting point is the nonlinear optimal decision problem for dynamic systems with multiple objectives under uncertainty. An optimal decision needs to take account of possible future uncertainties. As the process unfolds, and the uncertainty becomes known, the decision is revised and new future uncertainties are considered. This approach to optimal decisions is formulated as a static nonlinear problem, and the question of multiple objective decision making within this framework is considered using quadratic programming, nonlinear programming, nonlinear constrained min–max, mean-variance optimization and noncooperative Nash games. Regarding uncertainty in multiperiod decision problems, the treatment of scenario optimization is omitted from this book. This approach utilises the probability of the scenarios at each period to evaluate the expected value of the objective and ensure the satisfaction of the constraints arising from each scenario. It is essentially the application of optimization algorithms to very large-scale problems. The size of the problem is in particular due to the scenarios and adds a further layer of complexity, often resolved by efficient problem formulation, decomposition and parallel computers. Other areas not covered in this book are network optimization, combinatorial optimization and integer programming. The algorithms and tools for these seem to be beyond the main focus of the book.

We consider the static optimization problem with a single criterion in Chapter 1 and study the optimality conditions under equality and inequality constraints. We also describe in Chapter 1 a simple and approximate algorithm for solving the nonlinear dynamic policy optimization problem. If an approximate search strategy is required, when the multiple objective decision is being formulated, the last two sections of Chapter 1 provide an algorithm that has been tried and tested in numerous macroeconomic decision making and engineering process control problems. Nevertheless, it must be pointed out that the rest of the book is devoted to the discussion of methods for which accuracy is of primary importance.

Chapter 2 is devoted to the solution of the quadratic programming problem, encountered in Chapters 3–5, for the specification of multiple objective problems and, as a subproblem, in Chapters 6, 7, 8 and 12. Three different algorithms are considered in this chapter.

The basic view of multiple criteria is that the decision making process is a cognitive one. It is in the course of this process that the decision maker gains an increasingly concrete knowledge of what can be done and determines the trade-offs and targets. Chapters 3–5 describe iterative methods, involving interactions with the decision maker, for the specification of the relative weights and targets in multiple objective problems. In Chapter 5, the convergence properties of these algorithms are considered.

Chapters 6–8 cover nonlinear programming algorithms required for the solution of the optimization of a single objective. Convex optimization is discussed in Chapter 6, and an efficient version of the Goldstein–Levitin–Polyak algorithm is studied in detail with convergence rate results. The general nonlinear programming problem is considered in Chapter 7 with a detailed study of sequential quadratic programming algorithms. For example, considerations such as convexifying the problem in order to enlarge the region of convergence of the algorithm and augmented Lagrangians are introduced. Techniques for augmenting the Lagrangian are discussed along with stepsize strategies that measure the progress of the algorithm at every iteration. The rates of convergence of these algorithms are dependent on the nature of the approximate Hessian used. These are discussed in Chapter 8, with results concerning the rates for the variable-Lagrange multiplier pair.

A competition model in the presence of multiple decision makers is considered in Chapter 9. This is the case when each objective corresponds to an agent, or player, whose actions affect the system and thereby the objectives of other players. The aim is the computation of Nash equilibria in games. The algorithms considered are an asynchronous relaxation of the best replay algorithm, as well as variants of the Newton algorithm for solving the equilibrium condition.

The mean-versus-variance multiple objective problem is discussed in Chapters 10 and 11. This arises in decision making under uncertainty where we consider the simultaneous optimization of the expected value of the objective and its variance, which represents the associated risk. These are usually conflicting objectives. An example is the classical investment portfolio problem of maximizing expected return and minimizing expected risk. This is studied in Chapters 10 and 11. In the former, an extension of the portfolio problem is discussed, combining the risk of investments and exchange rates. In the latter, a study of the nonlinear case is given for dynamic decision problems with feedbacks.

The final approach to the multiple objective problem, discussed in Chapter 12, is the min–max formulation. The optimization of the worst-case objective requires an algorithm to solve the nonlinearly constrained min–max problem. The algorithm and its convergence rate properties are considered in detail.

The ideas presented in this book are based on experience in designing solutions to optimal decision problems in economics, finance and engineering. These were developed over a period during which I was privileged to have the opportunity to discuss, debate and argue related questions with Robin Becker, Jeremy Bray and Kumaraswamy Velupillai. Without their input, parts of the book would have been considerably weaker. On nonlinear programming, I am indebted to Laurence Dixon and David Mayne for numerous informative discussions and to Ioannis Akrotirianakis for proof reading the manuscripts of most of the related chapters. On various aspects of uncertainty, I am grateful for the comments and advice of Gregory Chow, David Kendrick and Martin Zarrop. Of course, none of the above bear responsibility for any remaining errors or misrepresentations.

The book was finished during a sabbatical year, and I am grateful to Imperial College for the sabbatical programme and to my colleagues in the Department of Computing for giving me this opportunity to finalize the project.

Notation

Sections within a chapter are referred to by their consecutive numbers; sections in other chapters are preceded by the chapter number. For example, in Chapter 1, the third section is referred to as Section 3, whereas within Chapter 2, the same section would be referred to as Section 1.3. The numbering of equations, theorems, etc. also follow the same rule. Within a chapter, an equation is referred to by the section number in which it occurs, followed by the equation or theorem number in the section. Outside the chapter, this is preceded by the chapter number. Although sometimes subsections are used to direct the discussion, these are not used in the referencing system.

SYMBOLS

$t = 1, 2, \ldots, \mathscr{T}$	index of discrete- time periods, starting at period 1 with final time period $\mathscr{T}$;
$u_t \in \mathbb{R}^u$	vector of controls, or decision variables at tth time period;
$y_t \in \mathbb{R}^y$	vector of output, or endogenous values, determined by the system, at tth time period;
$\epsilon_t \in \mathbb{R}^\epsilon$	vector of uncertainties, or random variables, at tth time period;
$\mathscr{E}(\cdot)$	expected value of $(\cdot)$;
$\text{var}\,(\cdot)$	variance of $(\cdot)$;
$U \equiv [u_1^{\mathrm{T}}, \ldots, u_t^{\mathrm{T}}, \ldots, u_{\mathscr{T}}^{\mathrm{T}}]^{\mathrm{T}} \in \mathbb{R}^{u \times \mathscr{T}}$	vector of controls of all time periods;
$Y \equiv [y_1^{\mathrm{T}}, \ldots, y_t^{\mathrm{T}}, \ldots, y_{\mathscr{T}}^{\mathrm{T}}]^{\mathrm{T}} \in \mathbb{R}^{y \times \mathscr{T}}$	vector of endogenous variables of all time periods;
$U^{\mathrm{d}}, Y^{\mathrm{d}}$	desired, or bliss, values of U, Y;
$\epsilon \equiv [\epsilon_1^{\mathrm{T}}, \ldots, \epsilon_t^{\mathrm{T}}, \ldots, \epsilon_{\mathscr{T}}^{\mathrm{T}}]^{\mathrm{T}} \in \mathbb{R}^{\epsilon \times \mathscr{T}}$	vector of random variables of all time periods;
$\mathbb{R}^n$	n-dimensional real vector space;
$x \in \mathbb{R}^n$	vector of optimization variables; sometimes we denote $x = [Y^{\mathrm{T}} \vdots U^{\mathrm{T}}]^{\mathrm{T}}$;
x^{d}	desired, or bliss, value of x;
x_{p}	preferred value of x (Chapters 3–5);
x_{c}	current optimal value of x (Chapters 3–5);
x_{n}	new optimal value of x (Chapters 3–5);

$\bar{x}$	unconstrained Newton step at x_k (Chapter 6);
x_k^{p}	projection of $\bar{x}$ (Chapter 6);
d_k	direction of search at x_k;
τ_k	stepsize along direction of search d_k;
c_k	penalty parameter value at x_k;
η_k	barrier parameter at iteration k;
$\mathscr{L}$	binary input length of quadratic programming problem;
$\mathbb{R}^{n\times m}$	set of real matrices of dimensions $n \times m$;
$\mathbb{R}_+^m \equiv \{\eta \in \mathbb{R}^m \vert \eta \geqslant 0\}$:	nonnegative orthant;
$1 \in \mathbb{R}^n$	$[1, 1, \ldots, 1]^{\mathrm{T}}$;
$\langle x, y\rangle$	inner product $x^{\mathrm{T}}y$ for $x, y \in \mathbb{R}^n$;
$\mathbb{E}_+^m \equiv \{\alpha \in \mathbb{R}^m \vert \alpha \geqslant 0; \langle 1, \alpha\rangle = 1\}$;	
$f(x)$	scalar objective function of $x \in \mathbb{R}^n$;
$f \in \mathbb{C}^1(\mathbb{R}^n)$ or $\in \mathbb{C}^1$	function f has continuous first partial derivatives with respect to $x \in \mathbb{R}^n$;
$f \in \mathbb{C}^2$	function f has continuous second partial derivatives with respect to x;
$\nabla_x f(x)$ or $\nabla f(x)$	gradient of f with respect to x; this is a column vector with jth element $\partial f(x)/\partial x^j$:

$$\nabla f(x) \equiv \begin{bmatrix} \dfrac{\partial f(x)}{\partial x^1} \\ \vdots \\ \dfrac{\partial f(x)}{\partial x^j} \\ \vdots \\ \dfrac{\partial f(x)}{\partial x^n} \end{bmatrix};$$

f_k and ∇f_k	$f(x_k)$ and $\nabla f(x_k)$;
$q(x)$ or $q_k(x)$	quadratic objective function or quadratic approximation to the objective function at $x_k j$;
$\mathscr{Y}(U)$	computational mapping between Y and U arising from model of the system $g(Y, U) = 0$ and the model solution algorithm (Chapters 1 and 11);
$\mathscr{F}(x)$ or $\mathscr{F}(U)$	$f(\mathscr{Y}(U), U)$, i.e. the objective function $f(Y, U)$ reduced using the mapping $Y = \mathscr{Y}(U)$;
$\mathfrak{Q} \equiv \nabla^2 f(x)$	Hessian, or second-derivative matrix, of f with respect to x; the j^{ℓ}th element of this matrix is given by $\partial^2 f(x)/\partial x^j\, \partial x^{\ell}$:

$$\nabla^2 f(x) \equiv \begin{bmatrix} \frac{\partial^2 f(x)}{\partial (x^1)^2} & \cdots & \frac{\partial^2 f(x)}{\partial x^j \, \partial x^1} & \cdots & \frac{\partial^2 f(x)}{\partial x^n \, \partial x^1} \\ \vdots & \ddots & \vdots & \vdots & \\ \frac{\partial^2 f(x)}{\partial x^1 \, \partial x^j} & \cdots & \frac{\partial^2 f(x)}{\partial (x^j)^2} & \cdots & \frac{\partial^2 f(x)}{\partial x^n \, \partial x^j} \\ \frac{\partial^2 f(x)}{\partial x^1 \, x^n} & \cdots & \frac{\partial^2 f(x)}{\partial x^j \, \partial x^n} & \cdots & \frac{\partial^2 f(x)}{\partial (x^n)^2} \end{bmatrix};$$

$\mathfrak{Q}_u, \mathfrak{Q}_y$	weighting matrics associated with $U - U^{\mathrm{d}}$, $Y - Y^{\mathrm{d}}$
$g(x) \in \mathbb{R}^e$	vector-valued function of equality constraints (i.e. $\{x \in \mathbb{R}^n \mid g(x) = 0\}$);
g_k	$g(x_k)$;
∇g_k	$\nabla g(x_k) \equiv [\nabla g^1(x_k) \vdots \nabla g^2(x_k) \vdots \ldots \vdots \nabla g^e(x_k)] \in \mathbb{R}^{n \times e}$;
$h(x) \in \mathbb{R}^i$	vector- valued function of inequality constraints (i.e. $\{x \in \mathbb{R}^n \mid h(x) \leqslant 0\}$);
h_k	$h(x_k)$
∇h_k	$\nabla h(x_k) \equiv [\nabla h^1(x_k) \vdots \nabla h^2(x_k) \vdots \ldots \vdots \nabla h^i(x_k)] \in \mathbb{R}^{n \times i}$;
$\mathscr{I}(x_k) \equiv \{i \mid h^i(x_k) = 0\}$	set of active inequality constraints at x_k;
$\mathscr{R}$	set of feasible points, usually given by $\mathscr{R} = \{x \in \mathbb{R}^n \mid g(x) = 0; h(x) \leqslant 0\}$; in Chapter 6, where we constrain $\mathscr{R}$ to be convex, we have $\mathscr{R} = \{x \in \mathbb{R}^n \mid h(x) \leqslant 0\}$; in Chapter 3, we take this set to be defined by linear equalities only: $\mathscr{R} = \{x \in \mathbb{R}^n \mid \mathscr{G}^{\mathrm{T}} x = g\}$;
Ω	set of policies acceptable to the decision maker (Chapters 3–5);
$L(x, \lambda, \mu)$	Lagrangian function for the constrained optimization problem;
$L^{\mathrm{a}}(x, \lambda, \mu, c, \alpha)$	augmented Lagrangian function for the constrained optimization problem;
$\alpha \in \mathbb{R}^i$	vector of offsets for inequality constraints, used in Chapters 7 and 8;
$\lambda \in \mathbb{R}^e$	multiplier vector for the equality constraints;
$\mu \in \mathbb{R}^i$	multiplier vector for the inequality constraints;
trace(A)	$\sum_{i=1}^{n} a_{ii}, A \in \mathbb{R}^{n \times n}$;
diag (x)	$\mathrm{diag}[x^1, x^2, \ldots, x^n] \equiv \begin{bmatrix} x^1 & 0 & \ldots & 0 \\ 0 & x^2 & \ldots & 0 \\ \vdots & \vdots & \ddots & \vdots \\ 0 & 0 & \ldots & x^n \end{bmatrix}; x \in \mathbb{R}^n$;
$\lceil \mathrm{x} \rceil$	smallest integer ι such that $\iota \geqslant x$;

□	end of a proof, an example or a particular train of thought;
policy maker	decision maker.

PART I

Introduction

CHAPTER 1

Introduction: Optimization of a Single Objective

Multiple objective decision making requires an understanding of optimization with a single objective. As most decisions affect the future state of the system, a certain awareness is expected about intertemporal implications and preferences. After formulating a static view of the basic dynamic problem, we embark on an exploration of the necessary and sufficient conditions for optimality. This is intended to provide the analytical tools for nonlinear optimization and the framework for augmented Lagrangians and algorithms in Chapters 6, 7, 8, 9 and 12.

We close this chapter with a simple, approximate yet effective optimization algorithm used in applications with macroeconomic decision models. The purpose of the approximate optimization algorithm is first to illustrate the interactive, iterative and adaptive nature of most decision processes, which sometimes require an indication of the neighbourhood of the optimum; and second to highlight the decision process as a current event that takes account of its future implications. The approximate approach to policy optimization is satisfactory for the intermediate steps required to formulate the decision problem. The steps represent a cognitive process in which the decision maker learns more about the system, the possibilities and the objectives. We discuss accurate methods for computing constrained optima in later chapters.

1 THE DYNAMIC DECISION PROBLEM AND STATIC OPTIMIZATION

Most decision making is inherently time-dependent: a decision made today affects the underlying system, and thereby any decision to be made tomorrow, or the next period. We study the dynamic decision problem in a *discrete-time* setting. The latter signifies that the problem is considered on discrete intervals of time, usually of equal length or duration. Each variable has a constant value during this interval. Examples of typical intervals are: for control systems, fractions of a second; for decision making (particularly in finance), hours, days or weeks; and for macroeconomic policies, months or quarters.

Let $y_t \in \mathbb{R}^y$ denote the vector (see CN.1) of the endogenous variables (see CN.2), $u_t \in \mathbb{R}^u$ denote the vector of control or decision variables of the system at time period t, and let $\varepsilon_t \in \mathbb{R}^\varepsilon$ denote the uncertainty or disturbance at time t. Thus, the basic dynamic system is represented by

$$y_{t+1} = g_t(y_t, u_t, \epsilon_t), \quad t = 1, \dots \mathscr{T},$$

where g_t is a time-dependent vector-valued nonlinear function and the initial state y_0 is given prior to any computation. The optimal decision problem may be formulated as the choice of variables that optimize a policy objective function subject to the model describing the system and other restrictions, throughout the decision period $[1, \mathcal{T}]$. For an uncertain system, this is expressed as

$$\min \mathcal{E}\{f(Y, U) \mid g(Y, U, \epsilon) = 0; h(Y, U) \leqslant 0\}, \tag{1.1a}$$

where $\mathcal{E}$ denotes expectations ϵ is a vector of random noise, Y and U are respectively the endogenous or output variables and decision variables, or controls, of the system (Rustem, 1990). f is the policy objective function and g is the model of the system, affected by the random disturbance vector ϵ. The equality indicates that each component of g is equal to zero. In general, g is a nonlinear dynamic relationship between Y, U and ϵ. Similarly, h represents the inequality constraints describing the feasible region and (1.1a) indicates that each component of h is nonpositive. Problem (1.1a) is essentially a static transcription of a dynamic optimization problem in discrete time, where

$$U = \begin{bmatrix} u_1 \\ \vdots \\ u_t \\ \vdots \\ u_{\mathcal{T}} \end{bmatrix}, \quad Y = \begin{bmatrix} y_1 \\ \vdots \\ y_t \\ \vdots \\ y_{\mathcal{T}} \end{bmatrix}, \quad \epsilon = \begin{bmatrix} \epsilon_1 \\ \vdots \\ \epsilon_t \\ \vdots \\ \epsilon_{\mathcal{T}} \end{bmatrix} \tag{1.1b}$$

The optimization covers the periods $t = 1, \ldots, \mathcal{T}$. Thus, $Y \in \mathbb{R}^{y \times \mathcal{T}}$, $U \in \mathbb{R}^{u \times \mathcal{T}}$, $\epsilon \in \mathbb{R}^{\epsilon \times \mathcal{T}}$, $g: \mathbb{G} \subset \mathbb{R}^{n + (\mathcal{T} \times \epsilon)} \to \mathbb{R}^e$, $h: \mathbb{H} \subset \mathbb{R}^n \to \mathbb{R}^i$ and $f: \mathbb{F} \subset \mathbb{R}^n \to \mathbb{R}^1$ $(n = \mathcal{T} \times (y + u))$. Thus, g represents, in static form, the dynamic system, for time periods $t = 1, \ldots, \mathcal{T}$. The size of the system is generally in the order of hundreds of equations per time period t.

In the *multiple objective* case, when the decision maker has a number of possibly conflicting targets for each Y and U, we can formulate the overall objective in terms of a general attainment measure of these targets. Thus, the objective may be formulated as the quadratic

$$f(Y, U) = \frac{1}{2} \sum_{i=1}^{y \times \mathcal{T}} q_y^i \left[y^i - (y^i)^{\mathrm{d}}\right]^2 + \frac{1}{2} \sum_{j=1}^{u \times \mathcal{T}} q_u^j \left[u^j - (u^j)^{\mathrm{d}}\right]^2, \tag{1.1c}$$

where y^i, u^j are elements of the vectors Y, U respectively. The target or desired values, characterising the *bliss point*, of the variables are indicated by the superscript d, and q_y^i, q_u^j are nonnegative weights indicating the importance associated with the achievement of each target. We invoke this special structure for $f(Y, U)$ in some chapters of this book. Algorithms for the optimization of the quadratic objective function, subject to linear equality and inequality constraints, are discussed in Chapter 2. Methods discussed in Chapters 6, 7, 8, 12 are able to handle general nonlinear $f(Y, U)$ structures, under certain continuity and differentiability assumptions.

Uncertainty problems are addressed in Chapters 10–12. The approaches to uncertainty, as well as other decision problems discussed in this book, entail the solution of a deterministic optimization problem (i.e. $\epsilon \equiv 0$). Indeed, central to most policy optimization considerations is a deterministic optimization algorithm. We discuss three such algorithms for solving deterministic problems of differing complexity.

The first is a Gauss–Newton (G–N)-type algorithm, ordinarily for problems constrained only by the model equations $g(Y, U) = 0$, given by

$$\min \{ f(Y, U) \mid g(Y, U) = 0 \}. \tag{1.2}$$

The G–N algorithm is discussed in Sections 4 and 5 below, and requires simple derivative information. It is quite effective for rough computations to aid the decision maker in the formulation or exploration of the policy objectives. This may entail the selection of $(y^i)^{\mathrm{d}}, (u^j)^{\mathrm{d}}, q_y^i, q_u^j$ in (1.1c), achieved by an interactive process between the decision maker and the optimization model, as discussed in Chapters 3–5. The G–N algorithm solves the unconstrained problem obtained by eliminating Y, using the system $g(Y, U) = 0$, from (1.2).

The second approach is the Goldstein–Levitin–Polyak (GLP) algorithm discussed in Chapter 6. This is concerned with a convex feasible region and a convex objective function.† Therefore, in practice, it addresses linear constraints, such as bounds, on the controls. For dynamic problems such as (1.2), the derivative information for this algorithm can be efficiently obtained using a backward evaluation approach. The algorithm is fast and involves a quadratic subproblem with inequality bounds. The formulation is particularly useful when $g(Y, U) = 0$ can be used to eliminate Y from the problem and when the remaining constraints on the decision, or control, variables are convex. A particular case is given by linear inequality constraints on the control variables

$$\min \{ f(Y, U) \mid g(Y, U) = 0; \mathscr{H}^{\mathrm{T}} U \leqslant h \}, \tag{1.3}$$

where $\mathscr{H}$ is a fixed matrix and h is a given vector. The general convex optimization problem can be written, for a convex feasible region described by $h(U)$, as

$$\min \{ \mathscr{F}(U) \mid h(U) \leqslant 0 \}, \tag{1.4}$$

having eliminated the equality constraints from the problem. We consider the GLP algorithm discussed in Rustem (1984) (see also Darlington *et al.*, 1996)

The third method is for optimization subject to general nonlinear equality and inequality constraints. It is a sequential quadratic programming (SQP) algorithm, discussed in Chapters 7 and 8, that admits general constraints. The derivative calculations are expensive, but the algorithm is fast. In this and

† A brief discussion of convexity is provided in Section 2.

several other decision problems such as multiple objective optimisation, min–max and equilibrium computations studied in Chapters 2–5, 7 and 8, and 10 and 11, the distinction between the decision (control) and endogenous (output) variables is not necessary. We therefore use the vector x to denote

$$x = \begin{bmatrix} Y \\ \dots \\ U \end{bmatrix}. \tag{1.5}$$

Consider, therefore, the problem

$$\min\{f(x) \mid g(x) = 0; h(x) \leqslant 0\}, \tag{1.6}$$

where no convexity requirement is associated with the region described by $g(x)$ and $h(x)$. In Chapters 7 and 8, we consider an SQP algorithm for (1.6).

For simplicity, we omit the time dimension from most of our discussion. Looking forward to other chapters in this book, multiple objective decisions that reflect the policy maker's preference are formulated in the form of an interactive algorithm for specifying the weighting matrices or desired (or bliss) values of a quadratic objective function. The algorithm and its properties are discussed in Chapters 3–5.

As an extension of the Newton and quasi-Newton algorithms for individual decisions, we consider, in Chapter 9, multiple decision makers engaged within a noncooperative framework. This requires algorithms for computing equilibrium points that achieve the optimality conditions for each individual.

In decision making under uncertainty, the optimization of the original objectives and the minimisation of risk is formulated as multiple objective decision making. The classical simultaneous optimisation of mean and variance is extended to nonlinear systems in Chapters 10 and 11.

A final approach to multiple objectives is provided by the minimization of the worst-case objective. The worst case and the decision are determined simultaneously. In the presence of rival objectives, a robust approach is formulated and an algorithm to solve the resulting discrete min–max problem is discussed in Chapter 12. The min–max decision yields the ultimate robust policy. If the price of such a robust approach is too high in policy terms then the formulation provides guidance for searching less robust, but more acceptable, decisions. The latter may be achieved by weighting the objectives, as discussed in Chapters 3–5, but using the weights implied by the min–max solution as the starting point for the search.

2 BASIC OPTIMALITY CONDITIONS

Most of the discussion in this book is focused on the formulation and solution of the optimization problem (1.6). We therefore discuss the optimality conditions of (1.6) for reference throughout this book. The satisfaction of these conditions is the primary aim of optimization algorithms. Hence, the terminations of algorithms are generally determined by testing for their satisfaction.

Let the set of *feasible points* be defined by

$$\mathscr{R} \equiv \{x \in \mathbb{R}^n \mid g(x) = 0;\ h(x) \leqslant 0\}. \tag{2.1}$$

Definition

A point x_* is a *local minimum point* of f over $\mathscr{R}$ if there is no better solution within an $\epsilon > 0$ neighbourhood of x_* such that

$$f(x) \geqslant f(x_*), \quad \forall x \in \mathscr{R}, \quad \|x - x_*\| < \epsilon$$

(see CN.3). If $f(x) > f(x_*)$, for all $x_* \neq x \in \mathscr{R}$, $\|x - x_*\| < \epsilon$, then x_* is a *strict local minimum point* of f over $\mathscr{R}$.

Definition

A point x_* is a *global minimum point* of f over $\mathscr{R}$ if $f(x) \geqslant f(x_*)$ for all $x \in \mathscr{R}$. If $f(x) > f(x_*)$, for all $x_* \neq x \in \mathscr{R}$, then x_* is a *strict global minimum point* of f over $\mathscr{R}$.

In practice, we can be reasonably confident in computing local minima only. The reason is that algorithms for solving the problem proceed on the basis of local information available to them. Global minima are restricted to special types of problems where convexity properties apply as discussed below. Specifically, Theorem 2.4 ensures that, for convex problems, the computed local minimum is also the global solution.

A constrained optimum is characterized by a set of inequalities that indicate that any *bounded feasible movement* away from the optimum leads to a deterioration of the objective function value. The next three results state this characterization.

First-order conditions involve the gradient, or first-derivative, vector of $f(x)$. Also, a condition is *necessary* if it is satisfied at an optimum point. However, the satisfaction of a necessary condition would not guarantee on its own that the point is an optimum. That guarantee is provided by *sufficiency* conditions.

Proposition 2.1 (First-order necessary condition)

Let $\mathscr{R} \subseteq \mathbb{R}^n$ and let $f \in \mathbb{C}^1$ be a function on $\mathscr{R}$ (see CN.4). If x_* is a local minimum point of f over $\mathscr{R}$, then, for any $x \in \mathscr{R}$, we have

$$\langle \nabla f(x_*), x - x_* \rangle \geqslant 0. \tag{2.2}$$

Proof

Let $d = x - x_*, x(\theta) = x_* + \theta d$ for any $\theta \in [0, \bar{\theta}]$, with upper bound $\bar{\theta}$, and consider the first-order Taylor expansion of $f(x)$ (with the term $\theta\langle \nabla f(x_*), \mathrm{d}\rangle$ added and subtracted on the right):

$$f(x(\theta)) = f(x_*) + \theta\langle\nabla f(x_*), d\rangle + \theta\int_0^1 \langle\nabla f(x(t\theta)) - \nabla f(x_*), d\rangle \mathrm{d}t$$

(see CN.5). Clearly, the last term on the right approaches zero faster than θ. If $\langle\nabla f(x_*), d\rangle < 0$, for sufficiently small values of $\theta > 0$, the last two terms on the right together will be negative and thus

$$f(x(\theta)) < f(x_*).$$

As $f(x(0)) = f(x_*)$, x_* cannot be a minimum. Thus, we have a contradiction. Hence, for a minimum, we have $\langle\nabla f(x_*), d\rangle \geqslant 0$. □

Inequality (2.2) indicates that, at the solution, there is no feasible direction along which the objective function can be improved. This inequality is the necessary condition upon which variational inequality methods for computing competitive equilibria are based. The implication of (2.2) is that each decision maker decides on the best strategy satisfying (2.2) (see e.g. Bertsekas and Gafni, 1982; Dafremos, 1983, 1990; Harker and Pang, 1990; Glowinski, Lions and Trémolières, 1981; Glowinski, 1984; Nagurney, 1993).

The second-order condition involves the second-derivative matrix of $f(x)$.

Proposition 2.2 (Second-order necessary conditions)
Let $\mathcal{R} \subseteq \mathbb{R}^n$ and let $f \in \mathbb{C}^2$ be a function on $\mathcal{R}$ (see CN.4). If x_* is a local minimum point of f over $\mathcal{R}$ then, for any $x \in \mathcal{R}$, we have (2.2) and if

$$\langle\nabla f(x_*), x - x_*\rangle = 0 \quad \text{then} \quad \langle x - x_*, \nabla^2 f(x_*)(x - x_*)\rangle \geqslant 0. \tag{2.3}$$

Proof
Using the same initial discussion as in the previous proof, we consider the second-order Taylor series expansion

$$\begin{aligned} f(x(\theta)) &= f(x_*) + \theta\langle\nabla f(x_*), d\rangle + \langle d, \nabla^2 f(x_*)d\rangle \\ &\quad + \theta^2\left(\int_0^1 (1-t)\langle d, [\nabla^2 f(x(t\theta)) - \nabla^2 f(x_*)]d\rangle \mathrm{d}t\right) \end{aligned}$$

(see CN.5). As $\langle\nabla f(x_*), d\rangle = 0$, if $\langle d, \nabla^2 f(x_*)d\rangle < 0$, the last two terms on the right become negative for sufficiently small $\theta > 0$. Thus, $f(x_*)$ cannot be a minimum. Hence, for a minimum, we have

$$\langle d, \nabla^2 f(x_*)d\rangle \geqslant 0.$$ □

Proposition 2.2 indicates that if there is a feasible direction $x - x_*$ that is orthogonal to $\nabla f(x_*)$ then the Hessian is positive-semidefinite (see CN.6) in this direction.

In Section 3 below, (1.2) is expressed as an unconstrained optimization problem by eliminating the Y variables, using $g(Y, U) = 0$, and formulating an unconstrained problem with respect to the decision or control variables. We therefore provide an outline of the above results for this case. A detailed discussion may be found in Luenberger (1984, Chap. 6).

Proposition 2.3 (The unconstrained case)
First-order necessary conditions. *Let the assumptions of Proposition 2.1 be satisfied and let the local minimum of f, x_* be an interior point of $\mathscr{R}$. Then* $\nabla f(x_*) = 0$.

Second-order necessary conditions. *Let the assumptions of Proposition 2.2 be satisfied and let the local minimum of f, x_*, be an interior point of $\mathfrak{R}$. Then* $\nabla f(x_*) = 0$ *and, for all* $v \in \mathbb{R}^n$,

$$\langle v, \nabla^2 f(x_*)v \rangle \geqslant 0.$$

Second-order sufficient conditions. *Let the assumptions of Proposition 2.2 be satisfied and let the local minimum of f, x_*, be an interior point of $\mathscr{R}$. Then x_* is a strict local minimum if* $\nabla f(x_*) = 0$ *and*

$$\langle v, \nabla^2 f(x_*)v \rangle > 0, \quad \forall v \in \mathbb{R}^n. \qquad \square$$

Thus, the strict positive-definiteness (see CN.6) of the Hessian of f is the additional requirement that ensures the sufficiency condition for a strict local minimum.

Convex optimization is the minimisation or maximisation of a convex function, subject to a convex feasible set. We provide a brief summary of related results. The omitted proofs can be found in Berge (1963, Chap. 8).

Definition
A set $\mathscr{C} \subseteq \mathbb{R}^n$ is *convex* if for every $x_1, x_2 \in \mathscr{C}$ and a real number $\alpha, 0 < \alpha < 1$, the point $\alpha x_1 + (1 - \alpha)x_2 \in \mathscr{C}$.

Definition
A function f defined on a convex set $\mathscr{C} \subseteq \mathbb{R}^n$ is *convex* if for every $x_1, x_2 \in \mathscr{C}$ and for all $\alpha, 0 \leq \alpha \leq 1$,

$$f(\alpha x_1 + (1 - \alpha)x_2) \leqslant \alpha f(x_1) + (1 - \alpha)f(x_2),$$

and f is *strictly convex* if the above inequality is strict for all $\alpha, 0 < \alpha < 1$.

Theorem 2.1
Let $f \in \mathbb{C}^1$. *Then f is convex over a convex set $\mathscr{C}$ if and only if*

$$f(x_1) \geqslant f(x_2) + \langle \nabla f(x_2), (x_1 - x_2) \rangle, \quad \forall x_1, x_2 \in \mathscr{C}.$$

Theorem 2.2
Let $f \in \mathbb{C}^2$. *Then f is convex over a convex set $\mathscr{C}$ if and only if the Hessian* $\nabla^2 f(x)$ *is positive-semidefinite for all* $x \in \mathscr{C}$.

Theorem 2.3
Let f be a convex function defined over the convex set $\mathcal{C}$. Then any local minimum of f is also a global minimum.

Proof
Suppose that x_1 is a local minimum. Suppose also that there is another point $x_2 \in \mathcal{C}, x_1 \neq x_2$, with $f(x_2) < f(x_1)$. Then, for $\alpha x_1 + (1 - \alpha)x_2, \alpha \in (0, 1)$, we have

$$f(\alpha x_1 + (1 - \alpha)x_2) \leqslant \alpha f(x_1) + (1 - \alpha)f(x_2) < f(x_1),$$

which contradicts the claim that x_1 is a local minimum point. □

Theorem 2.4 (Necessary and sufficient optimality for global optima in convex problems)
Let $f \in \mathbb{C}^1$ be convex on the convex set $\mathcal{C}$. If there is an $x_* \in \mathcal{C}$ such that, for all $x \in \mathcal{C}$, (2.2) is satisfied then x_* is a global minimum of f over $\mathcal{C}$.

Proof
The proof follows from Theorem 2.1, since

$$f(x) \geqslant f(x_*) + \langle \nabla f(x_*), x - x_* \rangle \geqslant f(x_*).$$ □

Convex optimization may arise in a number of formulations in this book. When (1.3) and (1.4) reduces to the minimization of $f(x)$ subject to convex constraints, eliminating Y may yield the minimization of a convex function subject to convex constraints on U only, or, alternatively, all the constraints, the model and other linear restrictions on Y *and* U may be convex. The convex constraints are typically linear inequalities. Other convex optimization problems arise, in practice, in the form of convex quadratic objective functions and convex, linear, constraints. We encounter these as the subproblems in Chapters 6–8 and in the quadratic optimization problems in Chapters 2–5.

3 NECESSARY AND SUFFICIENT CONDITIONS FOR NONLINEARLY CONSTRAINED OPTIMIZATION

We turn to the optimality conditions associated with the general nonlinear optimization problem (1.6). Similar to the definition of local and global minima in Section 2, these conditions are essentially based on the fact that at a minimum (maximum) there is no feasible descent (ascent) direction (see CN.12).

Definition
An inequality constraint $h^j(x) \leqslant 0$ is said to be *active* at a point x' if $h^j(x') = 0$, and *inactive* at a feasible point x' if $h^j(x') < 0$.

Definition
Let x' be a point satisfying the constraints

$$g(x) = 0, \qquad h(x) \leqslant 0,$$

and let $\mathscr{I}$ or $\mathscr{I}(x')$ be the set of active inequality constraints at x' (i.e. $\mathscr{I}(x') = \{i \mid h^i(x') = 0\}$). Then, x' is a *regular point* of the constraints if the gradient vectors

$$\nabla g^j(x'), \quad 1 \leqslant j \leqslant e; \qquad \nabla h^i(x'), \quad i \in \mathscr{I}(x')$$

are linearly independent (see CN.7).

Definition
Let the matrices $\nabla g(x)$ and $\nabla h(x)$ be defined as

$$\nabla g(x) \equiv [\nabla g^1(x) \vdots \ldots \vdots \nabla g^e(x)], \quad \nabla h(x) \equiv [\nabla h^1(x) \vdots \ldots \vdots \nabla h^i(x)]. \tag{3.1}$$

We characterize an optimal point as a point at which there is no feasible direction of improvement. A feasible direction for a problem with nonlinear constraints needs to be defined with some care. Consider $x' \in \mathscr{R}$, and any infinite sequence $\{x_k\} \to x'$ such that $x_k \neq x'$ and $x_k \in \mathscr{R}$. Then, we can write

$$x_k = x' + \tau_k d_k, \tag{3.2}$$

$\forall k$, with d_k a vector of any fixed length $\|d_k\| = \delta > 0, \{d_k\} \to d$, and $\tau_k > 0$ a scalar. It follows that $\{\tau_k\} \to 0$. The limiting vector d is referred to as a *feasible direction*, and $\mathscr{R}(x')$ denotes the *set of feasible directions at* x'. Taking the limit in (3.2), we consider a feasible incremental step along d. Intuitively, $\mathscr{R}(x')$ is the set of directions d that, for small $\tau > 0$, yields $x' + \tau d \in \mathscr{R}$. Next, we consider the linearized constraints at $x' \in \mathscr{R}, \langle \nabla g^j(x'), d\rangle + g^j(x') = 0$; $\langle h^i(x'), d\rangle + h^i(x') \leqslant 0$ and the set of feasible directions for these constraints

$$\mathscr{R}^{\mathrm{L}}(x') \equiv \{d \in \mathbb{R}^n \mid d \neq 0; \langle \nabla g^j(x'), d\rangle = 0; \langle \nabla h^i(x'), d\rangle \leqslant 0; 1 \leqslant j \leqslant e; i \in \mathscr{I}(x')\}.$$

For any $d \in \mathscr{R}^{\mathrm{L}}(x')$, it is not necessarily true that $x' + \tau d \in \mathscr{R}$, for all $\tau \geqslant 0$. This is because $\mathscr{R}^{\mathrm{L}}(x')$ is a linear system whereas $\mathscr{R}(x')$ consists of the set of original nonlinear constraints. Clearly, in general, $\mathscr{R}(x') \neq \mathscr{R}^{\mathrm{L}}(x')$. What may be a feasible direction for the linearized system may not feasible for the original nonlinear problem. Indeed, Kuhn and Tucker (1951) give the following example to illustrate this.

Example
Consider the system

$$h(x_1, x_2) = x_2 - (1 - x_1)^3 \leqslant 0, \qquad x_1 \geqslant 0, \qquad x_2 \geqslant 0$$

and the feasible point

$$\begin{bmatrix} \hat{x}_1 \\ \hat{x}_2 \end{bmatrix} = \begin{bmatrix} 1 \\ 0 \end{bmatrix}.$$

A move, from this point, in the direction

$$d = \begin{bmatrix} 1 \\ 0 \end{bmatrix},$$

is given by

$$\begin{bmatrix} x_1 \\ x_2 \end{bmatrix} = \begin{bmatrix} \hat{x}_1 \\ \hat{x}_2 \end{bmatrix} + \tau \begin{bmatrix} 1 \\ 0 \end{bmatrix}$$

for $\tau > 0$. Although such a move is not feasible with respect to the nonlinear inequality, it is feasible for its linear approximation at this point given by $\langle \nabla h(\hat{x}_1, \hat{x}_2), d \rangle \leqslant 0$.

Another example is $(x_1 - 1)^2 + x_2^2 - 1 = 0$, $(x_1 + 1)^2 + x_2^2 - 1 = 0$. At the origin, any vector

$$\begin{bmatrix} 0 \\ \alpha \end{bmatrix}$$

satisfies the linearized constraints. For the nonlinear constraints, however, the origin is the only feasible point. □

It is shown in Fletcher (1981, Lemma 9.2.1) that $\mathscr{R}^{\mathrm{L}}(x') \supseteq \mathscr{R}(x')$. Lemma 3.1 below establishes the conditions of equivalence between $\mathscr{R}(x')$ and $\mathscr{R}^{\mathrm{L}}(x')$.

The plan for establishing the first-order necessary conditions for optimality

First, we discuss the conditions under which $\mathscr{R}^{\mathrm{L}}(x') = \mathscr{R}(x')$. This is important since $\mathscr{R}^{\mathrm{L}}(x')$ is a more convenient set to consider in practice. Then, given the set of descent directions at x',

$$\mathscr{D}(x') \equiv \{d \mid \langle \nabla f(x'), d \rangle < 0\},$$

we establish that at a local optimal solution $\mathcal{R}^{\mathrm{L}}(x') \cap \mathcal{D}(x') = \varnothing$. Finally, we discuss Farkas' Lemma to relate this result to the first-order necessary conditions for x_* to be a local minimum of (1.6).

Lemma 3.1 (Fletcher, 1981)

Let $x' \in \mathscr{R}$. Then $\mathscr{R}^{\mathrm{L}}(x') = \mathscr{R}(x')$ if all the constraints are linear or x' is a regular point.

Proof

By definition, we have $\mathscr{R}^{\mathrm{L}}(x') = \mathscr{R}(x')$ for linear constraints. To establish the case for a regular point x', let $d \in \mathscr{R}^{\mathrm{L}}(x')$ and consider the nonlinear system of n equations

$$0 = r(x, \theta) = \begin{cases} g^j(x) - \theta\langle \nabla g^j(x'), d\rangle & \text{for } 1 \leqslant j \leqslant e, \\ h^i(x) - \theta\langle \nabla h^i(x'), d\rangle & \text{for } i \in \mathcal{I}(x'), \\ \langle z^\iota, x - x' - \theta d\rangle & \text{for } e + |\mathcal{I}(x')| + 1 \leqslant \iota \leqslant n, \end{cases} \tag{3.3}$$

where $|\mathcal{I}(x')|$ denotes the number of active constraints in the set $\mathcal{I}(x')$ and z^ι is a vector chosen below. The system (3.3) is solved by x' when $\theta = 0$. For sufficiently small $\theta \geqslant 0$, x solving (3.3) also satisfies the original constraints; hence $x \in \mathcal{R}$. Consider the $n \times n$ matrix

$$J(x) = [\nabla g^1(x) \vdots \ldots \vdots \nabla g^e(x) \vdots \nabla h^{i_1}(x) \vdots \ldots \vdots \nabla h^{i_{|\mathcal{I}(x')|}} \vdots z^{[e+|\mathcal{I}(x')|+1]} \vdots \ldots \vdots z^n],$$

where the indices $i_1, \ldots, i_{|\mathcal{I}(x')|}$ refer to the $|\mathcal{I}(x')|$ elements of $\mathcal{I}(x')$. As x' is a regular point, the vectors z^ι can be chosen such that the columns of $J(x')$ are linearly independent and hence $J(x')$ is non-singular. Thus, by the Implicit Function Theorem (see e.g. Apostol, 1981), there exist numbers $\kappa_1 > 0$, $\kappa_2 > 0$ and a unique solution $x(\theta)$ to (3.3) such that

$$x(\theta) \in \{x \mid \|x - x'\| < \kappa_1\},$$

with

$$\theta \in \{\theta \mid | \theta | < \kappa_2\},$$

is a once continuously differentiable function of θ. Using (3.3) and the chain rule, we get

$$0 = \frac{\mathrm{d}r}{\mathrm{d}\theta} = \nabla_x r(x, \theta)\frac{\mathrm{d}x}{\mathrm{d}\theta} + \nabla_\theta r(x, \theta) = J^{\mathrm{T}}(x)\frac{\mathrm{d}x}{\mathrm{d}\theta} - J^{\mathrm{T}}(x')d.$$

Thus, $\mathrm{d}x/\mathrm{d}\theta = d$ at $\theta = 0$ and, if θ_k is a sequence decreasing to zero, $x_k = x(\theta_k)$ is a sequence of feasible points defined by the feasible direction d. □

Lemma 3.2 (No feasible descent directions at a local minimum: Fletcher, 1981) *Let $f(x)$ be once differentiable with Lipschitz-continuous first derivatives (see CN.8) and let x_* be a local minimum of (1.6). Then $\mathcal{R}^{\mathrm{L}}(x_*) \cap \mathcal{D}(x_*) = \varnothing$.*

Proof

Let $d \in \mathcal{R}(x_*)$ so that there exists a feasible sequence $x_k = x_* + \tau_k d_k$, $\|d_k\| = \delta > 0$, $\{x_k\} \to x_*$, such that $\{d_k\} \to d$. Consider a first-order Taylor series expansion of $f(x)$ about x_*:

$$\begin{aligned} f(x_k) &= f(x_*) + \tau_k\langle d_k, \nabla f(x_*)\rangle + \tau_k \int_0^1 \langle \nabla f(x_* + t\tau_k d_k) - \nabla f(x_*), d_k\rangle \,\mathrm{d}t \\ &\leqslant f(x_*) + \tau_k\langle d_k, \nabla f(x_*)\rangle + c\tau_k^2\|d_k\|^2 \end{aligned}$$

for some $c \geqslant 0$. As x_* is a local minimum, we have $f(x_*) \leqslant f(x_k)$ for k sufficiently large. Thus,

$$0 \leqslant f(x_k) - f(x_*) \leqslant \tau_k\langle d_k, \nabla f(x_*)\rangle + c\tau_k^2\|d_k\|^2.$$

Dividing by $\tau_k > 0$, we have

$$0 \leqslant \langle \mathrm{d}_k, \nabla f(x_*) \rangle + c\tau_k \|\mathrm{d}_k\|^2.$$

In the limit, $\{d_k\} \to d$ and $\{\tau_k\} \to 0$, so $d \notin \mathcal{D}(x_*)$ and hence $\mathcal{R}^{\mathrm{L}}(x_*) \cap \mathcal{D}(x_*) = \emptyset$. □

The next lemma is required to establish Farkas' Lemma, and, geometrically, it states that given a point outside a closed (see CN.9) and convex set, a hyperplane (see CN.10) can be passed through the point that does not touch the convex set.

Lemma 3.3 (Separating hyperplane)
Let $\mathcal{C}$ be a closed convex set given by

$$\mathcal{C} = \left\{ v \mid v = \sum_{j=1}^{i} \nabla h^j \mu^j ; \mu^j \geqslant 0 \right\}, \tag{3.4}$$

for any vectors ∇h^i and nonnegative multipliers μ^i, and let $\nabla f(x')$ be a nonzero vector exterior to $\mathcal{C}$†*. Then there exists a hyperplane $\langle x, d \rangle = 0$ that separates $\mathcal{C}$ from $\nabla f(x')$.*

Proof
The result applies to all closed convex sets but is stated in relation to (3.4) for use in Lemma 3.4. Let

$$x_0 = \arg \min_{x \in \mathcal{C}} \|x - \nabla f(x')\|_2 > 0.$$

x_0 exists since the solution satisfies $\|x_0 - \nabla f(x')\|_2 \leqslant \|x - \nabla f(x')\|_2, \forall x \in \mathcal{C}$; thus x_0 is bounded, and, by continuity of $\|.\|_2$, a minimizing point exists. x_0 is also on the boundary of $\mathcal{C}$, since it is only necessary to consider x in the intersection of $\mathcal{C}$ and the sphere centred at $\nabla f(x')$, with radius given by $\|x_0 - \nabla f(x')\|_2$. We show that $d = x_0 - \nabla f(x')$ satisfies the lemma. Let $x \in \mathcal{C}$. By convexity, we have

$$x_0 + \alpha(x - x_0) \in \mathcal{C}, \quad \alpha \in [0, 1],$$

and thus

$$\|x_0 + \alpha(x - x_0) - \nabla f(x')\|_2^2 \geqslant \|x_0 - \nabla f(x')\|_2^2.$$

Expanding,

$$2\alpha \langle x_0 - \nabla f(x'), x - x_0 \rangle + \alpha^2 \|x - x_0\|_2^2 \geqslant 0.$$

Considering this as $\alpha \to 0+$, we obtain

$$\langle x_0 - \nabla f(x'), x - x_0 \rangle \geqslant 0. \tag{3.5}$$

† Such a set is also known as a *polyhedral cone*.

Given (3.4), it is clear that, since $x_0 \in \mathcal{C}, \beta x_0 \in \mathcal{C}, \forall \beta \geqslant 0$. Furthermore,

$$\|\beta x_0 - \nabla f(x')\|_2^2$$

has a minimum at $\beta = 1$. It follows that

$$0 = \left.\frac{\mathrm{d}\|\beta x_0 - \nabla f(x')\|_2^2}{\mathrm{d}\beta}\right|_{\beta=1} = \langle x_0, x_0 - \nabla f(x') \rangle. \tag{3.6}$$

Using this in (3.5), we have

$$\langle x_0 - \nabla f(x'), x \rangle \geqslant 0.$$

Thus, for $d = x_0 - \nabla f(x')$, we have

$$\langle d, x \rangle \geqslant 0, \quad \forall x \in \mathcal{C}.$$

From (3.6), we also have

$$0 = \langle x_0 - \nabla f(x') + \nabla f(x'), x_0 - \nabla f(x') \rangle = \langle d, d \rangle + \langle d, \nabla f(x') \rangle.$$

Since $\nabla f(x') \notin \mathcal{C}, d \neq 0$. Thus,

$$\langle d, \nabla f(x') \rangle < 0,$$

and the hyperplane $\langle d, x \rangle$ separates $\mathcal{C}$ and the vector $\nabla f(x')$. □

Lemma 3.4 (Farkas' Lemma: Farkas, 1901)
Given the vectors x' and $\nabla f(x'), \nabla g^j(x')$, $1 \leqslant j \leqslant e, \nabla h^i(x')$, $1 \leqslant i \leqslant i$, then

$$\{d \mid \langle d, \nabla f(x') \rangle < 0; \langle \nabla g^j(x'), d \rangle = 0; \langle \nabla h^i(x'), d \rangle \leqslant 0; 1 \leqslant j \leqslant e; 1 \leqslant i \leqslant i\} = \varnothing \tag{3.7}$$

if and only if there exist multipliers $\lambda^j, \mu^i (1 \leqslant j \leqslant e; 1 \leqslant i \leqslant i)$, with $\mu^i \geqslant 0$, such that

$$\nabla f(x') = -\left[\sum_{j=1}^{e} \nabla g^j(x') \lambda^j + \sum_{i=1}^{i} \nabla h^i(x') \mu^i\right].$$

Proof
First, we shall dispense with the equality constraints by observing that if the lemma holds for inequality constraints, we can reformulate the equality constraints as two sets of inequality constraints

$$\langle \nabla g^j(x'), d \rangle \leqslant 0, \qquad -\langle \nabla g^j(x'), d \rangle \leqslant 0,$$

and rewrite the relationship as

$$\nabla f(x') = -\left[\sum_{j=1}^{e} \nabla g^j(x') \lambda_+^j - \sum_{j=1}^{e} \nabla g^j(x') \lambda_-^j + \sum_{i=1}^{i} \nabla h^i(x') \mu^i\right]$$

for $\lambda^j_+, \lambda^j_- \geqslant 0$. Defining $\lambda^j = \lambda^j_+ - \lambda^j_-$ yields the original relationship. Thus, we ignore the equality constraints and only consider the case for inequality constraints, and establish the result

$$\{d \mid \langle d, \nabla f(x') \rangle < 0; \langle \nabla h^i(x'), d \rangle \leqslant 0; 1 \leqslant i \leqslant \mathcal{I}\} = \varnothing \tag{3.8}$$

if and only if there exist multipliers $\mu^i \geqslant 0$, $1 \leqslant i \leqslant \mathcal{I}$ such that

$$\nabla f(x') = -\sum_{i=1}^{\mathcal{I}} \nabla h^i(x') \mu^i. \tag{3.9}$$

Suppose there exist such multipliers. Then, by (3.9),

$$\langle d, \nabla f(x') \rangle = -\left\langle d, \sum_{i=1}^{\mathcal{I}} \nabla h^i(x') \mu^i \right\rangle.$$

If $\langle d, \nabla f(x') \rangle < 0$ then $\langle \nabla h^i(x'), d \rangle > 0$, for some i or if $\langle \nabla h^i(x'), d \rangle \leqslant 0$, $\forall i$, then $\langle d, \nabla f(x') \rangle \geqslant 0$. Thus, (3.8), and consequently (3.7), is satisfied.

Conversely, suppose (3.9) with $\mu^i \geqslant 0$ does not hold. It can then be shown that (3.8) or (3.7) do not hold, since there exists a vector d satisfying the inequalities $\langle d, \nabla f(x') \rangle < 0$, $\langle \nabla h^i(x'), d \rangle \leq 0$; $1 \leqslant i \leqslant \mathcal{I}$. If $\nabla f(x') \notin \mathcal{C}$ then, by Lemma (3.3), there exists a hyperplane $\langle d, x \rangle$ that separates $\mathcal{C}$ and $\nabla f(x')$ such that $\langle d, \nabla f(x') \rangle < 0$, $\langle \nabla h^i(x'), d \rangle \leqslant 0, 1 \leqslant i \leqslant \mathcal{I}$. Thus, a vector d exists. □

Theorem 3.1 (Karush–Kuhn–Tucker first-order necessary conditions: Karush, 1939; Kuhn and Tucker, 1951)
Let x_* be a local minimum for problem (1.6) and let x_* be a regular point of the constraints. Then, there is a vector $\lambda \in \mathbb{R}^{\mathcal{E}}$ and a vector $\mu \in \mathbb{R}^{\mathcal{I}}$ and $\mu \geqslant 0$ such that

$$\nabla f(x_*) + \nabla g(x_*)\lambda + \nabla h(x_*)\mu = 0, \tag{3.10}$$

$$\langle \mu, h(x_*) \rangle = 0. \tag{3.11}$$

Proof
The result is a consequence of Lemmas (3.1)–(3.4). We note that since $\mu \geqslant 0$ and $h(x_*) \leqslant 0$, (3.11) simply states that a component of μ may be nonzero if the corresponding constraint is active. x_* is a local minimum of the original problem as well as that obtained by treating the active constraints as equality constraints. Consequently, (3.10) holds with $\mu^i = 0$ if $h(x_*) \neq 0$. □

Definition
Condition (3.11) signifies that, at the optimum, $\mu^i h^i = 0$. This is known as the *complementarity condition*, which ensures that both μ^i and h^i cannot both be nonzero, or, equivalently, that inactive constraints have a zero multiplier. Furthermore, if there is no constraint such that $\mu^i = h^i = 0$ then *strict complementarity* holds.

The case $\mu^i = h^i = 0$ occurs when the constraint in question is not necessary to characterize the optimum point but happens to pass through the optimum. A simple example is the constraint passing through the unconstrained optimum of an objective function, making the latter the solution of the constrained problem as well.

Example
Consider the problem

$$\min\{(x^1-1)^2+(x^2-2)^2 \mid x^1+2x^2 \leqslant 5\}.$$

The solution $(1,2)$ also coincides with the unconstrained optimum of the objective function. From the optimality condition (3.10), we have

$$\nabla f(x_*) + \nabla h(x_*)\mu = 0 = \begin{bmatrix} 0 \\ 0 \end{bmatrix} + \begin{bmatrix} 1 \\ 2 \end{bmatrix}\mu$$

and

$$\mu = 0 = x_*^1 + 2x_*^2 - 5.$$

In the presence of equality constraints only, we can introduce the Lagrangian defined as

$$L(x,\lambda) = f(x) + \langle g(x), \lambda \rangle$$

and express the first-order conditions as

$$\nabla_x L(x,\lambda) = 0 \quad \text{and} \quad \nabla_\lambda L(x,\lambda) = 0.$$

To express the second-order conditions, we introduce the Lagrangian for the inequality-constrained problem

$$L(x,\lambda,\mu) = f(x) + \langle g(x), \lambda \rangle + \langle h(x), \mu \rangle \tag{3.12}$$

and let $\nabla^2 L(x,\lambda,\mu)$ denote the Hessian of (3.12) with respect to x.

Theorem 3.2 (Second-order necessary conditions)
Let f, g, $h \in \mathbb{C}^1$ and let x_* be a regular point of the constraints (2.1). If x_* is a local minimum of problem (1.6) then there is a $\lambda \in \mathbb{R}^e$ and a $\mu \in \mathbb{R}^i$, $\mu \geq 0$ such that (3.10) and (3.11) hold and that the Hessian

$$\nabla^2 L(x_*,\lambda,\mu) = \nabla^2 f(x_*) + \sum_{j=1}^{e} \nabla^2 g^j(x_*)\lambda^j + \sum_{i=1}^{i} \nabla^2 h^i(x_*)\mu^i \tag{3.13}$$

is positive-semidefinite on the subspace (see CN.2.2)

$$\mathscr{V} = \{v \mid v \neq 0; \nabla g^{\mathrm{T}}(x_*)v = 0; \langle \nabla h^i(x_*), v\rangle = 0, i \in g(x_*)\}.$$

Proof
Consider only the equality constraints and inequality constraints active at x_*. For inequality constraints not active, i.e. $i \notin \mathcal{G}(x_*)$, we have $\mu^i = 0$. Hence, their Hessian does not influence $\nabla^2 L(x_*, \lambda, \mu)$. Let $x(t)$ be a differentiable curve on the surface of the equality and active inequality constraints at x_*. From calculus, for every twice-differentiable curve on the surface defined by these constraints, passing through x_*, with $x(0) = x_*$, we have

$$\left.\frac{\mathrm{d}^2 f(x(t))}{\mathrm{d}t^2}\right|_{t=0} \geqslant 0. \tag{3.14}$$

By definition,

$$\left.\frac{\mathrm{d}^2 f(x(t))}{\mathrm{d}t^2}\right|_{t=0} = \left\langle \frac{\mathrm{d}x(0)}{\mathrm{d}t}, \nabla^2 f(x_*)\frac{\mathrm{d}x(0)}{\mathrm{d}t} \right\rangle + \left\langle \nabla f(x_*), \frac{\mathrm{d}^2 x(0)}{\mathrm{d}t^2} \right\rangle. \tag{3.15}$$

Furthermore, differentiating $\langle g(x_*), \lambda\rangle + \langle h(x_*), \mu\rangle = 0$ twice, and remembering that only the equality and active inequality constraints are influential, we obtain

$$\left\langle \frac{\mathrm{d}x(0)}{\mathrm{d}t}, \left[\sum_{j=1}^{e} \nabla^2 g^j(x_*)\lambda^j + \sum_{i \in \mathcal{J}(x_*)} \nabla^2 h^i(x_*)\mu^i\right] \frac{\mathrm{d}x(0)}{\mathrm{d}t} \right\rangle$$
$$+ \left\langle \nabla g(x_*)\lambda + \nabla h(x_*)\mu, \frac{\mathrm{d}^2 x(0)}{\mathrm{d}t^2} \right\rangle = 0. \tag{3.16}$$

Adding (3.15) to (3.16) yields, in view of (3.10) and (3.14),

$$\frac{\mathrm{d}^2 f(x(t))}{\mathrm{d}t^2}\Big|_{t=0} = \left\langle \frac{\mathrm{d}x(0)}{\mathrm{d}t}, \nabla^2 L(x_*, \lambda, \mu)\frac{\mathrm{d}x(0)}{\mathrm{d}t} \right\rangle \geqslant 0.$$

Since $\mathrm{d}x(0)/\mathrm{d}t$ is arbitrary in $\mathcal{V}$, we have the required result. □

Similar to the second-order condition for a minimum in ordinary calculus, the second-order sufficiency condition essentially requires (2.9) to be positive-definite. However, there is a complication due to inequality constraints that are active at the solution (i.e. $h^i(x_*) = 0$) but have *zero multipliers* (i.e. $\mu^i = 0$). To handle such *degenerate* constraints, we require that the Hessian be positive-definite in a subspace that only includes active constraints with strictly positive multipliers.

Theorem 3.3 (Second-order sufficiency conditions)
Let f, g, $h \in \mathbb{C}^2$ and let x_* be a regular point of the constraints (2.1). Sufficient conditions that x_* is a local minimum of problem (1.6) is that there is a $\lambda \in \mathbb{R}^e$ and a $\mu \in \mathbb{R}^i$, $\mu \geqslant 0$ such that (3.10) and (3.11) hold and that the Hessian (3.13) is positive-definite on the subspace

$$\mathscr{V}_0 = \{v \mid v \neq 0; \nabla g^{\mathrm{T}}(x_*)v = 0; \langle \nabla h^i(x_*), v\rangle = 0, i \in \mathscr{I}_0\}$$

where

$$\mathscr{I}_0 = \{i \mid h^i(x_*) = 0, \mu^i > 0\}.$$

Proof

If x_* is not a local minimum point, as in Lemma 3.2, there exists a sequence of feasible points $\{x_k\} \to x_*$, such that $f(x_k) \leqslant f(x_*)$. Let $\|d_k\| = 1$, and, as in (3.2) $x_k = x_* + \tau_k d_k, \tau_k > 0$. We have $\{\tau_k\} \to 0$, and, since $\{d_k\}$ is bounded, the sequence $\{d_k\}$ has a subsequence converging to d. We simply assume for convenience that $\{d_k\} \to d$. For the feasible sequence $\{x_k\} \to x_*$, we have

$$g(x_k) - g(x_*) = 0,$$

and for each active inequality constraint

$$h^i(x_k) - h^i(x_*) \leqslant 0.$$

By a similar argument to Lemma (3.2), dividing by τ_k and letting $k \to \infty$, we obtain

$$\langle \nabla f(x_*), d\rangle \leqslant 0, \quad \nabla_g^{\mathrm{T}}(x_*)d = 0, \quad \langle \nabla h^i(x_*), d\rangle \leqslant 0.$$

The second-order Taylor series expansions of $f(x), g^j(x), h^i(x), 1 \leqslant j \leqslant e$, $i \in \mathscr{I}_0$, about x_*, are given by

$$0 \geqslant f(x_k) - f(x_*) = \tau_k\langle \nabla f(x_*), d_k\rangle + \tau_k^2\left[\int_0^1 (1-t)\langle d_k, \nabla^2 f(x_k + t\tau_k d_k)(d_k)\rangle \mathrm{d}t\right], \tag{3.17}$$

$$0 = g^j(x_k) = g^j(x_*) + \tau_k\langle \nabla g^j(x_*), d_k\rangle + \tau_k^2\left[\int_0^1 (1-t)\langle d_k, \nabla^2 g^j(x_k + t\tau_k d_k)(d_k)\rangle \mathrm{d}t\right], \tag{3.18}$$

$$0 \geqslant h^i(x_k) - h^i(x_*) = \tau_k\langle \nabla h^i(x_*), d_k\rangle + \tau_k^2\left[\int_0^1 (1-t)\langle d_k, \nabla^2 h^i(x_k + t\tau_k d_k)(d_k)\rangle \mathrm{d}t\right]. \tag{3.19}$$

We need to consider two cases, both of which lead to a contradiction.

(i) If $\langle \nabla h^i(x_*), d\rangle = 0, \forall i \in \mathscr{I}_0$, and thence $d \in \mathscr{V}_0$, multiplying (3.18) by λ^j and (3.19) by μ^i, adding these to (3.17) and using (3.10), leads to

$$0 \geqslant \tau_k^2 \int_0^1 (1-t)\left\langle d_k, \left[\nabla^2 f(x_k + t\tau_k d_k) + \sum_{j=1}^{e} \nabla^2 g^j(x_k + t\tau_k d_k)\lambda^j \right.\right.$$

$$\left.\left. + \sum_{i \in \mathscr{I}_0} \nabla^2 g^i(x_k + t\tau_k d_k)\mu^j \right](d_k)\right\rangle \mathrm{d}t$$

which, as $k \to \infty$, contradicts the strict positive definiteness condition on the Hessian of (3.12).

(ii) If $\langle \nabla h^i(x_*), d\rangle < 0$, for some $i \in \mathscr{I}_0$, and thence $d \notin \mathscr{V}_0$, then

$$0 \geqslant \langle \nabla f(x_*), d\rangle = -\langle \lambda, \nabla_g^{\mathrm{T}}(x_*)d\rangle - \sum_{i \in \mathscr{I}_0} \langle \nabla h^i(x_*), d\rangle > 0.$$

which is also a contradiction. □

4 A SIMPLE GAUSS–NEWTON ALGORITHM FOR OPTIMAL DECISION PROBLEMS

We return to the dynamic decision problem (1.2) to discuss a simple algorithm that can be used for approximate solutions of large-scale problems. The approximate solution is often used as part of a *search procedure to aid the cognitive process in multiple-objective optimization*. It is in the course of this process that the decision maker gains an increasingly concrete knowledge of what can be done and formulates the trade-offs and targets.

The approach adopted uses the model of the system, $g(Y, U) = 0$, to eliminate the output or endogenous variables if, associated with $g(Y, U) = 0$, there is a model solver that, given U, computes the value of Y. We discuss algorithms for this computational mapping in Chapter 9 and denote the mapping by

$$Y = \mathscr{Y}(U). \tag{4.1}$$

We use (4.1) to eliminate the output variables from the problem and express (1.2) as

$$\min\{\mathscr{F}(U)\}, \tag{4.2a}$$

where

$$\mathscr{F}(U) = f(\mathscr{Y}(U), U). \tag{4.2b}$$

In optimal decision making, the objective function is chosen to reflect the aims and preferences of the decision maker. A quadratic function has often been considered sufficient for this purpose (see Chapters 3–5; Bray, 1972; 1982; Chow, 1975; Holbrook, 1974).

For the *multiple-objective decision problem*, we assume that the objective function is a quadratic such as (1.1c) and write this in a slightly more general form as

$$\frac{1}{2}\langle Y - Y^{\mathrm{d}}, (Y - Y^{\mathrm{d}})\rangle + \frac{1}{2}\langle U - U^{\mathrm{d}},\ \mathfrak{Q}_u(U - U^{\mathrm{d}})\rangle, \tag{4.3}$$

where $\mathfrak{Q}_y, \mathfrak{Q}_u$ are symmetric matrices, with $\mathfrak{Q}_y$ is positive-semidefinite and $\mathfrak{Q}_u$ positive-definite; $Y^{\mathrm{d}}, U^{\mathrm{d}}$ are given desired or bliss values of the multiple objectives. The diagonal elements of $\mathfrak{Q}_y$, $\mathfrak{Q}_u$ penalize the deviation of a variable from its bliss value. The off-diagonal elements measure the importance of the deviation in one variable versus the deviation in another. The determination of the matrices and the bliss values are discussed in Chapters 3–5 in the context of multiple-objective decision making.

When the model is used to eliminate Y from the objective function (4.3), we have

$$\mathcal{F}(U) = \frac{1}{2}\langle \mathcal{Y}(U) - Y^{\mathrm{d}}, \mathfrak{Q}_y(\mathcal{Y}(U) - Y^{\mathrm{d}})\rangle + \frac{1}{2}\langle U - U^{\mathrm{d}}, \mathfrak{Q}_u(U - U^{\mathrm{d}})\rangle. \quad (4.4)$$

Starting from an initial point U_0, the algorithm generates the sequence $\{U_k\}, k = 0, 1, \ldots,$ where

$$U_{k+1} = U_k + \tau_k d_k^0. \quad (4.5)$$

The direction d_k^0 is defined by

$$\mathfrak{Q}_k^0 d_k^0 = -\nabla\mathcal{F}(U_k), \quad \text{with} \quad \mathfrak{Q}_k^0 = N_k^{\mathrm{T}}\mathfrak{Q}_y N_k + \mathfrak{Q}_u, \quad (4.6)$$

$$\nabla\mathcal{F}(U_k) = N_k^T\mathfrak{Q}_y(\mathcal{Y}(U_k) - Y^{\mathrm{d}}) + \mathfrak{Q}_u(U_k - U^{\mathrm{d}})$$

$$N_k = \left.\frac{\partial\mathcal{Y}}{\partial U}\right|_{U=U_k} = \left.\frac{\partial Y}{\partial U}\right|_{U=U_k} \quad (4.7)$$

Newton's algorithm for the same problem would compute d_k as

$$\mathfrak{Q}_k d_k = -\nabla\mathcal{F}(U_k), \quad (4.8)$$

$$\mathfrak{Q}_k = \mathfrak{Q}_k^0 + B_k, \quad (4.9)$$

$$B_k = \sum_{j=1}^{y\times\mathcal{F}} \nabla^2\mathcal{Y}^j(U_k)\left[\mathfrak{Q}_y(\mathcal{Y}(U_k) - Y^d)\right]^j, \quad (4.10)$$

where j indicates the jth element of the vector and $\nabla^2\mathcal{Y}^j(U_k)$ is the Hessian of $\mathcal{Y}^j$ with respect to U, at U_k. This term is not generally available for some models, and ignoring it leads to the choice $\mathfrak{Q}_k^0$ and thence to the generalized *damped Gauss–Newton method* discussed below (Ortega and Rheinboldt, 1970).

In nonlinear least squares problems, with small $\|\mathcal{Y}(U_*) - Y^{\mathrm{d}}\|, B_k$ is expected to vanish at the solution. Thus, the Gauss–Newton algorithm works well for these cases. However, in the optimal decision making problem above, the term

$$\mathfrak{Q}_y[(\mathcal{Y}(U_k) - Y^{\mathrm{d}})]^j$$

does not normally vanish at the solution. Therefore, G_k^0 cannot be regarded as an accurate Hessian approximation near the solution of the problem. This may cause problems when the omitted B_k is important relative to G_k^0 (Dennis and Schnabel, 1983). However, as an inaccurate, yet inexpensive, method it has been widely used (see e.g. Rustem and Zarrop, 1979, 1981; Karakitsos and Rustem, 1985). Another advantage of the method is that it is easily implementable for complex models, $g(Y, U) = 0$, with which the simplest interaction is through the mapping $Y = \mathscr{Y}(U)$.

The stepsize $\tau_k \in [0, 1]$ is determined using an Armijo (1966) strategy to satisfy

$$\mathscr{F}(U_{k+1}) - \mathscr{F}(U_k) \leqslant \rho\tau_k\langle\nabla\mathscr{F}(U_k), d_k^0\rangle, \qquad \rho \in (0, 1) \tag{4.11}$$

(see Armijo, 1966). This algorithm is a generalization of Holbrook's approach of linearizing the model and solving the quadratic optimization subject to the linearized model (Holbrook, 1974; Chow, 1975).

5 A QUASI-NEWTON IMPLEMENTATION

The evaluation of the Jacobian N_k is computationally expensive for large systems. As a feasible, but not very accurate, alternative it may be approximated by using Broyden's rank-one formula

$$N_{k+1} = N_k + \frac{\left[\mathscr{Y}(U_{k+1}) - \mathscr{Y}(U_k) - \tau_k N_k d_k^0\right]\left[d_k^0\right]^{\mathrm{T}}}{\tau_k\langle d_k^0, d_k^0\rangle}. \tag{5.1}$$

Causality in a dynamic system such as $g(Y, U) = 0$ requires that decision or control variables affect the present and future, but not the past (see CN.11). Hence, the Jacobian N_k is required to be of block lower-triangular structure. To preserve this causality–time model structure, Schubert's (1970) modification to the above formula can be used. This updating is accomplished as follows:

Definition
The row vector n_k^j is defined to be the jth row of the matrix N_k.

Definition
For the jth row of N_k, define the column vector d_k^j derived from d_k^0 by setting equal to zero those elements in d_k^0 that correspond to constant (e.g. zero) values in n_k^j.

The resulting updating formula for those elements of N_k that are not constant is given for each row of N_{k+1}, since d_k^j depends on the row being calculated. Hence,

$$n_{k+1}^j = n_k^j + \frac{[\mathscr{Y}^j(U_{k+1}) - \mathscr{Y}^j(U_k) - \tau_k n_k^j d_k^0][d_k^j]^{\mathrm{T}}}{\tau_k\langle d_k^j, d_k^j\rangle} \tag{5.2}$$

may be used to compute the jth row of N_{k+1}, $j = 1, \ldots, y \times \mathscr{T}$. The convergence properties of Schubert's modification have been discussed in Broyden (1971) and Marvill (1978).

The approximation of the model Jacobian N_k is an important issue for the algorithm below. Even for solutions of systems of equations with a Newton-type algorithm, discussed in Chapter 9, the Jacobian is required to be periodically evaluated (Moré and Trangenstein, 1976). Experiments have shown that a computationally acceptable strategy is to reevaluate N_k only when a descent direction (see CN.12) cannot be obtained using the current N_k (Becker and Rustem, 1993; Li, 1989; Rustem and Zarrop, 1979, 1981).

Algorithm 5.1

Step 0: Given $Y^d, U^d, \mathfrak{Q}_y, \mathfrak{Q}_u, U_0$, compute $Y_0 = \mathscr{Y}(U_0)$. Set $k = 0, q_{\max}$ to the largest element of $\mathfrak{Q}_y, \phi = q_{\max}\epsilon_0$, where ϵ_0 is the accuracy of the computational mapping $Y = \mathscr{Y}(U)$. Set the small positive numbers ν, δ and $\rho \in (0, 1), \bar{\tau} \in (0, 1)$.

Step 1: Compute N_k by numerical differentiation.

Step 2: Compute $\nabla\mathscr{F}(U_k)$.

Step 3: Optimality check: if

$$\begin{aligned} \|\nabla\mathscr{F}(U_k)\| \leqslant \phi, \quad \text{or if} \quad k > 0, \quad \|U_{k-1} - U_k\| \leqslant \nu \\ \text{and} \quad \mathscr{F}(U_{k-1}) - \mathscr{F}(U_k) \leqslant \phi, \end{aligned} \tag{5.3}$$

compute $N_k, \nabla\mathscr{F}(U_k)$ in Steps 1–3, and if (5.3) is still satisfied, stop.

Step 4: Solve (4.6) to compute d_k^0.

Step 5: Compute the stepsize τ_k: set τ_k to the first member of the sequence $\{(\bar{\tau})^j; j = 0,1,\ldots\}$ such that (4.11) is satisfied. If

$$\tau_k < \delta, \tag{5.4}$$

compute Steps 1–5 and if (5.4) is again satisfied, stop.

Step 6: Set $U_{k+1} = U_k + \tau_k d_k^0$.

Step 7: If $\|Y_{k+1} - Y_k\| \leqslant \epsilon_0$, go to Step 8. Otherwise, update N_k using (5.1) or (5.2) and compute $\mathfrak{Q}_{k+1}^0$.

Step 8: Set $k = k + 1$, go to Step 2.

In numerical experiments discussed in Rustem and Zarrop (1981), Karakitsos and Rustem (1985) and Westcott *et al.* (1981), Algorithm 5.1 was used with $\phi = \nu = 10^{-4}, \tau = \rho = 0.1, \epsilon_0 = 5 \times 10^{-4}, \rho = 10^{-4}$. The restriction $\|Y_{k+1} - Y_k\| > \epsilon_0$ is imposed to avoid updating N_k when the difference $Y_{k+1} - Y_k$ is dominated by the errors in computing $Y_{k+1} = \mathscr{Y}(U_{k+1})$ and $Y_k = \mathscr{Y}(U_k)$.

Algorithm 5.1 uses N_k computed by (5.1) or (5.2) when a descent direction can be obtained. If (4.11) is not satisfied for $\tau_k > \delta$, with the approximation given by (5.1) or (5.2), N_k is re-evaluated by numerical differentiation. If, furthermore, N_k is re-evaluated periodically after a number of iterations, the convergence of the $\{U_k\}$ to the solution follow from the convergence of the damped Gauss–Newton theorem (Ortega and Rheinboldt, 1970, Theorem 14.4.4; Dennis and Schnabel, 1983).

The aim of the algorithm is to avoid computing N_k numerically for as long as progress towards the solution is possible. Using the updating formulae (5.1) and (5.2) is one option. Another is to leave N_k unaltered as long as the algorithm is able to progress. Most of the computation involved is inexpensive relative to the cost of numerically evaluating N_k. Hence, leaving N_k unchanged is a desirable strategy in practice.

6 COMMENTS AND NOTES

CN.1 Vectors and matrices

Let $\mathbb{R}^n$ be the set of n-dimensional vectors. Any $x \in \mathbb{R}^n$ has n components, or coordinates. Generally, we refer to the ith component of x as x^i. A *null vector* refers to $x^i = 0, \forall i$, and is denoted by $x = 0$. Vectors in $\mathbb{R}^n$ are viewed as column vectors, unless the contrary is stated explicity. The transpose of x is a row vector and is denoted by x^{T}.

A matrix A is a rectangular array of numbers. For any A, we use a_{ij}, or $[A]_{ij}$ as the ijth element of A. For $i = 1, \ldots, n$ and $j = 1, \ldots, m$, A is referred to as an $(n \times m)$ matrix, with n rows and m columns. When $n = m$ then the matrix A is referred to as a *square* matrix, and if $a_{ij} = 0, \forall\, i, j$, then $A = 0$. The transpose of the $(n \times m)$ matrix A is denoted by A^{T} and refers to the $(m \times n)$ matrix whose elements are given by a_{ij}. A matrix is *symmetric* if $A = A^{\mathrm{T}}$. The sum of two $(m \times n)$ matrices A and B is written as $A + B$ and is the $(m \times n)$ matrix whose elements are the sum of the corresponding elements in A and B. The product of a matrix A and scalar τ, $\tau A = A\,\tau$, is obtained by multiplying each element of A by τ. The product of a $(m \times n)$ matrix A and a $(n \times q)$ matrix B is written as AB and is the $(m \times q)$ matrix F whose elements are given by $[F]_{ij} = \sum_{l=1}^{n} [A]_{il}[B]_{lj}$. Finally, as in the vector case, a $(m \times n)$ matrix A is considered to be a member of the space $\mathbb{R}^{n \times m}$ such that $A \in \mathbb{R}^{n \times m}$.

CN.2 Endogenous variables

Variables determined by the system are endogenous. In contrast to decision, or control, variables chosen by the decision maker and the exogenous variables such as random effects or variables determined outside the system, endogenous variables are determined within the system. World energy prices are generally considered exogenous. Inflation and unemployment are considered to be endogenous variables. In the present setting, the endogenous variables can also be interpreted as the state, as well as output, variables of the dynamic system. The subsequent discussion does not make a distinction between these concepts.

CN.3 Inner product, vector and matrix norms

For any two vectors $x, y \in \mathbb{R}^n$, their *inner product* is defined by

$$\langle x, y \rangle = x^{\mathrm{T}} y = \sum_{i=1}^{n} x^i y^i.$$

Any two vectors x, $y \in \mathbb{R}^n$ satisfying $\langle x, y \rangle = 0$ are called *orthogonal*.

A vector *norm* $\|\cdot\|$ on $\mathbb{R}^n$ is a mapping that assigns a real scalar to every $x \in \mathbb{R}^n$, and has the properties:

(i) $\|x\| \geqslant 0, \quad \forall x \in \mathbb{R}^n$;
(ii) $\|\tau x\| = |\tau| \, \|x\|$ for every $\tau \in \mathbb{R}^1$ and every $x \in \mathbb{R}^n$;
(iii) $\|x\| = 0$, if and only if $x = 0$;
(iv) $\|x + z\| \leqslant \|x\| + \|y\|, \quad \forall x, y \in \mathbb{R}^n$ (i.e. the triangle inequality).

The *Euclidean norm* is defined by

$$\|x\|_2 = \langle x, x \rangle^{1/2} = \left(\sum_{i=1}^{n} (x^i)^2 \right)^{1/2}$$

and satisfies the Cauchy–Schwartz inequality

$$|\langle x, y \rangle| \leqslant \|x\|_2 \|y\|_2.$$

The *matrix norms* we consider belong to the class of norms that are induced by corresponding vector norm. Given any norm $\|\cdot\|$ for a vector x, the induced norm for a $(n \times m)$ matrix A is given by

$$\|A\| = \max_{\{x \in \mathbb{R}^n | \, \|x\| = 1\}} \|Ax\|.$$

We note that

$$\|A\| = \max_{\|x\| \neq 0} \frac{\|Ax\|}{\|x\|},$$

which follows because for any $x \neq 0$, x can be scaled so that its norm is equal to unity without changing the value of $\|Ax\|/\|x\|$. Furthermore, we have

$$\|Ax\| \leqslant \|A\| \, \|x\| \quad \text{and} \quad \|AB\| \leqslant \|A\| \, \|B\|,$$

which follows from the above for B of appropriate dimension. A detailed discussion on matrix norms can be found in Golub and van Loan (1983) and Horn and Johnson (1993).

CN.4 Differentiable function definitions

If the function f has continuous first partial derivatives with respect to x, this is denoted by $f \in \mathbb{C}^1$. If f has continuous second partial derivatives with respect to x, this is denoted by $f \in \mathbb{C}^2$.

CN.5 First- and second-order Taylor expansions

The *first-order Taylor series expansion* of a function f: $\mathbb{R}^n \to \mathbb{R}^1$, and $f \in \mathbb{C}^1$, for any $x, d \in \mathbb{R}^n$, $\theta \in \mathbb{R}^1$, is given by

$$f(x+\theta d) = f(x) + \theta \int_0^1 \langle \nabla f(x + t\theta d), d \rangle \, \mathrm{d}t.$$

The *second-order Taylor series expansion* of a function f: $\mathbb{R}^n \to \mathbb{R}^1$, and $f \in \mathbb{C}^2$, for any $x, d \in \mathbb{R}^n$, $\theta \in \mathbb{R}^1$, is given by

$$f(x+\theta d) = f(x) + \theta \langle \nabla f(x), d \rangle + \theta^2 \left(\int_0^1 (1-t) \langle d, \nabla^2 f(x + t\theta d) d \rangle \, \mathrm{d}t \right).$$

CN.6 Positive-semidefinite and -definite matrices

The matrix $\nabla^2 f$ is *positive-semidefinite* if, for all $0 \neq v \in \mathbb{R}^n$, $\langle v, [\nabla^2 f] v \rangle \geqslant 0$. The matrix $\nabla^2 f$ is *positive-definite* if, for all $0 \neq v \in \mathbb{R}^n$, $\langle v, [\nabla^2 f] v \rangle > 0$.

CN.7 Linear independence of vectors

A set of vectors $v_1, v_2, \cdots, v_s$ is said to be *linearly dependent* if there are scalars $\omega_1, \omega_2, \cdots, \omega_s$, not all zero, such that $\sum_{i=1}^{s} v_i \omega_i = 0$. If no such $\omega_1, \omega_2, \cdots, \omega_s$ exist, the vectors $v_1, v_2, \cdots, v_s$ is said to be *linearly independent*. A further discussion of this subject is given in Chapter 2.

CN.8 Lipschitz continuity of a function

A function $f(x) : \mathscr{D} \subset \mathbb{R}^n \to \mathbb{R}^p$ is said to be *Lipschitz-continuous* on $\mathscr{D}_0 \subset \mathscr{D}$ if there is a constant $c \geqslant 0$ such that, for all $x, y \in \mathscr{D}$,

$$\|f(x) - f(y)\| \leqslant c \|x - y\|$$

CN.9 Closed sets

A set $\mathscr{C}$ is said to be *closed* if every point arbitrarily close to $\mathscr{C}$ is a member of $\mathscr{C}$. Equivalently, $\mathscr{C}$ is closed if it contains each of its limit points. Thus, for example, the set

$$\{x \mid \|x\| \leqslant 1\}$$

is closed. A further discussion on this is given in Chapter 6.

CN.10 Hyperplanes

A *hyperplane* is essentially the n-dimensional generalization of a three-dimensional plane. Let the column vector $a \in \mathbb{R}^n$ and let c be a scalar. Then, the set $\{x \in \mathbb{R}^n \mid \langle a, x \rangle = c\}$ is a hyperplane in $\mathbb{R}^n$.

CN.11 Dynamic multipliers

In dynamic models, the possibility of approximating N_k with the model dynamic multipliers has been used as a successful strategy (see Holbrook, 1974; Rustem and Zarrop, 1979; 1981). The elements of $N = \partial Y/\partial U|_{U=U_0}$ are

$$\frac{\partial y_{il}}{\partial u_{jt}}, \quad 1 \leqslant t, \ell \leqslant \mathscr{T}, \quad 1 \leqslant i \leqslant y, \quad 1 \leqslant j \leqslant u.$$

It is assumed that

$$\frac{\partial y_{i\ell}}{\partial u_{jt}} = 0, \quad \forall \ell < t,$$

which implies that changes in the decision variable values can only affect current and future endogenous endogenous values. For constructing N with the dynamic multipliers, it is further assumed that

$$\frac{\partial y_{i\ell}}{\partial u_{jt}} = M_{ij}(\ell - t), \quad \forall t \leqslant \ell.$$

The matrix $M(\tau)$ is the dynamic multiplier at lag τ, and the above equality is only true for a time-invariant model. However, it avoids the numerical differentiation of the model equations, since N given by

$$N = \begin{bmatrix} M(0) & 0 & \dots & 0 \\ M(1) & M(0) & \dots & 0 \\ \vdots & \ddots & & 0 \\ M(\mathscr{T}-1) & M(\mathscr{T}-2) & \dots & M(0) \end{bmatrix}$$

can be numerically constructed by solving the model (i.e. computing $\mathscr{Y}(U)$) u times, rather than $u \times \mathscr{T}$ times.

CN.12 Motivation for descent

In general, at a point x_k, d_k is a descent direction such that the objective function decreases for small steps taken along d_k. Consider a linear approximation of $f(x)$,

$$f(x_k + \tau_k d_k) \cong f(x_k) + \tau_k \langle \nabla f(x_k), d_k \rangle.$$

For small τ_k, this approximation is fairly accurate and

$$f(x_k + \tau_k d_k) \leqslant f(x_k)$$

provided

$$\langle \nabla f(x_k), d_k \rangle \leqslant 0.$$

Consider the problem of determining the direction d_k that satisfies the last inequality. Let

$$d_k = \arg\min\{f(x_k) + \langle \nabla f(x_k), d\rangle \mid \|d\|_2 = 1\},$$

where d is required to be of unit, or fixed, length to ensure a well-defined solution. The first-order optimality conditions for this problem are

$$\nabla f(x_k) + 2\lambda d_k = 0, \quad \|d_k\|_2^2 = 1.$$

Premultiplying the first by d_k and applying the constraint, we have

$$\lambda = -\frac{1}{2}\langle \nabla f(x_k), d_k\rangle.$$

Since $2\lambda\langle \nabla f(x_k), d_k\rangle = -\langle \nabla f(x_k), \nabla f(x_k)\rangle$,

$$\lambda^2 = \frac{1}{4}\langle \nabla f(x_k), \nabla f(x_k)\rangle,$$

and, since the minimization of $f(x_k) + \langle \nabla f(x_k), d\rangle$ is being considered, this yields

$$\lambda = \frac{1}{2}\|\nabla f(x_k)\|_2 \quad \text{and} \quad d_k = -\frac{\nabla f(x_k)}{\|\nabla f(x_k)\|_2},$$

where d_k is known as the *steepest-descent* direction.

If, instead of the linear we assume a quadratic approximation to $f(x)$, we have

$$f(x_k) + \langle \nabla f(x_k), d\rangle + \frac{1}{2}\langle d, \nabla^2 f(x_k), d\rangle.$$

For $\nabla^2 f(x_k)$ positive-definite, the quadratic is minimized by d_k satisfying the first-order optimality condition

$$\nabla f(x_k) + \nabla^2 f(x_k) d_k = 0,$$

or $d_k = -[\nabla^2 f(x_k)]^{-1}\nabla f(x_k)$. This is the *Newton direction*, whose descent property,

$$\langle d_k, \nabla f(x_k)\rangle \leqslant 0,$$

can easily be verified.

REFERENCES

Apostol, T. (1981). *Mathematical Analysis*, 2nd Edition. Addison-Wesley, Reading, MA.

Armijo, L. (1966). Minimization of functions having Lipschitz-continuous first partial derivatives. *Pacific J. Math.* **16**, 1–3.

Becker, R. and B. Rustem (1993). Algorithms for solving nonlinear dynamic decision models. *Ann. Oper. Res.* **44**, 117–142.

Berge, C. (1963). *Topological Spaces.* Macmillan, New York.
Bertsekas, D. and E. Gafni (1982). Projection methods for variational inequalities with application to the traffic assignment problem. *Math. Programming Study* **17**, 139–159.
Bray, J. (1972). Some considerations in the practice of economic management. *PREM Discussion Paper* No. 2.
Bray, J. (1982). *Production, Purpose and Structure.* Frances Pinter, London.
Broyden, C. G. (1971). The convergence of an algorithm for solving sparse nonlinear systems, *Mathematics of Computation*, **25**, 285–294.
Chow, G. C. (1975). *Analysis and Control of Dynamic Economic Systems.* Wiley, New York.
Dafremos, S. (1983). An iterative scheme for variational inequalities. *Math. Programming* **26**, 40–47.
Dafremos, S. (1990). Exchange price equilibria and variational inequalities. *Math. Programming* **46**, 391–402.
Darlington, J., C. C. Pantelides, B. Rustem and B.A. Tanyi (1996). An efficient approximate algorithm for optimal robust decisions under uncertainty. Department of Computing, Imperial College.
Dennis, J. B. and R. B. Schnabel (1983). *Numerical Methods for Unconstrained Optimization and Nonlinear Equations.* Prentice- Hall, Englewood Cliffs, NJ.
Farkas, J. (1901). Uber die Theorie der einfachen Ungleichungen. *J. Reine Angew. Math.* **124**, 1–24.
Fletcher R., (1981). *Practical Methods of Optimization*, Vol. 2, Wiley, Chichester.
Glowinski, R. (1984). *Numerical Methods for Variational Problems.* Springer-Verlag, Berlin.
Glowinski, R., J. L. Lions and R. Trémolières (1981). *Numerical Analysis of Variational Inequalities.* North Holland, Amsterdam.
Golub, G. H. and C. F. van Loan (1983). *Matrix Computations.* North Oxford Academic, Oxford.
Harker, P. and J. S. Pang (1990). Finite dimensional variational inequality and nonlinear complementarity problems: a survey of theory and algorithms. *Math. Programming* **48**, 161–220.
Holbrook, R. S. (1974). A practical method for controlling a large nonlinear stochastic system. *Ann. Econ. Social Measurement* **3**, 796–811.
Horn, R. A. and C. R. Johnson (1993). *Matrix Analysis.* Cambridge University Press.
Karakitsos E. and B. Rustem (1985). Optimal fixed rules and simple feedback laws for nonlinear econometric models. *Automatica* **21**, 169–180.
Karush, W. (1939). Minima of Functions of Several Variables with Inequalities as Side Constraints. MS Thesis, Department of Mathematics, University of Chicago.
Kuhn, H. W. and A. W. Tucker (1951). Nonlinear programming. In *Proceedings of the Second Berkeley Symposium on Mathematical Statistics and Probability* (J. Neyman, ed.). University of California Press, Berkeley.
Li, G. (1989). Successive column correction algorithms for solving sparse nonlinear systems of equations. *Math. Programming* **43**, 187–208.
Luenberger, D. G. (1984). *Linear and Nonlinear Programming.* Addison-Wesley, Reading, MA.
Marvill E. S., (1978). Exploiting sparsity in Newton-type methods, PhD Thesis Cornell University.
Moré J. and J. A. Trangenstein (1976). On the global convergence of Broyden's method. *Math. Comput.* **30**, 523–540.
Nagurney, A. (1993). *Network Economics: A Variational Inequality Approach.* Kluwer, Dordrecht.

Ortega J. M. and W. C. Rheinboldt (1970). *Iterative Solution to Nonlinear Equations in Several Variables*. Academic Press, London.

Rustem, B. (1981). *Projection Methods in Constrained Optimisation and Applications to Optimal Policy Decisions*. Springer–Verlag, Berlin.

Rustem, B. (1984). A class of superlinearly convergent projection algorithms with relaxed stepsizes. *Appl. Math. Optim.* **12**, 29–43.

Rustem, B. (1990). Methods for optimal policy design. In *Advances in Control and Dynamic Systems*, Vol. 36 (C.T. Leondes, ed.), pp. 17–74. Academic Press, Orlando.

Rustem, B. and M. Zarrop (1979). Newton-type methods for the optimization and control of nonlinear econometric models. *J. Econ. Dynam. Control* **1**, 288–300.

Rustem, B. and M. Zarrop (1981). A quasi-Newton algorithm for the control of large nonlinear econometric models. *Large Scale Systems* **2**, 105–111.

Schubert, L. K. (1970). Modification of a quasi-Newton algorithm for nonlinear equations with a sparse Jacobian. *Math. Comput.* **24**, 27–30.

Westcott, J., E. Karakitsos, B. Rustem and R. G. Becker (1981). Memorandum of Evidence on Monetary Policy to the Select Committee on the Treasury and Civil Service. *House of Commons: Third Report from the Treasury and Civil Service Committee (1980–81) Monetary Policy*. HMSO, HC 1980–81, 163–III, 22–67.

PART II

Quadratic Programming Algorithms and Multi-Objective Optimization

CHAPTER 2

Quadratic Programming Algorithms

Quadratic programming is the optimization of a quadratic function subject to linear equality and inequality constraints. It arises in several problems covered in this book.

In multiple objective decision making, quadratic programming can be used to evaluate the decision in terms of a single quadratic objective function. This is essentially a weighted quadratic norm, measuring the departure of the actual decisions from their corresponding ideal, or bliss, values. The decision can be computed by minimizing this norm, subject to linear equality and inequality constraints. The question of determining the weights, the bliss values and related issues, are discussed in Chapters 3–5.

The formulation of mean-variance optimization of uncertain systems also leads to quadratic programming. An important application of mean variance is in simple optimal portfolio problems where the constraints are linear and the objective function is quadratic (Markowitz, 1959). We discuss this in Chapter 10 and its nonlinear formulation in Chapter 11. The multiple objective aspect is that the decision maker has to reconcile the conflicting desires of maximizing expected portfolio return, represented by the linear term, and minimizing the portfolio risk, represented by the quadratic (variance) term, in the objective function.

In general nonlinear optimization problems, discussed in Chapters 6, 7 and 12, sequential quadratic programming algorithms require the solution of quadratic programming subproblems to generate descent directions.

We consider three approaches to quadratic programming. The first is based on generating a basis for the null space of the constraints active at a point. The basis is used to obtain feasible directions of improvement. The second is based on the corresponding range space. The third is a model interior point algorithm that aims to solve directly the optimality conditions of the inequality constrained problem, including the complementarity conditions.

1 THE QUADRATIC PROGRAMMING PROBLEM

Consider the quadratic program

$$\min \{\langle x - x^{\mathrm{d}}, \mathfrak{Q}(x - x^{\mathrm{d}})\rangle \mid \mathscr{H}^{\mathrm{T}} x \leqslant h\} \tag{1.1}$$

where x^{d} is the desired, or bliss, value of $x \in \mathbb{R}^n$, $\mathfrak{Q}$ is a symmetric positive-semidefinite matrix penalizing the deviations of x from x^{d}. The matrix $\mathscr{H} \in \mathbb{R}^{n \times i}$ and the column vector $h \in \mathbb{R}^i$ form the linear inequality constraints. We study the multiple objective optimization problem (1.1) in Chapters 3–5. The matrix $\mathfrak{Q}$ is sometimes assumed to be positive-definite. Throughout this chapter, the assumption regarding $\mathfrak{Q}$ is explicitly stated.

In Chapters 6 and 7 we study nonlinear optimization algorithms that require the sequential solution of quadratic programming subproblems of the type

$$\min\left\{\langle a, x\rangle + \frac{1}{2}\langle x, \mathfrak{Q}x\rangle \,\middle|\, \mathscr{H}^{\mathrm{T}}x \leqslant h\right\}, \tag{1.2}$$

where $a \in \mathbb{R}^n$ is a constant vector. When $\mathfrak{Q}$ is positive-definite, it is clear that (1.1) is equivalent to (1.2) and that (1.2) can be formulated as (1.1) using $x^{\mathrm{d}} = -\mathfrak{Q}^{-1}a$, which is the unconstrained optimum of the objective function of (1.2).

There are numerous methods for solving quadratic programming problems. In this chapter, we only consider a subset. The first, and historically most important, class of methods are those relying on an active set strategy. These methods identify a feasible point that lies on some of the inequality constraints. They generate a descent direction that lies in the intersection of all the constraints currently satisfied as equalities. A step along this direction is either restricted by another inequality constraint or eventually reaches the optimum within the intersection of the constraints currently satisfied as equalities. If a constraint is encountered then the constraint is added to the active set and the optimum is sought in the intersection of the new set of constraints. If the optimum in the intersection of currently active constraints is attained then the multipliers at this optimum indicate which of the currently active constraints can be dropped from the active set in order to generate a direction of further progress. Thus, a sequence of equality constraints is solved, each considering a subset of the constraints as active, or satisfied as equalities. For non- degenerate problems (see CN.1), each equality-constrained problem leads to an improvement in the quadratic objective function. Thus, it is not possible to return to the same active set. As there are a finite number of active sets, convergence is ensured in a finite number of steps.

There exist two types of active set methods, depending on the subspace in which the direction of progress is generated (see CN.2). One is based on the range space of the matrix of active constraints and the other on the null space (see CN.3). We discuss the former in Section 2 and the latter in Section 3. The basic operation and progress of both algorithms are similar, and are highlighted in Algorithm 2.1. In Section 4, we discuss more recent methods based on interior point barrier functions. These introduce a logarithmic barrier function to penalize the transgression of the inequality constraints. The algorithms then solve the optimality condition of the overall problem by applying a nonlinear equation solver.

2 ACTIVE SET ALGORITHM BASED ON NULL SPACE OF THE CONSTRAINTS

Using the terminology introduced in Section 1.2, let the subset of the inequality constraints active at point x_k be represented by

$$\mathscr{I}_k = \{i \mid \langle \nabla h^i, x_k\rangle = h^i\}, \tag{2.1a}$$

and, in matrix form, we represent these active constraints as

$$\mathscr{H}_k^{\mathrm{T}} x_k = h_k, \tag{2.1b}$$

where $\mathscr{H}_k \in \mathbb{R}^{n \times i_k}$ is the coefficient matrix, $h_k \in \mathbb{R}^{i_k}$ is the corresponding column vector and i_k is the number of active constraints. Methods in this class generate a basis of the null space of $\mathscr{H}_k^{\mathrm{T}}$. Starting from a feasible point, the direction of progress towards the solution is then defined as a linear combination of the vectors spanning this null space. This approach was pioneered by Gill and Murray (1978).

Let $Z \in \mathbb{R}^{n \times (n-i_k)}$ be a matrix whose columns span the subspace orthogonal to the columns of $\mathscr{H}_k$; hence, $\mathscr{H}_k^{\mathrm{T}} Z = 0$.

Example: Upper and lower bounds

For upper and lower bound constraints such as $x_\ell \leqslant x \leqslant x_{\mathrm{u}}$, where x_ℓ, x_{u} are fixed vectors, we define a vector $J(x_k) \in \mathbb{R}^n$ as

$$J^i(x_k) = \begin{cases} 0, & x_k^i \neq x_\ell^i \text{ or } x_{\mathrm{u}}^i, \\ 1, & x_k^i = x_{\mathrm{u}}^i, \\ -1, & x_k^i = x_\ell^i. \end{cases}$$

Let i_k denote the number of active bounds, or equivalently the number of nonzero elements of $J(x_k)$. $Z \subset \mathbb{R}^{n \times (n-i_k)}$ is the matrix consisting of the non null columns of the diagonal matrix whose ith diagonal element is $1 - |J^i(x_k)|$. Thus, Z defines the subspace of the non-fixed (free) variables (see Gill and Murray, 1976).

There are two ways of representing the matrix Z in general. The first involves the partitioning of the matrix $\mathscr{H}_k$ as

$$\mathscr{H}_k - \begin{bmatrix} \mathscr{H}_{\mathrm{B}} \\ \mathscr{H}_{\mathrm{N}} \end{bmatrix}$$

where $\mathscr{H}_{\mathrm{B}}$ is a $i_k \times i_k$ nonsingular matrix and x is conformably partitioned as $x = [x_{\mathrm{B}}, x_{\mathrm{N}}]$. Then, the constraints (2.1) can be written as

$$\mathscr{H}_{\mathrm{B}}^{\mathrm{T}} x_{\mathrm{B}} + \mathscr{H}_{\mathrm{N}}^{\mathrm{T}} x_{\mathrm{N}} = h_{\mathrm{k}},$$

and, since $\mathscr{H}_{\mathrm{B}}$ is nonsingular, we have

$$x_{\mathrm{B}} = \mathscr{H}_{\mathrm{B}}^{-\mathrm{T}} \left(h_k - \mathscr{H}_{\mathrm{N}}^{\mathrm{T}} x_{\mathrm{N}} \right).$$

The i_k variables x_{B}, corresponding to the columns of $\mathscr{H}_{\mathrm{B}}^{\mathrm{T}}$, are the endogenous variables dependent on the remaining $n - i_k$ exogenous, or independent, variables. Thus, any $x_{\mathrm{N}}, x_{\mathrm{B}}$ pair given as above satisfy the linear equality constraints $\mathscr{H}_k x = h_k$. As we express the variables x_{B} in terms of x_{N}, we are able to eliminate x_{B} from the objective function. This technique is called the *variable reduction method*, and Z given by

$$Z = \begin{bmatrix} -\mathscr{H}_{\mathrm{B}}^{-\mathrm{T}}\mathscr{H}_{\mathrm{N}}^{\mathrm{T}} \\ I \end{bmatrix}$$

satisfies the orthogonality condition

$$\mathscr{H}_k^{\mathrm{T}} Z = [\mathscr{H}_{\mathrm{B}}^{\mathrm{T}} \vdots \mathscr{H}_{\mathrm{N}}^{\mathrm{T}}] \begin{bmatrix} -\mathscr{H}_{\mathrm{B}}^{-\mathrm{T}}\mathscr{H}_{\mathrm{N}}^{\mathrm{T}} \\ I \end{bmatrix} = 0.$$

The second way of obtaining the matrix Z is based on the QR decomposition of $\mathscr{H}_k$, where $Q \in \mathbb{R}^{n\times n}$ is an orthonormal matrix such that $Q^{\mathrm{T}}Q = I$ and $R \in \mathbb{R}^{i_k \times i_k}$ is a nonsingular upper-triangular matrix such that

$$\mathscr{H}_k = Q\begin{bmatrix} R \\ 0 \end{bmatrix} \tag{2.2}$$

Let Q be partitioned into two submatrices

$$Q = [Y \vdots Z],$$

where $Y \in \mathbb{R}^{n\times i_k}$ and $\mathrm{Z} \in \mathbb{R}^{n\times (n-i_k)}$ and $\mathscr{H}_k^{\mathrm{T}} Z = 0$. As Q is orthonormal, we have

$$Z^{\mathrm{T}}Z = I \in \mathbb{R}^{(n-i_k)\times(n-i_k)}.$$

The QR decomposition of the matrix can be obtained using either Householder transformations or Givens rotations (see CN.4).

To generate the basic search direction, let x_k be a feasible point, with active constraints (2.1), and let the direction be given by $d_k = \tilde{x} - x_k$. We consider the first-order necessary conditions of the equality-constrained problem

$$\min\left\{ \langle a, x\rangle + \frac{1}{2}\langle x, \mathfrak{Q}x\rangle \,\middle|\, \mathscr{H}_k^{\mathrm{T}} x = h_k \right\} \tag{2.3}$$

given by

$$a + \mathfrak{Q}\tilde{x} + \mathscr{H}_k\,\mu_{k+1} = 0, \tag{2.4a}$$

$$\mathscr{H}_k^{\mathrm{T}}\tilde{x} = h_k. \tag{2.4b}$$

These can be written as

$$a + \mathfrak{Q}x_k + \mathfrak{Q}d_k + \mathscr{H}_k\,\mu_{k+1} = 0, \tag{2.5}$$
$$\mathscr{H}_k^{\mathrm{T}} d_k = 0,$$

and d_k satisfying $\mathscr{H}_k^{\mathrm{T}} d_k = 0$ can be expressed as a linear combination of the columns of Z (which span the null space of $\mathscr{H}_k^{\mathrm{T}}$). Thus, we have $d_k = Z\omega$ and $\mathscr{H}_k^{\mathrm{T}} Z\omega = 0$. Hence, premultiplying (2.5) by Z^{T} yields

$$Z^{\mathrm{T}}\mathfrak{Q}Z\omega = -Z^{\mathrm{T}}(a + \mathfrak{Q}x_k). \tag{2.6}$$

Thus, d_k is given by

$$d_k = -Z(Z^{\mathrm{T}}\mathfrak{Q}Z)^{-1}Z^{\mathrm{T}}(a + \mathfrak{Q}x_k). \tag{2.7}$$

The matrix $Z^{\mathrm{T}}\mathfrak{Q}Z$ is referred to as the projected Hessian, or the projection of $\mathfrak{Q}$ on the intersection of the constraints $\mathscr{H}_k^{\mathrm{T}}x = h_k$.† Thus, $Z^{\mathrm{T}}\mathfrak{Q}Z$ is positive-definite if the Hessian $\mathfrak{Q}$ is positive-definite in the intersection of the constraints. The positive-definiteness of $\mathfrak{Q}$ clearly ensures that $Z^{\mathrm{T}}\mathfrak{Q}Z$ is positive-definite as well. However, $Z^{\mathrm{T}}\mathfrak{Q}Z$ may be positive-definite even when $\mathfrak{Q}$ is not (see Gill, Murray and Wright, 1981). Also, since $Z^{\mathrm{T}}\mathfrak{Q}Z \in \mathbb{R}^{(n-i_k)\times(n-i_k)}$, the dimension of the linear equations that need to be solved in (2.6), or equivalently size of the matrix that needs to be inverted to compute d_k, reduces as the number of the active constraints increases.

To establish that d_k is a *descent direction*, that is, starting from x_k, a positive step along direction d_k produces a reduction in the quadratic objective function, we consider the objective function at

$$x(\tau) = x_k + \tau d_k, \tag{2.8}$$

where $\tau \in (0, 1]$ is the stepsize to be determined. From (2.8), we have

$$\begin{aligned}\langle a, x(\tau)\rangle &+ \frac{1}{2}\langle x(\tau), \mathfrak{Q}x(\tau)\rangle \\ &= \langle a, x_k\rangle + \frac{1}{2}\langle x_k, \mathfrak{Q}x_k\rangle + \tau_k\langle d_k, a + \mathfrak{Q}x_k\rangle + \frac{1}{2}\tau_k^2\langle d_k, \mathfrak{Q}d_k\rangle.\end{aligned}$$

It follows from (2.7) that

$$\begin{aligned}\langle d_k, \mathfrak{Q}x_k + a\rangle &= -\langle (a + \mathfrak{Q}x_k), Z(Z^{\mathrm{T}}\mathfrak{Q}Z)^{-1}Z^{\mathrm{T}}(a + \mathfrak{Q}x_k)\rangle, \\ \langle d_k, \mathfrak{Q}d_k\rangle &= \langle (a + \mathfrak{Q}x_k), Z(Z^{\mathrm{T}}\mathfrak{Q}Z)^{-1}Z^{\mathrm{T}}(a + \mathfrak{Q}x_k)\rangle.\end{aligned}$$

As the columns of Z are linearly independent and $Z^{\mathrm{T}}\mathfrak{Q}Z$ is positive-definite, we have

$$\langle a, x(\tau)\rangle + \frac{1}{2}\langle x(\tau), \mathfrak{Q}x(\tau)\rangle < \langle a, x_k\rangle + \frac{1}{2}\langle x_k, \mathfrak{Q}x_k\rangle. \tag{2.9}$$

As the objective function is quadratic, the unconstrained optimum along the direction d_k, in the intersection of the constraints in the active set $\mathscr{I}_k$, is located where the derivative

$$\frac{\mathrm{d}(\langle a, x(\tau)\rangle + \frac{1}{2}\langle x(\tau), \mathfrak{Q}x(\tau)\rangle)}{\mathrm{d}\tau} = 0. \tag{2.10}$$

For d_k given by (2.7), this is achieved at $\tau = 1$. Thus, proceeding on the line segment from x_k to $x(1) = x_k + d_k$, if a constraint, not in $\mathscr{I}_k$, is encountered then the stepsize has to be shortened to ensure that this constraint is not

† A discussion of projections can be found in Chapter 3.

transgressed. The distance from x_k, along d_k, to all constraints not in $\mathcal{I}_k$ is given by the value of τ that satisfies $\langle \nabla h^j, x(\tau)\rangle = h^j$, $j \notin \mathcal{I}_k$. To ensure feasibility, the smallest value of τ, over all those constraints, is required, and hence

$$\tau_k = \min_{j \notin \mathcal{I}_k} \left\{ 1, \frac{h^j - \langle \nabla h^j, x_k\rangle}{\langle \nabla h^j, d_k\rangle} \middle| \langle \nabla h^j, d_k\rangle > 0 \right\}. \tag{2.11}$$

The condition $\langle \nabla h^j, d_k\rangle > 0$ indicates the constraints likely to be violated, so that only those are considered. From (2.11), we can set $x_{k+1} = x_k + \tau_k d_k$. If $\tau_k < 1$, the constraints that have been encountered are added to the active set, thus forming $\mathcal{I}_{k+1}$. If further descent is possible in the intersection of the constraints in $\mathcal{I}_{k+1}$, this is done by generating descent directions similar to (2.7).

If the optimum in the intersection of a given set of active constraints is attained then the multipliers at that point indicate if one of the currently active constraints can be dropped from $\mathcal{I}_k$ in order to generate a further descent direction.

To establish the basic procedure and results for dropping a constraint from the active set, let $\mathcal{I}_k, \mathcal{H}_k, h_k$ be obtained by *dropping the* ith *constraint from* $\mathcal{I}_{k-1}$, and hence the ith column and element of $\mathcal{H}_{k-1}$ and h_{k-1} respectively. We write the first-order necessary conditions (2.4a,b) for the equality-constrained problem with active set $\mathcal{I}_{k-1}$ as

$$a + \mathfrak{Q}x_k = -\mathcal{H}_{k-1}\mu_k, \tag{2.12a}$$

$$\mathcal{H}_{k-1}^{\mathrm{T}} x_k = h_{k-1}. \tag{2.12b}$$

The same conditions for the equality-constrained problem (2.3), with active set $\mathcal{I}_k$, are

$$a + \mathfrak{Q}x_{k+1} = -\mathcal{H}_k\, \mu_{k+1}, \tag{2.12c}$$

$$\mathcal{H}_k^{\mathrm{T}} x_{k+1} = h_k. \tag{2.12d}$$

The multipliers μ_{k+1} are evaluated as the least squares solution to (2.5). When the columns of $\mathcal{H}_k$ are linearly independent, this is given by

$$\mu_{k+1} = -(\mathcal{H}_k^{\mathrm{T}} \mathcal{H}_k)^{-1} \mathcal{H}_k^{\mathrm{T}} (a + \mathfrak{Q}x_k + \mathfrak{Q}d_k). \tag{2.13}$$

We show in Section 3 below that this choice of μ_{k+1} is consistent with the direction d_k and (2.12c) (see CN.5).

The following is a summary of the basic null-space algorithm. The statement of the range space algorithm discussed in Section 3 is basically the same, except that the direction d_k and the multiplier μ_{k+1} are evaluated using an alternative formulation.

Algorithm 2.1

Step 0: Given an initial feasible point x_0, identify the active set $\mathcal{I}_0$. Set $k = 0$.

Step 1: Solve the equality-constrained quadratic programming problem (2.3) to obtain $d_k = \tilde{x} - x_k$. If $\|d_k\| = 0$, **go to Step 3**.

Step 2: Set $x_{k+1} = x_k + \tau_k d_k$, where τ_k is determined by (2.11). If $\tau_k = 1$, **go to Step 3**. Otherwise, add the new constraint determined by (2.11) to the active set $\mathscr{I}_k$ to form $\mathscr{I}_{k+1}$. Set $k = k+1$ and **go to Step 1**.

Step 3: Compute the multipliers μ_{k+1} using (2.13). Determine the minimum element of μ_{k+1}. If this element $\mu^i_{k+1} < 0$, drop the ith constraint from the active set $\mathscr{I}_k$ to obtain $\mathscr{I}_{k+1}$, set $k = k+1$, **go to Step 1**. If $\mu^i \geqslant 0, x_{k+1}$ is optimal; **stop**.

Clearly, $d_k = 0$ occurs when Algorithm 2.1 has reached the optimum $\tilde{x} = x_k$ in the intersection of the currently active constraints. A realistic implementation would also need to consider the possibility of linear dependence of the columns of $\mathscr{H}_k$ (despite our earlier assumption). Also, it needs to check if new constraints have become active even with $\tau_k = 1$ and add such constraints to $\mathscr{I}_{k+1}$ in **Step 3**. The multipliers need to be checked to see if, by dropping any of the currently active constraints, a further feasible descent direction can be generated. Finally, as the active set changes, it is necessary to compute Z as well as other matrices involved in the computation of d_k and μ_{k+1}. Gill *et al.* (1974) discuss numerically stable methods for modifying these matrices.

Proposition 2.1
Let $\mathfrak{Q}$ be positive-definite. If there exists a negative element of μ_k, say $\mu^i_k < 0$, and $\mathscr{I}_{k+1}, \mathscr{H}_{k+1}, h_{k+1}$, are obtained by deleting the constraint corresponding to that μ^i_k from the active set, then

(i) *d_k solving the equality constrained problem in **Step 1** with active set $\mathscr{I}_{k+1}$ determined in **Step 3**, is a descent direction at x_k and hence $\langle d_k, (a+ \mathfrak{Q}x_k)\rangle < 0$;*

(ii) *d_k in **Step 1** satisfies the ith constraint dropped from $\mathscr{I}_k$ in **Step 3**, and hence x_{k+1} is feasible for the* ith *constraint;*

(iii) *if in **Step 3**, μ_k does not have a negative element then x_k satisfies the first-order necessary conditions of the quadratic programming problem.*

Proof
To establish (i), we write

$$\begin{aligned}\langle d_k, (a + \mathfrak{Q}x_k)\rangle &- \langle d_k, (a + \mathfrak{Q}x_k - \mathfrak{Q}x_{k+1} + \mathfrak{Q}x_{k+1})\rangle \\ &= \langle d_k, ((a + \mathfrak{Q}x_{k+1}) - \mathfrak{Q}d_k)\rangle \\ &< \langle d_k, (a + \mathfrak{Q}x_{k+1})\rangle \qquad (2.14)\\ &= 0, \qquad (2.15)\end{aligned}$$

where the last equality is due to (2.12c, d).

To establish (ii), consider descent property (2.14) and (2.12a)

$$0 > \langle d_k, (a + \mathfrak{Q}x_k)\rangle = -\langle d_k, \mathscr{H}_k \mu_k\rangle. \tag{2.16}$$

From (2.12b, d), we note that $\mathscr{H}_{k+1}^T d_k = 0$, and $\mathscr{H}_k$ only has one additional column corresponding to the ith constraint, dropped in $\mathscr{H}_k$. Thus, (2.16) becomes

$$0 > -\mu^i \langle \nabla h^i, d_k\rangle, \tag{2.17}$$

and to ensure (2.17), with $\mu^i < 0$, we must have $\langle \nabla h^i, d_k\rangle < 0$.

When μ_k does not have any negative elements, and hence $\mu_k \geqslant 0$, we note from the optimality conditions of Chapter 1 that the first-order necessary conditions of the quadratic programming problem are satisfied at (x_k, μ_k). □

Corollary
Let $\mathfrak{Q}$ be positive-semidefinite. Then

(i) in general, the strict descent condition is replaced by

$$\langle d_k, (a + \mathfrak{Q}x_k)\rangle \leqslant 0;$$

(ii) if $Z^T \mathfrak{Q} Z$ is positive-definite, strict descent is ensured.

Proof
For (i), inequality (2.14) becomes

$$\langle d_k, (a + \mathfrak{Q}x_k)\rangle \leqslant \langle d_k, (a + \mathfrak{Q}x_{k+1})\rangle = 0.$$

For (ii), $d_k = Z\omega$ and $\langle d_k, \mathfrak{Q}d_k\rangle = \langle \omega, Z^T \mathfrak{Q} Z\omega\rangle > 0$. Thus, we have the strict inequality in (2.14):

$$\langle d_k, (a + \mathfrak{Q}x_k)\rangle < 0. \qquad □$$

In practice, it is usual to choose the constraint to be dropped from $\mathscr{I}_k$ as the one corresponding to the negative multiplier with the largest magnitude. This ensures the largest decrease in the objective function per unit step along d_k.

We have assumed that each $\mathscr{I}_k$ is characterized by a set of constraints that are linearly independent. Linear dependence can be taken into account by specialized techniques beyond the scope of this chapter. One simple way of handling such degeneracy is to eliminate linearly dependent constraints from the initial formulation. Barring degeneracy, the algorithm proceeds from active set to active set, by either adding further constraints or dropping a constraint. As the objective function is reduced by every such move, return to the same active set is not possible. The number of constraints is finite, and therefore so is the number of possible active sets. Thus, the algorithm converges in a finite number of steps.

3 ACTIVE SET ALGORITHM BASED ON RANGE SPACE OF THE CONSTRAINTS

If the number of linear constraints is large then the null space of $\mathscr{H}_k^T$ is relatively small, and therefore characterizing the solution to the problem in terms of this null space is of special importance. This is discussed in Section 2. If the constraints are relatively few then it is convenient to use the range space of $\mathscr{H}_k^{\mathrm{T}}$ to construct the solution of the equality-constrained problem. To do this, we write the optimality conditions (2.4) as

$$\mathfrak{Q}d_k + \mathscr{H}_k\,\mu_{k+1} = -(a + \mathfrak{Q}x_k), \tag{3.1}$$

$$\mathscr{H}_k^{\mathrm{T}} d_k = 0. \tag{3.2}$$

If $\mathfrak{Q}$ is positive-definite then d_k in (3.1) is given by

$$d_k = -\mathfrak{Q}^{-1}[\mathscr{H}_k\,\mu_{k+1} + (a + \mathfrak{Q}x_k)]. \tag{3.3a}$$

Moreover, if the columns of $\mathscr{H}_k$ are linearly independent, premultiplying (3.3a) by $\mathscr{H}_k^{\mathrm{T}}$ and using (3.2), μ_{k+1} is given by

$$\mu_{k+1} = -\left(\mathscr{H}_k^{\mathrm{T}}\mathfrak{Q}^{-1}\mathscr{H}_k\right)^{-1}\mathscr{H}_k^{\mathrm{T}}\mathfrak{Q}^{-1}(a + \mathfrak{Q}x_k). \tag{3.3b}$$

Expression (3.3) can also be obtained by considering (3.1), (3.2) as a system

$$\begin{bmatrix} \mathfrak{Q} & \mathscr{H}_k \\ \mathscr{H}_k^{\mathrm{T}} & 0 \end{bmatrix}\begin{bmatrix} d_k \\ \mu_{k+1} \end{bmatrix} = -\begin{bmatrix} (a + \mathfrak{Q}x_k) \\ 0 \end{bmatrix},$$

with the inverse of the coefficient matrix on the left given by

$$\begin{bmatrix} \mathfrak{Q}^{-1} - \mathfrak{Q}^{-1}\mathscr{H}_k\left(\mathscr{H}_k^{\mathrm{T}}\mathfrak{Q}^{-1}\mathscr{H}_k\right)^{-1}\mathscr{H}_k^{\mathrm{T}}\mathfrak{Q}^{-1} & \mathfrak{Q}^{-1}\mathscr{H}_k\left(\mathscr{H}_k^{\mathrm{T}}\mathfrak{Q}^{-1}\mathscr{H}_k\right)^{-1} \\ \left(\mathscr{H}_k^{\mathrm{T}}\mathfrak{Q}^{-1}\mathscr{H}_k\right)^{-1}\mathscr{H}_k^{\mathrm{T}}\mathfrak{Q}^{-1} & \left(\mathscr{H}_k^{\mathrm{T}}\mathfrak{Q}^{-1}\mathscr{H}_k\right)^{-1} \end{bmatrix}.$$

In practice, the orthogonal factorization of $\mathscr{H}_k$ can be used to devise a version of the range-space algorithm that involves orthogonal matrices, thereby injecting an element of numerical stability (Stewart, 1973; Golub and van Loan, 1983). Consider the QR factorization of $\mathscr{H}_k$ in (2.2) and the submatrices of Q,

$$\mathscr{H}_k = [Y \vdots Z]\begin{bmatrix} R \\ 0 \end{bmatrix} = YR.$$

Substituting this in (3.3), we have

$$d_k = -\mathfrak{Q}^{-1}[Y\theta + (a + \mathfrak{Q}x_k)], \tag{3.4a}$$

$$\theta = -\left(Y^{\mathrm{T}}\mathfrak{Q}^{-1}Y\right)^{-1}Y^{\mathrm{T}}\mathfrak{Q}^{-1}(a + \mathfrak{Q}x_k). \tag{3.4b}$$

The multiplier μ_{k+1} can be recovered using the relation $R\mu_{k+1} = \theta$.

We now establish the equivalence of d_k given by (2.7) and (3.3a) and μ_{k+1} given by (2.13) and (3.3b)

Proposition 3.1
Let $\mathfrak{Q}$ be positive-definite and the columns of $\mathscr{H}_k$ linearly independent. Then d_k given by (2.7) and (3.3a) and μ_{k+1} given by (2.13) and (3.3b) are equivalent.

Proof
Since d_k given by (3.3a) satisfies $\mathscr{H}_k^{\mathrm{T}} d_k = 0$, it follows that d_k can be expressed as a linear combination of the columns of Z, which form a basis of the null space of $\mathscr{H}_k^{\mathrm{T}}$. Thus, $d_k = Z\omega$, for some ω. Then, using (3.3a), we have

$$Z^{\mathrm{T}}\mathfrak{Q}Z\omega = Z^{\mathrm{T}}\mathfrak{Q}d_k = -Z^{\mathrm{T}}[\mathscr{H}_k\mu_{k+1} + (a + \mathfrak{Q}x_k)] = -Z^{\mathrm{T}}(a + \mathfrak{Q}x_k),$$

and thus

$$d_k = Z\omega = -Z\left(Z^{\mathrm{T}}\mathfrak{Q}Z\right)^{-1}Z^{\mathrm{T}}(a + \mathfrak{Q}x_k),$$

which establishes the equivalence with (2.7).

Having established the equivalence of d_k, we use d_k given by (3.3a), which satisfies the optimality condition (3.1), with μ_{k+1} given by (3.3b), to show that this multiplier is equivalent to the one given by (2.13). We start with μ_{k+1} given by (2.13) and use d_k given by (3.3a) to establish equivalence with (2.13):

$$\begin{aligned}\mu_{k+1} &= -\left(\mathscr{H}_k^{\mathrm{T}}\mathscr{H}_k\right)^{-1}\mathscr{H}_k^{\mathrm{T}}(a + \mathfrak{Q}x_k + \mathfrak{Q}d_k) \\ &= \left(\mathscr{H}_k^{\mathrm{T}}\mathscr{H}_k\right)^{-1}\mathscr{H}_k^{\mathrm{T}}\left(\mathscr{H}_k\mu_{k+1}\right) = \mu_{k+1}.\end{aligned}$$

□

The basic algorithm, due to Fletcher (1971) and Goldfarb (1972), is the same as the null-space method, discussed in Algorithm 2.1, with the exception that d_k, μ_{k+1} are evaluated using (3.3) or (3.4).

4 INTERIOR POINT ALGORITHM FOR QUADRATIC PROGRAMMING

The active set strategies discussed in Section 2 and 3 are essentially similar to the strategy employed by the simplex algorithm, developed by G. Dantzig in 1947, for linear programming (Dantzig, 1963). These algorithms proceed on the surface of the polygon described by the linear constraints, generating a sequence of active sets until the active set corresponding to the optimal solution is reached. The main difference between linear and quadratic programming problems is that the optimal solution of the latter may lie in the interior of the linear inequalities, in which case the solution is the unconstrained optimum of the quadratic function, or on one of the sides of the constraint polygon. By contrast, the solution of a linear programming problem is generally at a vertex. Although the simplex algorithm performs quite satisfactorily in solving linear programming problems, there is the theoretical danger that in searching for the optimum vertex, it may visit all, or almost all feasible vertices of the problem. Indeed, Klee and Minty (1972) construct examples in which this occurs. In such cases, the number of vertices visited, and consequently the time taken, by the algorithm grows exponentially with the number of variables and constraints. In an effort to justify the observed satisfactory behaviour of the simplex algorithm, its average speed has been studied using a probabilistic framework (Smale, 1983). Attempts to design a linear programming algorithm to solve the problem in polynomial time (i.e. the solution time for any linear programming problem bounded above by a polynomial function of the size of the problem) have resulted in the ellipsoid algorithm of Khachian (1979) and the projective method of Karmarkar (1984). We describe Khachian's algorithm briefly in Chapter 5 (Section 4). In this section, we discuss a simple extension of Karmarkar's algorithm to solve quadratic programming problems.

Karmarkar's projective algorithm has been shown by Gill *et al.* (1986) to be an application of a barrier function method to linear (and quadratic) programming. The inequality-constrained problem is reformulated as an optimization of a penalty function constrained by equality constraints only. The optimality conditions of the reformulation can be interpreted as the perturbed optimality conditions of the original problem. Karmarkar's algorithm can then be seen as the application of the Newton algorithm to the system of nonlinear equations formed by the perturbed optimality conditions. The perturbation is gradually allowed to disappear, leading to the solution of the original optimality conditions.

Let the objective function be given by

$$q(x) = \langle a, x \rangle + \frac{1}{2}\langle x, \mathfrak{Q}x \rangle, \quad \mathcal{G} \in \mathbb{R}^{n \times e}, \quad g \in \mathbb{R}^{e}, \quad e < n,$$

and consider the quadratic programming problem

$$\min \{ q(x) \mid \mathcal{G}^{\mathrm{T}} x = g; x \geqslant 0 \}. \tag{4.1}$$

The original quadratic programming problem (1.2) can be reformulated as (4.1) by simply introducing nonnegative slack variables to transform the inequalities in (1.2) to equalities (see CN.6). For some $\eta > 0$, we define the *barrier function*

$$\ell_\eta(x) \equiv q(x) - \eta \sum_{j=1}^{n} \ln x^j \tag{4.2}$$

and the equality-constrained optimization problem

$$\min \left\{ \langle a, x \rangle + \frac{1}{2}\langle x, \mathfrak{Q}x \rangle - \eta \sum_{j=1}^{n} \ln x^j \,\middle|\, \mathcal{G}^{\mathrm{T}} x = g \right\}. \tag{4.3}$$

The function $\ell_\eta(x)$ is well defined for $x^j > 0$. It is referred to as a barrier function since it has a positive singularity at the constraint boundary (as x^j approaches the boundary of the feasible region). Many barrier functions have been proposed. We consider the logarithmic barrier function, first suggested by Frisch (1955). As the weight η assigned to the singularities approaches zero, the solution of (4.3) approaches the solution of the original problem (4.1). Methods based on barrier functions are referred to as *interior point methods*, since they require a strictly feasible initial point for each minimization, and generate a sequence of strictly feasible iterates. We introduce the properties of barrier method in this section. Detailed discussions of the general method are given by Fiacco and McCormick (1968) and Fiacco (1979).

Example: Upper and lower bounds

The barrier transformation of problem (4.1) and the associated solution algorithm can also be defined for quadratic programs with upper and lower bounds on the variables, of the form

$$\min\{q(x) \mid \mathcal{G}^{\mathrm{T}}x = g; l \leqslant \, < x \leqslant u\}.$$

The problem analogous to (4.3) is

$$\min\left\{ \langle a, x\rangle + \frac{1}{2}\langle x, \mathfrak{Q}x\rangle - \eta \sum_{j=1}^{n} \ln\,(x^j - l^j\,) + \eta \sum_{j=1}^{n} \ln\,(u^j - x^j) \,\middle|\, \mathcal{G}^{\mathrm{T}}x = g \right\}$$

(see e.g. Gill *et al.* 1986; Lustig, Marsten and Shanno, 1989, 1992).

To ensure the existence of a global constrained minimum for (4.2), we assume that $\mathfrak{Q}$ is positive-definite on the null space of $\mathcal{G}^{\mathrm{T}}$. In this case, (4.3) is a convex problem, and, by invoking the discussion in Chapter 1 (Section 2), the first-order conditions become both necessary and sufficient for optimality. Convexity also ensures that we are concerned with global optima of the barrier function ℓ_η. The main question and the underlying motivation for interior point methods is the convergence to the solution of the original problem as the barrier parameter η approaches zero. Let $\{\eta_k\}$ be a nonnegative sequence, strictly decreasing to zero, and define x_k such that

$$x_k = x(\eta_k) = \arg\min\,\{\ell_{\eta_k}(x) \mid \mathcal{G}^{\mathrm{T}}x = g\}. \tag{4.4}$$

The existence of x_k is guaranteed by the continuity of the quadratic objective and the barrier function for $x > 0$. Also, each x_k satisfies the equality constraints in (4.4) and $x > 0$. Let $x*$ denote a limit point of the uniformly bounded sequence $\{x_k\}$. Clearly, we have

$$x_* \in \{x \mid \mathcal{G}^{\mathrm{T}}x = g; x \geqslant 0\}.$$

Let

$$q_* = \min\{q(x) \mid \mathcal{G}^{\mathrm{T}}x = g; x \geqslant 0\},$$

and suppose that x_* is *not* the solution of (4.1). Then, by the basic property of q_*, we have $q(x_*) > q_*$. There is a $x_0 \in \{x \mid \mathscr{G}^{\mathrm{T}} x = g; x > 0\}$ where

$$q_* < q(x_0) < q(x_*).$$

Then

$$\liminf_{k\to\infty} \ell_{\eta_k}(x_k) \geqslant q(x_*) > q(x_0) = \lim_{k\to\infty} \ell_{\eta_k}(x_0).$$

This contradicts the assumption that x_k minimizes ℓ_η, subject to the equality constraints, for large k. Hence, $\lim_{k\to\infty} q(x_k) = q_*$, and x_* solves (4.1).

Example

Consider the problem

$$\min\{(x^1 - 1)^2 + (x^2 - 1)^2 \mid x^1 + x^2 \leqslant 1\}$$

and the barrier function formulation

$$\min\{(x^1 - 1)^2 + (x^2 - 1)^2 - \eta_k \ln(x^1 + x^2 - 1)\}.$$

The first-order conditions for an unconstrained minimum of the barrier function yield

$$(x^1 - 1) - \frac{\eta_k}{x^1 + x^2 - 1} = 0, \qquad (x^2 - 1) - \frac{\eta_k}{x^1 + x^2 - 1} = 0.$$

The solution of the above system of nonlinear equations is

$$x^1(\eta_k) = x^2(\eta_k) = \frac{3 \pm \sqrt{9 - 8(1 - \eta_k)}}{4}.$$

The negative term corresponds to the minimum. As $\eta_k \to 0$, the constrained minimum

$$x^1 = x^2 = \frac{1}{2}$$

is achieved. □

It can also be shown that as $k \to \infty$, the sequences $\{q(x_k)\}$ and $\{-\eta_k \sum_{j=1}^{n} \ln x_k^j\}$ are monotonically decreasing and increasing, respectively. Let

$$q(x_k) = q(x(\eta_k)) = q_k.$$

Because each $x_k = x(\eta_k)$ is a global minimum of ℓ_η, we have,

$$q_k - \eta_k \sum_{j=1}^{n} \ln x_k^j \leqslant q_{k+1} - \eta_k \sum_{j=1}^{n} \ln x_{k+1}^j, \tag{4.5}$$

$$q_{k+1} - \eta_{k+1} \sum_{j=1}^{n} \ln x_{k+1}^j \leqslant q_k - \eta_{k+1} \sum_{j=1}^{n} \ln x_k^j.$$

Multiplying (4.5) by η_{k+1}/η_k, adding to the second inequality yields

$$\left(1 - \frac{\eta_{k+1}}{\eta_k}\right) q_{k+1} \leqslant \left(1 - \frac{\eta_{k+1}}{\eta_k}\right) q_k.$$

Since $\eta_{k+1} < \eta_k$, the monotonic decrease of $\{q(x_k)\}$ follows. Using this result and (4.5) leads to the monotonic increase of $\{-\eta_k \sum_{j=1}^{n} \ln x_k^j\}$.

Interior point methods for quadratic programming depart slightly from the above framework by applying the Newton algorithm for solving the optimality condition of (4.3) while allowing η_k to decrease. The Newton algorithm is a general method for solving systems of nonlinear equations in many variables. It is discussed in detail in Chapter 9. In this chapter, we consider the application of the algorithm to the particular problem of solving the optimality conditions of (4.3).

The first-order conditions of (4.3) are given by the nonlinear system

$$\begin{bmatrix} \mathcal{G}^{\mathrm{T}} x - g \\ \mathcal{G}\lambda + a + \mathfrak{Q}x - \eta_k X^{-1} 1 \end{bmatrix} = \begin{bmatrix} 0 \\ 0 \end{bmatrix}, x > 0 \tag{4.6}$$

where $1 \in \mathbb{R}^n$ is the vector with all unit elements,

$$X \equiv \operatorname{diag}(x) = \begin{bmatrix} x^1 & 0 & \dots & 0 \\ 0 & x^2 & \dots & 0 \\ \vdots & \vdots & \ddots & \vdots \\ 0 & 0 & \dots & x^n \end{bmatrix}$$

and λ is the Lagrange multiplier of $\mathcal{G}^{\mathrm{T}} x = g$. By introducing the transformation $y = \eta_k X^{-1} 1$, $Y = \operatorname{diag}(y)$ and then expressing this relation in the form

$$XY1 = \eta 1,$$

we obtain the perturbed first-order conditions of the original quadratic program:

$$\begin{bmatrix} \mathcal{G}^{\mathrm{T}} x - g \\ \mathcal{G}\lambda + a + \mathfrak{Q}x - y \\ XY1 \end{bmatrix} = \begin{bmatrix} 0 \\ 0 \\ \eta_k 1 \end{bmatrix}, \quad x, \quad y > 0. \tag{4.6}.$$

With $x \geqslant 0, y \geqslant 0$ and $\eta_k = 0$, it can be verified that (4.6) is the first-order optimality condition for (4.1). It should be noted that (4.6) is not the optimality condition for (4.3) or any other smooth unconstrained or equality constrained optimization problem. The reason is that (4.6), as a first-order condition, would be the gradient of some Lagrangian function. It is easy to verify that the gradient of (4.6) itself (i.e. the Jacobian matrix) is not symmetric. If (4.6) were the gradient of a Lagrangian, its gradient (i.e. Jacobian) would be a Hessian and therefore a symmetric matrix (El-Bakry *et al.*, 1996).

To solve (4.6), as in Section 2.1, we choose the columns of $Z \in \mathbb{R}^{n\times(n-e)}$ to form a basis for the null space of $\mathcal{G}^{\mathrm{T}}$. We can eliminate the Lagrange multipliers λ from (4.6) by premultiplying the second equation by the nonsingular matrix $[\mathcal{G}\vdots Z]^{\mathrm{T}}$. Noting that $Z^{\mathrm{T}}\mathcal{G} = 0$, we obtain

$$0 = \begin{bmatrix} \mathcal{G}^{\mathrm{T}} \\ Z^{\mathrm{T}} \end{bmatrix} [\mathcal{G}\lambda + a + \mathfrak{Q}x - y] = \begin{bmatrix} \mathcal{G}^{\mathrm{T}}\mathcal{G}\lambda + \mathcal{G}^{\mathrm{T}}(a + \mathfrak{Q}x - y) \\ Z^{\mathrm{T}}(a + \mathfrak{Q}x - y) \end{bmatrix}.$$

Since $\mathcal{G}^{\mathrm{T}}\mathcal{G}$ is nonsingular, once x and y are known, λ is uniquely determined by

$$\lambda = (\mathcal{G}^{\mathrm{T}}\mathcal{G})^{-1}\left[\mathcal{G}^{\mathrm{T}}(a + \mathfrak{Q}x - y)\right].$$

Removing the equation for λ, we have the following $2n$-dimensional nonlinear system with nonnegativity constraints for x, y:

$$\begin{bmatrix} \mathcal{G}^{\mathrm{T}}x - g \\ Z^{\mathrm{T}}(a + \mathfrak{Q}x - y) \\ XY1 \end{bmatrix} = \begin{bmatrix} 0 \\ 0 \\ \eta_k 1 \end{bmatrix}, \quad x \geqslant 0, \quad y \geqslant 0. \tag{4.7}$$

We note that

$$F(x,y) \equiv \begin{bmatrix} \mathcal{G}^{\mathrm{T}}x - g \\ Z^{\mathrm{T}}(a + \mathfrak{Q}x - y) \\ XY1 \end{bmatrix} = \begin{bmatrix} 0 \\ 0 \\ 0 \end{bmatrix}, \quad x \geqslant 0, \quad y \geqslant 0, \tag{4.8}$$

denotes the first-order optimality conditions of the original quadratic program, and, as $\{\eta_k\} \to 0$, the solution of (4.7) approaches to that of (4.8).

Algorithm 4.1 is essentially aimed at solving (4.8) by successive steps taken using (4.7) while η_k is allowed to approach zero. It can be stated in compact form as follows.

Algorithm 4.1

Step 0: Given a pair $(x_0, y_0) > 0$, set $k = 0$.

Step 1: Choose $\sigma_k \in [0,1), \alpha_k \in (0,1)$ and set $\eta_k = \sigma_k\langle x_k, y_k\rangle/n$.

Step 2: Solve for d_k^x, d_k^y the system

$$\nabla F(x_k, y_k)\begin{bmatrix} d^x \\ d^y \end{bmatrix} = -F(x_k, y_k) + \begin{bmatrix} 0 \\ \eta_k 1 \end{bmatrix}. \tag{4.9}$$

Step 3: Compute the step length

$$\tau_k = \frac{-\alpha_k}{\min\{X_k^{-1}d^x, Y_k^{-1}d^y, -\alpha_k\}}.$$

Step 4: Set

$$\begin{bmatrix} x_{k+1} \\ y_{k+1} \end{bmatrix} = \begin{bmatrix} x_k \\ y_k \end{bmatrix} + \tau_k \begin{bmatrix} d_k^x \\ d_k^y \end{bmatrix},$$

$k = k + 1$ and go to Step 1.

This algorithm is due to Zhang, Tapia and Potra (1993). Among many variants of interior point methods, it is one of the simplest to describe. For a survey of this area, we refer the reader to den Hertog (1994), Ye (1990, 1991) and Wright (1997).

Algorithm 4.1 depends on the specification of the sequences $\{\alpha_k\}$ and $\{\sigma_k\}$. It is shown in Zhang, Tapia and Potra (1993, Theorem 3.1) that, provided $\{\alpha_k\} \to 1$ and $\{\sigma_k\} \to 0$, the sequence $\{F(x_k, y_k)\}$ is convergent at a Q-superlinear rate, i.e.

$$\lim_{k\to\infty} \sup \frac{\|F(x_{k+1}, y_{k+1})\|}{\|F(x_k, y_k)\|} = 0,$$

and that

$$\lim_{k\to\infty} \tau_k = 1.$$

It should nevertheless be noted that the Newton and quasi-Newton algorithms for computing the solution of nonlinear systems of equations, such as $F(x, y) = 0$, require a merit function to ensure the global progress of the algorithm. This aspect is discussed further in Chapter 9.

Without the perturbation term $(\eta_k I)$ on the right-hand side of (4.9), the search direction (d^x, d^y) is the Newton step for solving $F(x, y) = 0$. The starting point (x_0, y_0) is not required to be feasible. Also, $\tau_k \in (0, 1]$ and $\tau_k = 1$ if and only if

$$\min\{X_k^{-1}d^x, Y_k^{-1}d^y\} \geqslant -\alpha_k.$$

In most cases, it is expected that $\tau_k < 1$. The choice $\alpha_k = 1$ corresponds to allowing steps to the boundary of the bounds $x \geqslant 0$ and thus to a loss of strict feasibility. Using a similar argument as in (2.11), we see that the maximum step Algorithm 4.1 may take, from a feasible point (x_k, y_k), to reach the boundary of the inequalities $x \geqslant 0$, $y \geqslant 0$, is given by

$$\frac{1}{\min\{X_k^{-1}d^x, Y_k^{-1}d^y\}}.$$

Thus, if (x_k, y_k) is strictly feasible, so is (x_{k+1}, y_{k+1}).

The above algorithm is closely related to a number of interior point algorithms, including those of Kojima, Meggido and Noma (1991), Kojima, Mizuno and Noma (1989), Kojima, Mizuno and Yoishe (1989) and Monteiro and Adler (1989a, b). The strategies defining the stepsize and the penalty parameter are particular characteristics that vary across algorithms. For example, Monteiro and Adler (1989b) use an alternative stepsize strategy and the penalty parameter defined by

$$\eta_{k+1} = \eta_k\left(1 - \frac{\delta}{\sqrt{n}}\right)$$

for some δ (set at the value $\delta \equiv 0.1$). In other studies, including the special case of linear programming (i.e. $\mathfrak{Q} \equiv 0$), varying stepsize strategies have been used to define x_{k+1}, y_{k+1}, with η depending on $\langle x_{k+1}, y_{k+1} \rangle / n$ (see e.g. Ye and Anstreicher, 1993; Ye *et al.*, 1993; see also Lustig, Marsten and Shanno, 1989, 1992).

The exciting aspect of interior point algorithms is their polynomial complexity: that they terminate in a number of iterations that can be bounded by a polynomial depending on the data used to describe the problem (Papadimitriou and Steiglitz, 1982). To give an example for the quantities involved, we assume that the entries of the vectors a, c and the matrices $\mathfrak{Q}, \mathscr{G}$ are integral and define the size of the quadratic programming problem (4.1) as†

$$\mathscr{L}(\mathfrak{Q}, \mathscr{G}, a, g) = \left\lceil \log \left[\begin{array}{c} \text{largest absolute value of determinant of any} \\ \text{square submatrix of} \begin{bmatrix} \mathfrak{Q} & \mathscr{G} \\ \mathscr{G}^{\mathrm{T}} & 0 \end{bmatrix} \end{array} + 1 \right] \right\rceil + \lceil \log (1 + \max_i |a^i|) \rceil + \lceil \log (1 + \max_j |h^j|) \rceil + \lceil \log (n + e) \rceil.$$

By starting from an appropriate initial position satisfying $\langle x_0, y_0 \rangle \leqslant 2^{v\mathscr{L}}$, where v is a positive constant, independent of n, Zhang and Tapia (1993, Theorem 6.1, Corollary 6.2) show that they ensure sufficient finite progress, at each iteration, to convergence in at most $O(n\mathscr{L})$ iterations for an interior point linear programming algorithm.‡ Monteiro and Adler (1989b) show that if $\eta_0 = 2^{O(\mathscr{L})}$, then their quadratic programming algorithm solves the problem in at most $O(\sqrt{n}\mathscr{L})$ iterations. Actually, these results are based on interior point algorithms solving the system $F(x, y) = 0$ such that $\langle \hat{x}, \hat{y} \rangle \leqslant 2^{-O(\mathscr{L})}$. At such a point, it can be shown that the exact solution of the system would require $O(n^3)$ iterations, using conventional solvers (see Papadimitriou and Steiglitz, 1982, Lemma 8.7). A detailed discussion of the polynomial complexity of interior point algorithms is given in Den Hertog (1994).

Other related interior point algorithms include that of Kojima, Mizuno and Yoshise (1991), who consider the reduction of the potential function

$$f(x, y) = \eta \ln \langle x, y \rangle - \sum_{j=1}^{n} \ln x^j y^j$$

where $\eta = n + \sqrt{n}$ (see also Kojima, Megiddo, Noma and Yoshise, 1991). Starting from a feasible point, the algorithm generates a sequence $\{x_k, y_k\}$ that converges in $O(\sqrt{n}\mathscr{L})$ steps using a stepsize τ_k that ensures the feasibility of

$$\begin{bmatrix} x_k \\ y_k \end{bmatrix} + \tau \begin{bmatrix} d_k^x \\ d_k^y \end{bmatrix}$$

† For an alternative definition of $\mathscr{L}$, see e.g. Kojima, Megiddo, Noma and Yoshise (1991).
‡ $O(\cdot)$ denotes the order of magnitude of $(\cdot)$.

and

$$f(x_k + \tau d_k^x, y + \tau d_k^y) - f(x_k, y_k) \leqslant \beta$$

for $\beta > 0$ (chosen as $\beta = 0.2$).

Another algorithm, similar in spirit to that of Kojima, Mizuno and Yoshise (1991), is due to Goldfarb and Liu (1993), who consider the minimization of the potential function

$$\Phi(x, y) = \eta \ln \langle x, y \rangle - \sum_{j=1}^{n} \ln x^j - \sum_{j=1}^{n} \ln y^j,$$

where $\eta = n + \sigma\sqrt{n}$, with $\sigma \geqslant 1$. The method generates a sequence $\{x_k, y_k, \lambda_k\}$ that converges in $O(n^3 \mathscr{L})$ steps using a stepsize τ_k determined (as also in Anstreicher and Bosch, 1992) using an Armijo (1966)-type strategy that satisfies

$$\Phi(x_k + \tau d_k^x, y + \tau d_k^y) - \Phi(x_k, y_k) \leqslant \rho\tau(\langle \nabla_x \Phi(x_k, y_k), d_k^x \rangle + \langle \nabla_y \Phi(x_k, y_k), d_k^y \rangle)$$

for $\rho \in (0, 1)$. Furthermore, a potential function reduction requirement is satisfied by τ_k:

$$\Phi(x_k + \tau d_k^x, y + \tau d_k^y) - \Phi(x_k, y_k) \leqslant -\beta$$

for $\beta > 0$.

An important issue with interior point methods is the numerical problem encountered while approaching the boundary of the feasible set; in other words, when the existence of X_k^{-1} or Y_k^{-1} becomes questionable. There are numerous heuristic approaches to deal with this problem. A discussion of these is given by El-Bakry, Tapia and Zhang (1994).

5 CONCLUDING REMARKS

The three approaches to the quadratic programming problem discussed in this chapter constitute the main solution concepts. The algorithm based on the range space discussed in Section 3 is the earliest approach, and is economical when the number of active constraints is relatively small compared with the number of variables. The null-space algorithm discussed in Section 2 is based on orthogonal matrices, and consequently numerically stable, and is economical for problems with a large number of active constraints, relative to the number of variables. The interior point algorithm, on the other hand, is a promising efficient tool for large-scale problems.

6 COMMENTS AND NOTES

CN.1 Degeneracy

Degeneracy is essentially caused by the columns of $\mathscr{H}_k$ being linearly dependent. We assume nondegeneracy by considering that $\mathscr{H}_k$ has only linearly independent columns.

CN.2 Notes on spaces

A subset $\mathbb{S}$ of the n-dimensional vector space $\mathbb{R}^n$ is called a *subspace* of $\mathbb{R}^n$ if $\alpha x + \beta y \in \mathbb{S}$ for every $x, y \in \mathbb{S}$ and every $\alpha, \beta \in \mathbb{R}^1$. A *linear combination* of a set of vectors $v_1, v_2, \ldots, v_s$ is a vector of the form $\sum_{i=1}^{s} v_i\omega_i$, where $\omega_1, \omega_2, \ldots, \omega_s \in \mathbb{R}^1$. A set of vectors $v_1, v_2, \ldots, v_s$ is said to be *linearly dependent* if there are scalars $\omega_1, \omega_2, \ldots, \omega_s$, not all zero, such that $\sum_{i=1}^{s} v_i\omega_i = 0$. If no such $\omega_1, \omega_2, \ldots, \omega_s$ exist, the set of vectors $v_1, v_2, \ldots, v_s$ is said to be *linearly independent*. For example, if the vectors $v_1, v_2, \ldots, v_s$ are nonzero and mutually orthogonal (i.e. $\langle v_i, v_j \rangle = 0, \forall i \neq j$) then they are linearly independent. The set of vectors y of the form $y = \sum_{i=1}^{s} v_i\,\omega_i$ is the set *spanned* by $v_1, v_2, \ldots, v_s$. This set of y is a subspace of $\mathbb{R}^n$. A linearly independent set of vectors that span $\mathbb{R}^n$ is said to be a *basis* of $\mathbb{R}^n$. Given a subspace $\mathbb{S}$ of $\mathbb{R}^n$, a *basis for* $\mathbb{S}$ is a collection of vectors that are linearly independent, with span equal to $\mathbb{S}$. For a given subspace, every basis has the same number of vectors. This number is called the *dimension* of $\mathbb{S}$. Thus, the dimension of $\mathbb{S}$ is equal to the maximum number of linearly independent vectors in $\mathbb{S}$. The dimension of $\mathbb{R}^n$ is equal to n. Also for every subspace, there exists a basis of mutually orthogonal vectors. If $\mathbb{S}$ is a subspace of $\mathbb{R}^n$, the *orthogonal complement* of $\mathbb{S}$, denoted by $\mathbb{S}^\perp$, is the collection of all vectors that are orthogonal to the vectors in $\mathbb{S}$. $\mathbb{S}^\perp$ is a subspace and $\mathbb{S}$ and $\mathbb{S}^\perp$ jointly span $\mathbb{R}^n$, since every vector $x \in \mathbb{R}^n$ can be written as $x = y + z, y \in \mathbb{S}, z \in \mathbb{S}^\perp$. In this case, y is the *orthogonal projection* of x onto $\mathbb{S}$ and z is the *orthogonal projection* of x onto $\mathbb{S}^\perp$.

CN.3 Range and null space

Let $\mathscr{H}$ be an $n \times m$ matrix. The *rank* of $\mathscr{H}$, denoted rank $(\mathscr{H})$, is the maximum number of linearly independent columns of $\mathscr{H}$, which is equal to the maximum number of linearly independent rows of $\mathscr{H}$. $\mathscr{H}$ is of *full rank* if rank $(\mathscr{H}) = \min\{n, m\}$. The set of all n-dimensional vectors that are linear combinations of the columns of $\mathscr{H}$ is called the *range space*, or simply the *range*, of $\mathscr{H}$. This is denoted by range $(\mathscr{H})$. Thus,

$$r \in \text{range } (\mathscr{H}) \quad \text{if and only if} \quad \exists \lambda \in \mathbb{R}^m \quad \text{such that} \quad r = \mathscr{H}\lambda.$$

We observe that range $(\mathscr{H})$ is a subspace of $\mathbb{R}^n$ by considering $r_1, r_2 \in$ range $(\mathscr{H})$. By definition of range $(\mathscr{H})$, for some λ_1, λ_2, we have $r_1 = \mathscr{H}\lambda_1$, $r_2 = \mathscr{H}\lambda_2$. For any $\alpha_1, \alpha_2 \in \mathbb{R}^1$, we have

$$\alpha_1 r_1 + \alpha_2 r_2 = \alpha_1 \mathscr{H}\lambda_1 + \alpha_2 \mathscr{H}\lambda_2 = \mathscr{H}(\alpha_1\lambda_1 + \alpha_2\lambda_2) = \mathscr{H}\lambda,$$

which satisfies the above definition of subspace. For example,

$$\mathscr{H} = \begin{bmatrix} 1 & 0 \\ 0 & 2 \\ 3 & -1 \end{bmatrix}, \quad \lambda = \begin{bmatrix} \ell_1 \\ \ell_2 \end{bmatrix} = \begin{bmatrix} 1 \\ 2 \end{bmatrix},$$

$$\mathscr{H}\lambda = \begin{bmatrix} \ell_1 \\ 2\ell_2 \\ 3\ell_1 - \ell_2 \end{bmatrix} = \begin{bmatrix} 1 \\ 4 \\ 1 \end{bmatrix}$$

and, for any $\ell_1, \ell_2, \ell_3 \in \mathbb{R}^1$,

$$\begin{bmatrix} \ell_1 \\ 2\ell_2 \\ 3\ell_1 - \ell_2 \end{bmatrix} \in \text{range}\,(\mathscr{H}).$$

The set of all m-dimensional vectors, orthogonal to vectors in range $(\mathscr{H})$ is the *null space* of $\mathscr{H}^{\mathrm{T}}$, denoted null$\,(\mathscr{H}^{\mathrm{T}})$. Thus,

$$y \in \text{null}\,(\mathscr{H}^{\mathrm{T}}) \quad \text{if and only if} \quad \mathscr{H}^{\mathrm{T}} y = 0.$$

To establish that null $(\mathscr{H}^{\mathrm{T}})$ is a subspace, consider $y_1, y_2 \in$ null $(\mathscr{H}^{\mathrm{T}})$. Then $\mathscr{H}^{\mathrm{T}} y_1 = 0, \mathscr{H}^{\mathrm{T}} y_2 = 0$. For any $\alpha_1, \alpha_2 \in \mathbb{R}^1$, we have

$$\mathscr{H}^{\mathrm{T}}(\alpha_1 y_1 + \alpha_2 y_2) = \alpha_1 \mathscr{H}^{\mathrm{T}} y_1 + \alpha_2 \mathscr{H}^{\mathrm{T}} y_2 = 0,$$

so that $(\alpha_1 y_1 + \alpha_2 y_2) \in$ null $(\mathscr{H}^{\mathrm{T}})$. For example, given the above $\mathscr{H}$, any $y \in$ null $(\mathscr{H}^{\mathrm{T}})$ satisfies

$$\mathscr{H}^{\mathrm{T}} y = \begin{bmatrix} 1 & 0 & 3 \\ 0 & 2 & -1 \end{bmatrix} \begin{bmatrix} v_1 \\ v_2 \\ v_3 \end{bmatrix} = \begin{bmatrix} v_1 + 3v_3 \\ 2v_2 - v_3 \end{bmatrix} = 0,$$

and therefore y must be of the form

$$y = \begin{bmatrix} -3v_3 \\ \frac{1}{2} v_3 \\ v_3 \end{bmatrix}$$

for some v_3, and null $(\mathscr{H}^{\mathrm{T}}) = 1$. For $v_3 = 1$, we have $y = [-3, \frac{1}{2}, 1]^{\mathrm{T}}$.

We have established that range $(\mathscr{H})$ is all vectors given by linear combinations of the columns of $\mathscr{H}$, and null $(\mathscr{H}^{\mathrm{T}})$ is given by all vectors that are orthogonal to those columns. Thus, any n-dimensional vector can be written as a linear combination of vectors in range $(\mathscr{H})$ and null $(\mathscr{H}^{\mathrm{T}})$. The intersection of range $(\mathscr{H})$ and null $(\mathscr{H}^{\mathrm{T}})$ is empty, except for the zero vector. Consequently, range $(\mathscr{H})$ and null $(\mathscr{H}^{\mathrm{T}})$ are orthogonal complements. The sum of their dimensions is n. If $\mathscr{H}$ has linearly independent columns (hence $n \geqslant m$), the dimensions of range $(\mathscr{H})$ and null $(\mathscr{H}^{\mathrm{T}})$ are m and $n - m$ respectively.

As range $(\mathscr{H})$ and null $(\mathscr{H}^{\mathrm{T}})$ are orthogonal complements that only have in common the zero vector, any nonzero n-dimensional vector x can be expressed as the projection of x onto these two subspaces:

$$x = y + r, \quad y \in \text{null}\,(\mathscr{H}^{\mathrm{T}}), \quad r \in \text{range}\,(\mathscr{H}),$$
$$\langle y, r \rangle = 0, \quad \langle x, x \rangle = \langle y, y \rangle + \langle r, r \rangle.$$

Consider, for example, the vector derived from above:

$$x = \begin{bmatrix} -2 \\ 4\frac{1}{2} \\ 1 \end{bmatrix} = y + r = \begin{bmatrix} -3 \\ \frac{1}{2} \\ 1 \end{bmatrix} + \begin{bmatrix} 1 \\ 4 \\ 1 \end{bmatrix},$$

where it can be verified that $\langle y, r \rangle = 0$.

CN.4 Householder and Givens transformations

QR decomposition of the matrix $\mathscr{H}_k$ is done using a sequence of either Householder or Givens transformations. Briefly, a *Householder transformation* is a matrix $H \in \mathbb{R}^{n \times n}$ of the form

$$H = I - 2\frac{\omega\omega^{\mathrm{T}}}{\langle \omega, \omega \rangle}.$$

H is symmetric and $H^{\mathrm{T}}H = \mathrm{I}$, and hence *orthogonal*. Householder matrices are important because of their ability to zero specified entries in a matrix or vector. In particular, given a nonzero vector $v \in \mathbb{R}^n, \omega$ can be constructed such that Hv is a multiple of e_1, the first column of I. As

$$Hv = v - 2\frac{\langle \omega, v \rangle}{\langle \omega, \omega \rangle}\omega,$$

the requirement for $H\,v$ to be a multiple of e_1 implies that ω is a combination of e_1 and v:

$$\omega \in \{x \in \mathbb{R}^n \mid x = \alpha_1 e_1 + \alpha_2 v; \alpha_1, \alpha_2 \in \mathbb{R}^1\}.$$

Setting $\omega = v + \alpha e_1$ yields

$$\langle \omega, v \rangle = \langle v, v \rangle + \alpha\langle v, e_1 \rangle,$$

where $\langle v, e_1 \rangle = v_1$, the first element of v, and

$$\langle \omega, \omega \rangle = \langle v, v \rangle + 2\alpha\langle \omega, v \rangle + \alpha^2.$$

Thus, we have

$$Hv = \left[1 - 2\frac{\langle v, v \rangle + \alpha v_1}{\langle v, v \rangle + 2\alpha v_1 + \alpha^2}\right]v - 2\alpha\frac{\langle \omega, v \rangle}{\langle \omega, \omega \rangle}e_1.$$

To ensure that the coefficient of v on the right is zero, we set $\alpha = \pm\|v\|_2$. Hence, ω is given by

$$\omega = v \pm \|v\|_2 e_1,$$

and $Hv = \pm\|v\|_2 e_1$. To choose the sign of α, it should be noted that if v is close to a multiple of e_1 then $\omega = v - \mathrm{sign}\,(v_1)\|v\|_2 e_1$ has a small norm. Consequently, the relative error in $2/\langle \omega, \omega \rangle$ may be large. This problem may be avoided by choosing the sign of α to be the same as the sign of the first component of v. Thus, we have

$$\omega = v + \operatorname{sign}(v_1)\,\|v\|_2 e_1.$$

The QR decomposition of a matrix can be computed using a sequence of Householder matrices that zero the all the elements below the diagonal element of the matrix. After the first column is thus evaluated, the procedure is repeated for the second and subsequent columns by concentrating on the submatrix obtained by deleting the first row and column, or the row and columns above the given diagonal element. The Householder matrix essentially operates on the column vector consisting of the diagonal element and all the elements below it, zeroing all but the diagonal element. Thus, we progress through the columns to construct the upper-triangular matrix R. In order to maintain the evaluated part of the triangular matrix R, a leading identity matrix block of the dimension just above the current diagonal element is augmented to the Householder matrices. A sequence of such matrices yields the upper-triangular matrix R. As the product of orthogonal matrices is also orthogonal, the product of this sequence is the orthogonal matrix Q.

The *Givens transformation* is an orthogonal matrix

$$T(p,q,\theta) = \begin{bmatrix} 1 & 0 & \dots & 0 & 0 \\ 0 & t_{pp} & \dots & t_{pq} & 0 \\ \vdots & \vdots & \ddots & \vdots & \vdots \\ 0 & t_{qp} & \dots & t_{qq} & 0 \\ 0 & 0 & \dots & 0 & 1 \end{bmatrix},$$

with $t_{pq} = -t_{qp} = \sin\theta$, and $t_{pp} = t_{qq} = \cos\theta$, all other diagonal elements being unity and nondiagonal elements being zero. The matrix T is used to introduce zeros in selected locations in a vector or matrix. Let $x = T(p,q,\theta)v$, where $x, v \in \mathbb{R}^n$. Then

$$x_p = v_p\cos\theta + v_q\sin\theta, \quad x_q = -v_p\sin\theta + v_q\cos\theta, \quad x_i = v_i, \quad i \neq p \text{ or } q,$$

and we can force $x_q = 0$ by setting

$$\cos\theta = \frac{v_p}{\sqrt{(v_p^2 + v_q^2)}}, \quad \sin\theta = \frac{v_q}{\sqrt{(v_p^2 + v_q^2)}}.$$

For example, let $\cos\theta = 1/\sqrt{5}$, $\sin\theta = 2/\sqrt{5}$ and

$$T(2,4,\theta)\begin{bmatrix} 1 \\ 2 \\ 3 \\ 4 \end{bmatrix} = \begin{bmatrix} 1 \\ \frac{1}{\sqrt{5}} \times 2 + \frac{2}{\sqrt{5}} \times 4 = \sqrt{20} \\ 3 \\ 0 \end{bmatrix}.$$

CN.5 Least squares solution of $\mathscr{H}_k \mu_{k+1} = -(a + \mathfrak{Q}x_k + \mathfrak{Q}d_k)$

When $\mathscr{H}_k$ has linearly independent columns, the least squares solution is commonly obtained using the $Q\ R$ factorization of $\mathscr{H}$ given by (2.2). Using the fact that Y is an orthonormal matrix and therefore $\|Yv\|_2 = \langle Yv, Yv\rangle^{1/2} = \|v\|_2$, for any v, we have

$$\begin{aligned}\|\mathscr{H}_k \mu_{k+1} + a + \mathfrak{Q}x_k + \mathfrak{Q}d_k\|_2 &= \left\| [Y \vdots Z] \begin{bmatrix} R \\ 0 \end{bmatrix} \mu_{k+1} + a + \mathfrak{Q}x_k + \mathfrak{Q}d_k \right\|_2 \\ &= \left\| \begin{bmatrix} R \\ 0 \end{bmatrix} \mu_{k+1} + Y(a + \mathfrak{Q}x_k + \mathfrak{Q}d_k) \right\|_2\end{aligned}$$

and the least squares solution is achieved for

$$\mu_{k+1} = R^{-1} Y(a + \mathfrak{Q}x_k + \mathfrak{Q}d_k).$$

CN. 6 Slack variables

For $\mathscr{H}^{\mathrm{T}} \hat{x} \leqslant \mathrm{h}$, for some $\mathscr{H}$ and $\hat{x}$, we define *slack variables* $x_{\mathrm{s}} \geqslant 0$, and

$$\mathscr{G} = \begin{bmatrix} \mathscr{H} \\ I \end{bmatrix}, \quad x = \begin{bmatrix} \hat{x} \\ x_{\mathrm{s}} \end{bmatrix}, \quad g = h.$$

REFERENCES

Anstreicher, K.M. and R.A. Bosh (1992). Long Steps in a $O(n^3 L)$ algorithm for linear programming. *Math. Programming*, **54**, 251–265.

Armijo, L. (1966). Minimization of functions having Lipschitz- continuous first partial derivatives. *Pacific J. Math.* **16**, 1–3.

Dantzig, G. (1963). *Linear Programming and Extensions.* Princeton University Press.

Den Hertog, D. (1994). *Interior Point Approach to Linear, Quadratic and Convex Programming*. Kluwer, Dordrecht.

El-Bakry, A.S., R.A. Tapia, T. Tsuchiya and Y. Zhang (1996). On the formulation of the Newton interior-point method for nonlinear programming. *J. Optim. Theory Applic.* **89**, 507–541.

El-Bakry, A.S., R.A. Tapia and Y. Zhang (1994). A study of indicators for identifying zero variables in interior point methods. *SIAM Rev.* **36**, 45–72.

Fiacco, A.V. (1979). Barrier methods for nonlinear programming. In *Operations Research Support Methodology* (A. Holzman, ed.). Marcel Dekker, New York.

Fiacco, A.V. and G.P. McCormick (1968). *Nonlinear Programming: Sequential Unconstrained Minimization Techniques*, Wiley, New York.

Fletcher, R. (1971). A general quadratic programming algorithm. *J. Inst. Math. Applic.* **7**, 76–91.

Frisch, R. (1955). Logarithmic potential method of convex programming. Institute of Economics, University of Oslo.

Gill, P.E., G.H. Golub, W. Murray and M.A. Saunders (1974). Methods for modifying matrix factorizations. *Math. Comput.* **28**, 505–535.

Gill, P.E., W. Murray, M.A. Saunders, J.A. Tomlin and M.H. Wright (1986). On projected Newton barrier methods for linear programming and an equivalence to Karmarkar's projective method. *Math. Programming* **36**, 183–209.
Gill, P.E. and W. Murray (1976). Minimization subject to bounds on the variables. Report NAC 72, National Physical Laboratory, England.
Gill, P.E. and W. Murray (1978). Numerically stable methods for quadratic programming. *Math. Programming* **14**, 349–372.
Gill, P.E., W. Murray and M.H. Wright (1981). *Practical Optimization.* Academic Press, New York.
Goldfarb, D. (1972). Extensions of Newton's method and simplex methods for solving quadratic programs. In *Numerical Methods for Non-Linear Optimization* (F.A. Lootsma, ed.). Academic Press, London.
Goldfarb, D. and S. Liu (1993). An $O(n^3L)$ primal–dual potential reduction algorithm for solving convex quadratic programs. *Math. Programming* **61**, 161–170.
Golub, G.H. and C.F. van Loan (1983). *Matrix Computations.* North Oxford Academic, Oxford.
Karmarkar, N. (1984). A new polynomial-time algorithm for linear programming. *Combinatorica* **4**, 373–395.
Khachian, L.G. (1979). A polynomial algorithm in linear programming. *Soviet Math. Dokl.* **20**, 191–194.
Klee, V. and G. Minty (1972). How good is the simplex algorithm? In *Inequalities–III* (O. Shisha, ed.). Academic Press, New York.
Kojima, M., N. Meggido and T. Noma (1991). Homotopy continuation methods for complimentarity problems. *Math. Oper. Res.* **16**, 754–774.
Kojima, M., N. Meggido, T. Noma and A. Yoshise (1991). *A Unified Approach to Interior Point Algorithms for Linear Complimentarity Problems.* Springer-Verlag, Berlin.
Kojima, M., S. Mizuno and T. Noma (1989). A new continuation method for complimentarity problems with uniform *p*-functions. *Math. Programming* **43**, 107–113.
Kojima, M., S. Mizuno and A. Yoshise (1989). A primal–dual algorithm for a class of linear complimentarity problems. *Math. Programming*, **44**, 1–26.
Kojima, M., S. Mizuno and A. Yoshise (1991). An $O(\sqrt{n}L)$ iteration potential reduction algorithm for linear complimentarity problems. *Math. Programming*, **50**, 331–342.
Lustig, I.J., R.E. Marsten and D.F. Shanno (1989). Computational experience with a primal–dual interior point method for linear programming'. *Linear Algebra Applic.* **152**, 191–122.
Lustig, I.J., R.E. Marsten and D.F. Shanno (1992). On implementing Mehrotra's predictor–corrector interior point method for linear programming. *SIAM J. Optim.* **2**, 435–449.
Markowitz, H. (1959). *Portfolio Selection: Efficient Diversification of Investment*, Wiley, New York.
Monteiro, R.C. and I. Adler (1989a). Interior path-following primal–dual algorithms. Part I: Linear programming. *Math. Programming* **44**, 27–41.
Monteiro, R.C. and I. Adler (1989b). 'Interior path-following primal–dual algorithms. Part II: Convex quadratic programming. *Math. Programming* **44**, 43–66.
Papadimitriou, C.R. and K. Steiglitz (1982). *Combinatorial Optimization: Algorithms and Complexity.* Prentice-Hall, Englewood Cliffs, NJ.
Smale, S. (1983). On the average number of steps of the simplex method of linear programming. *Math. Programming.* **27**, 241–262.
Stewart, G.W. (1973). *Introduction to Matrix Computations.* Academic Press, London.
Wright, S. J. (1997). *Primal-dual Interior Point Methods*, SIAM, Philadelphia.

Ye, Y. (1990). Interior point methods for global optimization. *Ann. Oper. Res.* **25**, 59–74.
Ye, Y. (1991). Interior point algorithms for quadratic programming. In *Recent Developments in Mathematical Programming* (S. Kumar, ed.). Gordon and Breach, New York.
Ye, Y. and K. Anstreicher (1993). On quadratic and $O(\sqrt{n}L)$ convergence of a predictor–corrector algorithm for LCP. *Math. Programming* **62**, 537–551.
Ye, Y., O. Güler, R.A. Tapia and Y. Zhang (1993). A quadratically convergent $O(\sqrt{n}L)$-iteration algorithm for linear programming. *Math. Programming* **59**, 151–162.
Zhang, Y. and R.A. Tapia (1993). A superlinearly convergent polynomial primal–dual interior point linear programming algorithm for linear programming. *SIAM J. Optim.* **3**, 118–133.
Zhang, Y., R.A. Tapia and J.E. Dennis (1992). On the superlinear and quadratic convergence of primal–dual interior point linear programming algorithms. *SIAM J. Optim.* **2**, 304–324.
Zhang, Y., R.A. Tapia and F. Potra (1993). On the superlinear convergence of interior point algorithms for a general class of problems. *SIAM J. Optim.* **3**, 413–422.

CHAPTER 3

Multiple-Objective Optimization 1 Interactive Search for Acceptable Decisions: Updating Quadratic Objective Weights

The existence of multiple, and often conflicting, objectives seems an integral part of most decision making. A traditional approach to multi-objective optimization is through goal programming in which the attainment of each objective is weighted by its importance to the decision maker. These are then used to express a single combined objective function to evaluate the decision.

As the weights assigned to the objectives need to reflect the preferences of the decision maker, their determination is central to multi-objective decision making. Assuming that, for practical purposes, the preferences of the decision maker cannot be easily specified, we discuss an interactive algorithm with desirable properties.

The interactive algorithm is an iterative scheme that allows a single decision maker to search for an acceptable decision that strikes the right balance between the achievement of multiple, and often conflicting, objectives.

The method uses a rank-one updating formula to update the Hessian, or the weights, of the preference function. Numerical results illustrating an earlier version of the method are discussed in Rustem, Velupillai and Westcott (1978). The issue of sensitivity is also considered in this chapter by evaluating the applicability of the objective function when the underlying decision model changes.

1 THE MULTI-OBJECTIVE DECISION PROBLEM

Let $x \in \mathbb{R}^n$ be the vector of target variables and $\mathscr{R} \subseteq \mathbb{R}^n$ be the feasible set of x. The set $\mathscr{R}$ includes the model of the underlying system as well as other operational constraints. We let the elements of x^{d} denote the *bliss point* (French *et al.*, 1983) or desired values of the corresponding variables in x. The decision maker would ideally like to achieve $x = x^{\mathrm{d}}$. However, we usually have

$$x^{\mathrm{d}} \notin \mathscr{R}.$$

The attainment of some elements of x^{d} *conflicts* with the same aim for other elements of x^{d} when feasibility is required. Given the restriction $\mathscr{R}$, it is possible to construct a feasible x such that a subset of its elements attain their desired values but the rest depart from the corresponding elements of x^{d} to ensure $x \in \mathscr{R}$. Clearly, all the elements of x^{d} cannot be attained simultaneously when x^{d} is infeasible. Equivalently, when $x^{\mathrm{d}} \in \mathscr{R}$, all the values desired by the decision maker can be attained and thus the optimal policy would simply be $x = x^{\mathrm{d}}$. This chapter and the next two are therefore concerned with the problems arising when $x^{\mathrm{d}} \notin \mathscr{R}$. In such cases, the *relative importance* of each element of x attaining its corresponding desired value has to be determined. The information is then used to compute the best feasible alternative to x^{d} on $\mathscr{R}$ (Dixon, 1981). In *goal programming*, this is resolved by defining the problem

$$\min \left\{ \sum_{i=1}^{n} \omega^i \mid x^i - x^{\mathrm{d}i} | \middle| x \in \mathscr{R} \right\}$$

where $x^i, x^{\mathrm{d}i}$ are respectively the ith elements of x and x^{d}, and $\omega^i \geqslant 0$ is the weight, or importance, attached by the decision maker to achieving $x^{\mathrm{d}i}$. This has a standard linear programming formulation obtained by introducing the variables $x^{i+} \geqslant 0, x^{i-} \geqslant 0$ such that $x^i - x^{\mathrm{d}i} = x^{i+} - x^{i-}$ and $|x^i - x^{\mathrm{d}i}| = x^{i+} + x^{i-}$ (see e.g. Hillier and Lieberman, 1996). The resulting linear programming problem is given by

$$\min \left\{ \sum_{i=1}^{n} \omega^i (x^{i+} + x^{i-}) \middle| x \in \mathscr{R}; x^i - x^{\mathrm{d}i} = x^{i+} - x^{i-}; x^{i+} \geqslant 0, x^{i-} \geqslant 0; i = 1, \ldots, n \right\}.$$

The above formulation serves the same purpose as the min-max approach considered by Pinto (1995), Pinto and Rustem (1996a; b), and in Chapter 12. This unity of purpose also applies to the quadratic objective function framework considered in this chapter and the next two. The latter has the additional desirable property that it relates the changes in the optimal solution to the direction of desired change as weights are altered and alternatives explored. To introduce the problem, consider the quadratic objective framework

$$\min \{ q_{\mathrm{c}}(x) \mid x \in \mathscr{R} \} \tag{1.1}$$

where (see CN.1)

$$q_{\mathrm{c}}(x) = \frac{1}{2} \langle x - x^{\mathrm{d}}, \mathfrak{Q}_{\mathrm{c}}(x - x^{\mathrm{d}}) \rangle \tag{1.2}$$

$$= \frac{1}{2} \| x - x^{\mathrm{d}} \|^2_{\mathfrak{Q}_{\mathrm{c}}} \tag{1.3}$$

and $\mathfrak{Q}_{\mathrm{c}}$ is a symmetric positive-semidefinite weighting matrix (see CN.2). $\mathfrak{Q}_{\mathrm{c}}$ specifies the relative importance of each element of x attaining its corresponding desired, or bliss, value. The diagonal element of $\mathfrak{Q}_{\mathrm{c}}$ penalize the departure of a variable from its desired value. The off-diagonal elements measure the importance attached to the deviation in one variable versus the deviation in another. Hence, these elements express trade-offs between variables. Clearly, by specifying a positive-definite $\mathfrak{Q}_{\mathrm{c}}$, a measure of distance from x to x^{d} is defined as indicated in (1.3).

It can be verified that the solution of the linear programming formulation of *goal programming* has the following property:$x^i > 0 \Rightarrow (x^{i+} > 0, x^{i-} = 0)$ and

$x^i > 0 \Rightarrow (x^{i+} > 0, x^{i-} = 0)$. However, this applies to linear programming prblems. For (1.1), the same can be achieved by ensuring explicitly the condition $(x^{i+} \times x^{i-}) = 0$. The latter is realised by augmenting (1.1) with an appropriate penalty term. To show this let $\mathfrak{Q}_{c_{ii}}$ denote the ith diagonal element of $\mathfrak{Q}_c$, $d_i \geqslant \mathfrak{Q}_{c_{ii}}$ and the diagonal matrix $D_{\mathfrak{Q}_c}$ be defined as

$$D_{\mathfrak{Q}_c} \equiv \begin{bmatrix} d_1 & 0 & \dots & 0 \\ 0 & d_2 & \dots & 0 \\ \vdots & \vdots & \ddots & \vdots \\ 0 & 0 & \dots & d_n \end{bmatrix}$$

Let the vectors x^+, x^- denote

$$x^+ \equiv [x^{1+}x^{2+}, \dots, x^{n+}]^{\mathrm{T}}; \; x^- \equiv [x^{1-}, x^{2-}, \dots, x^{n-}]^{\mathrm{T}}$$

Consider the problem

$$\min_{x^+, x^-} \{\| x^+ - x^- \|^2_{\mathfrak{Q}_c} + 2\langle x^+, D_{\mathfrak{Q}_c} x^- \rangle \,|\, x \in \Re; \; x - x^{\mathrm{d}} = x^+ - x^-; \; x^+, \; x^- \geqslant 0\}$$

which penalises $(x^{i+} \times x^{i-})$ sufficiently to ensure that either one or both of $x^{i+}, x^{i-}, i = 1, \dots, n$, are null. If $(x^{i+} \times x^{i-}) \neq 0$, then the objective in can be further reduced, without altering $x^i = x^{i+} - x^{i-}$. Thus, the solution satisfies $(x^{i+} \times x^{i-}) = 0$.

The iterative method of this chapter is aimed at the tailoring of $\mathfrak{Q}_c$ to meet the requirements of the decision maker regarding the target variables. The method is not concerned with a 'best' set of weights *independent* of the desired value x^{d}. A solution, optimally generated via (1.1), acceptable to the decision maker, is the main aim. Once the initial optimal solution is determined using an initial $\mathfrak{Q}_c$, the method provides a systematic way in which the decision maker can specify her dissatisfaction with the various elements of the optimal solution and leaves it to the method to alter $\mathfrak{Q}_c$ to generate a more acceptable optimal solution. The method translates the decision maker's dissatisfaction with the initial optimal solution into a rank-one correction to $\mathfrak{Q}_c$. It is proposed that an 'optimal' weighting matrix will be obtained by repeated updating, so that, at the end of the iterative procedure, the 'final' optimal solution will be totally acceptable to the decision maker.

1.1 Projections in $\mathbb{R}^n$

When $\mathfrak{Q}_c$ is positive-definite and $\mathscr{R}$ is convex, the solution of (1.1) has a simple geometric interpretation. Simply stated, the projection of any vector $x^{\mathrm{d}} \notin \mathscr{R}$ is the point $x \in \mathscr{R}$ nearest to x^{d}. We measure distance with respect to the norm

$$\|x - x^{\mathrm{d}}\|^2_{\mathfrak{Q}_c} = \langle x - x^{\mathrm{d}}, \mathfrak{Q}_c(x - x^{\mathrm{d}}) \rangle.$$

As (1.1) minimizes the norm defined in (1.3), the solution is the projection of x^{d} onto $\mathscr{R}$, with respect to the norm (1.3). This is the extension of the three-dimensional situation that the the projection of a point x, not on a given plane, is the point y, on the plane, nearest to x. Furthermore, the line joining x and y is perpendicular to the plane. In general, as $\mathscr{R}$ is convex and $\mathfrak{Q}_{\mathrm{c}}$ positive-definite, by the discussion on convex problems in Chapter 1, there is a unique minimizing vector that solves (1.1).

When $\mathscr{R}$ denotes a set of linear equality constraints, i.e.

$$\mathscr{R} = \{x \in \mathbb{R}^n \mid \mathscr{G}^{\mathrm{T}} x = g\} \tag{1.4}$$

where g is an $(e \times 1)$ constant vector and $\mathscr{G}$ is an $(n \times e)$ matrix whose columns are assumed (for simplicity) to be linearly independent, the projection of x^{d}, and the solution of (1.1), is given by the following lemma.

Lemma 1.1 (Projection of x^{d} onto $\mathscr{R}$)

(i) Let $\mathfrak{Q}_{\mathrm{c}}$ be symmetric positive-definite, let the feasible region $\mathscr{R}$ be given by (1.4) and let the columns of the matrix $\mathscr{G}$ be linearly independent. Then the solution of (1.1) is given by

$$x_{\mathrm{c}} = x^{\mathrm{d}} - \mathfrak{Q}_{\mathrm{c}}^{-1}\mathscr{G}(\mathscr{G}^{\mathrm{T}}\mathfrak{Q}_{\mathrm{c}}^{-1}\mathscr{G})^{-1}(\mathscr{G}^{\mathrm{T}}x^{\mathrm{d}} - g) \tag{1.5}$$

and the Lagrange multipliers (or shadow prices) associated with (1.1) are given by

$$\lambda_{\mathrm{c}} = -(\mathscr{G}^{\mathrm{T}}\mathfrak{Q}_{\mathrm{c}}^{-1}\mathscr{G})^{-1}(\mathscr{G}^{\mathrm{T}}x^{\mathrm{d}} - g). \tag{1.6}$$

(ii) Let $\mathfrak{Q}_{\mathrm{c}}$ be positive-semidefinite and let x_k be any feasible point such that $\mathscr{G}^{\mathrm{T}}x_k = g$ and $Z \in \mathbb{R}^{n\times(n-e)}$ be an orthogonal matrix such that $Z^{\mathrm{T}}\mathscr{G} = 0$. Then the optimal solution and Lagrange multipliers can be written as

$$x_{\mathrm{c}} = x_k - Z(Z^{\mathrm{T}}\mathfrak{Q}_{\mathrm{c}}Z)^{-1}Z^{\mathrm{T}}\left(\mathfrak{Q}_{\mathrm{c}}(x_{\mathrm{k}} - x^{\mathrm{d}})\right) \tag{1.7a}$$

$$\lambda_{\mathrm{c}} = (\mathscr{G}^{\mathrm{T}}\mathscr{G})^{-1}\mathscr{G}^{\mathrm{T}}\mathfrak{Q}_{\mathrm{c}}\left(x_{\mathrm{c}} - x^{\mathrm{d}}\right) \tag{1.7b}$$

Remark

There are a number of ways of defining Z. These are discussed briefly in Chapter 2. The numerically stable method, arising from the orthogonal decomposition of $\mathscr{G}$ (see CN.3), can also be extended in order to update Z in the presence of linear inequality constraints (Gill and Murray, 1974, 1978).

Remark

The equivalence of (1.5) and (1.7a) is discussed in Lemma 3.1.

Proof

Writing the Lagrangian associated with (1.1),

$$L_c(x, \lambda) = q_c(x) - \langle (\mathscr{G}^T x - g), \lambda \rangle, \tag{1.8}$$

both (1.5) and (1.6) can be derived straightforwardly from the following first-order optimality conditions:

$$\mathfrak{Q}_c(x_c - x^d) - \mathscr{G}\lambda_c = 0, \tag{1.9}$$

$$\mathscr{G}^T x_c - g = 0. \tag{1.10}$$

To establish (ii), we define the subspace

$$\mathscr{R}_0 = \{d \in \mathbb{R}^n \mid \mathscr{G}^T \mathrm{d} = 0\}, \tag{1.11}$$

and note that

$$x_c - x_k \in \mathscr{R}_0.$$

As $\mathscr{G}^T Z = 0$, any vector $x_c - x_k$ can be written as a linear combination of the columns of Z. Thus, we have

$$x_c - x_k = Z\omega,$$

for some vector $\omega \in \mathbb{R}^{n-m}$. From the first-order optimality condition of (1.1), we can write

$$Z^T \mathfrak{Q}_c(Z\omega + x_k - x^d) = Z^T \mathscr{G} \lambda_c = 0,$$

and this yields the required relationship

$$x_c - x_k = Z\omega = -Z(Z^T \mathfrak{Q}_c Z)^{-1} Z^T \mathfrak{Q}_c (x_k - x^d).$$

Finally, the multiplier in (1.7) is recovered as the least squares solution of (1.9). □

As x_c is the projection of x^d, with respect to the norm $\|x - x^d\|^2_{\mathfrak{Q}_c}$, for any x_k satisfying $\mathscr{G}^T x_k = g$, the vector $x_k - x_c$, which lies in the intersection of all the constraints, is orthogonal to $x^d - x_c$. Thus, we have

$$\langle x_k - x_c, \mathfrak{Q}_c(x^d - x_c) \rangle = \langle x_k - x_c \mathscr{G}(\mathscr{G}^T \mathfrak{Q}_c^{-1} \mathscr{G})^{-1} (\mathscr{G}^T x^d - g) \rangle = 0,$$

since $\mathscr{G}^T(x_k - x_c) = 0$.

The operator

$$P = I - \mathfrak{Q}_c^{-1} \mathscr{G}(\mathscr{G}^T \mathfrak{Q}_c^{-1} \mathscr{G})^{-1} \mathscr{G}^T \tag{1.12}$$

projects the vector $x \in \mathbb{R}^n$ onto the subspace $\mathscr{R}_0$ and those vectors of the form $\mathfrak{Q}_c^{-1} \mathscr{G} \lambda$ into 0 (see e.g. Goldfarb, 1969; Rustem, 1981). Also, the projection operation is idempotent:

$$PP = [I - \mathfrak{Q}_c^{-1} \mathscr{G}(\mathscr{G}^T \mathfrak{Q}_c^{-1} \mathscr{G})^{-1} \mathscr{G}^T][I - \mathfrak{Q}_c^{-1} \mathscr{G}(\mathscr{G}^T \mathfrak{Q}_c^{-1} \mathscr{G})^{-1} \mathscr{G}^T] = P.$$

Hence, we have $P(I - P) = 0$.

The operator $I - P$ projects $x \in \mathbb{R}^n$ onto the subspace of vectors spanned by the columns of $\mathfrak{Q}_c^{-1}\mathcal{G}$ or, equivalently, the subspace of vectors that can be expressed in the form $\mathfrak{Q}_c^{-1}\mathcal{G}\lambda$ and only those vectors in $\mathcal{R}_0$ into 0 (see CN.2 of chapter 2). Thus, for $x \in \mathbb{R}^n$,

$$x = Px + (I - P)x = Zw + \mathfrak{Q}_c^{-1}\mathcal{G}\lambda.$$

Since Z spans the null space of $\mathcal{G}$, we have $Zw \in \mathcal{R}_0$.

In subsequent sections, the positive-definiteness of $\mathfrak{Q}_c$ and the restrictive structure of (1.4) are relaxed. Expressions (1.5)–(1.10) are nevertheless needed to study projection properties of the method.

Lemma 1.2 (Projection property)
Let x_c be the projection of $x^d \in \mathbb{R}^n$ on the convex set $\mathcal{R}$, i.e.

$$x_c = \arg\min\{\|x - x^d\|^2_{\mathfrak{Q}_c} \mid x \in \mathcal{R}\}. \tag{1.13}$$

For any symmetric positive-definite $\mathfrak{Q}_c$ and any vector $y \in \mathcal{R}$, the following inequality holds:

$$\langle x_c - y, \mathfrak{Q}_c(x_c - x^d)\rangle \leqslant 0. \tag{1.14}$$

Proof
Suppose, contrary to (1.14), that there exists $u \in \mathcal{R}$ such that

$$\langle x_c - u, \mathfrak{Q}_c(x_c - x^d)\rangle > 0.$$

Consider the vector

$$\omega(\alpha) = (1 - \alpha)x_c + \alpha u, \quad 0 \leqslant \alpha \leqslant 1.$$

Since $\mathcal{R}$ is convex, $\omega(\alpha) \in \mathcal{R}$. Also, we have

$$\begin{aligned}\|\omega(\alpha) - x^d\|^2_{\mathfrak{Q}_c} &= \|(1 - \alpha)(x_c - x^d) + \alpha(u - x^d)\|^2_{\mathfrak{Q}_c}\\ &= (1 - \alpha)^2\|x_c - x^d\|^2_{\mathfrak{Q}_c} + 2\alpha(1 - \alpha)\langle u - x^d, \mathfrak{Q}_c(x_c - x^d)\rangle\\ &\quad + \alpha^2\|u - x^d\|^2_{\mathfrak{Q}_c}.\end{aligned}$$

The norm $\|\omega(\alpha) - x^d\|^2_{\mathfrak{Q}_c}$ is a differentiable function of α, and its derivative at $\alpha = 0$ is given by

$$\begin{aligned}\left.\frac{\mathrm{d}\|\omega(\alpha) - x^d\|^2_{\mathfrak{Q}_c}}{\mathrm{d}\alpha}\right|_{\alpha=0} &= -2\,\|x_c - x^d\|^2_{\mathfrak{Q}_c} + 2\langle u - x^d, \mathfrak{Q}_c(x_c - x^d)\rangle\\ &= -2\langle x_c - u, \mathfrak{Q}_c(x_c - x^d)\rangle\\ &= -2\epsilon\\ &< 0.\end{aligned}$$

Thus, for small positive α, we have

$$\|\omega(\alpha) - x^{\mathrm{d}}\|^2_{\mathfrak{Q}_\mathrm{c}} < \|x_\mathrm{c} - x^{\mathrm{d}}\|^2_{\mathfrak{Q}_\mathrm{c}},$$

which is a contradiction since x_c is, by hypothesis, the projection of x^d and thereby the vector that minimizes this norm. Hence, no such u can exist.

Conversely, suppose that $x_\mathrm{c} \in \mathscr{R}$ is such that (1.14) holds $\forall y \in \mathscr{R}$. Then, for any $y \in \mathscr{R}$ and $y \neq x_\mathrm{c}$, we have

$$\begin{aligned} \|y - x^{\mathrm{d}}\|^2_{\mathfrak{Q}_\mathrm{c}} &= \|y - x_\mathrm{c} + x_\mathrm{c} - x^{\mathrm{d}}\|^2_{\mathfrak{Q}_\mathrm{c}} \\ &= \|y - x_\mathrm{c}\|^2_{\mathfrak{Q}_\mathrm{c}} + 2\langle y - x_\mathrm{c}, \mathfrak{Q}_\mathrm{c}(x_\mathrm{c} - x^{\mathrm{d}})\rangle + \|x_\mathrm{c} - x^{\mathrm{d}}\|^2_{\mathfrak{Q}_\mathrm{c}} \\ &> \|x_\mathrm{c} - x^{\mathrm{d}}\|^2_{\mathfrak{Q}_\mathrm{c}} \end{aligned}$$

Hence, x_c is the unique minimizing vector of (1.13). □

1.2 Motivation for updating the weighting matrix

Before discussing the iterative specification of $\mathfrak{Q}_\mathrm{c}$, it would be appropriate to introduce the underlying motivation, which stems from concepts in numerical optimization. Suppose the decision maker has a globally valid true weighting matrix $\mathfrak{Q}_{\mathrm{true}}$ that is not known. The corresponding true quadratic objective function is given by

$$q_{\mathrm{true}}(x) = \frac{1}{2}\langle x - x^{\mathrm{d}}, \mathfrak{Q}_{\mathrm{true}}(x - x^{\mathrm{d}})\rangle,$$

and, consequently,

$$\nabla q_{\mathrm{true}}(x) = \mathfrak{Q}_{\mathrm{true}}(x - x^{\mathrm{d}}).$$

The problem of determining $\mathfrak{Q}_{\mathrm{true}}$ can be resolved by generating a finite sequence of matrices that converge to $\mathfrak{Q}_{\mathrm{true}}$. The following algorithm generates such a sequence.

Algorithm 1.1 (Model for specification of the quadratic objective)

Step 0: A current positive-definite matrix $\mathfrak{Q}_\mathrm{c}$ and a current value $x_\mathrm{c} \in \mathbb{R}^n$ are given.

Step 1: Ask the decision maker to specify the vectors, $\delta, \gamma \in \mathbb{R}^n$ such that the new value

$$x_\mathrm{n} = x_\mathrm{c} + \tau\delta,$$

where $\delta = -\mathfrak{Q}_\mathrm{c}^{-1}\nabla q_{\mathrm{true}}(x_c)$ is a descent direction such that $\langle \nabla q_{\mathrm{true}}(x_\mathrm{c}), \delta\rangle < 0$ and τ is determined to minimize q_{true} (x) along δ (signifying that the direction indicated by δ has been chosen for exploration, as discussed further in Section 2) and

$$\gamma = \nabla q_{\mathrm{true}}(x_\mathrm{n}) - \nabla q_{\mathrm{true}}(x_\mathrm{c}).$$

Step 2: Update $\mathfrak{Q}_c$ using the rank-one formula

$$\mathfrak{Q}_n = \mathfrak{Q}_c + \frac{(\gamma - \tau\mathfrak{Q}_c\delta)(\gamma - \tau\mathfrak{Q}_c\delta)^T}{\langle\gamma - \tau\mathfrak{Q}_c\delta, \tau\delta\rangle};$$

set $c = n$, **go to Step 1.**

The termination of the algorithm is discussed in the theorem below. The rank-one formula in **Step 2** is a member of the variable-metric family widely used for computing approximations to second derivative matrices in function optimization (see Broyden, 1967; Davidon, 1968). The vector γ represents the change in the gradient vector of q_{true} evaluated at x_n and x_c respectively. Algorithm 1.1 is stated in terms of generating a sequence of matrices that suppresses the original purpose of the rank-one formula, which is to solve

$$\min\{q_{\text{true}}(x) \mid x \in \mathbb{R}^n\}.$$

When $\mathfrak{Q}_{\text{true}}$ is not known, increasingly accurate approximations are provided by the sequence of $\mathfrak{Q}_c$ generated by the rank-one formula. In this case, ∇q_{true} and δ are known. The vector δ is a descent direction at x_c, implying that q_{true} is reduced along δ. The latter is usually given by

$$\delta = -\mathfrak{Q}_c^{-1}\nabla q_{\text{true}}(x_c). \tag{1.15}$$

In a multiple-objective decision making environment ∇q_{true} is generally not available and, as discussed in Algorithm 1.1, δ is specified by the decision maker. In this case, assigning values to γ requires further interaction with the decision maker (Bray, 1972, 1975). In Section 2, the approximation

$$\gamma \simeq \phi\mathfrak{Q}_c\delta$$

is considered for some scalar $\phi \geqslant 1$.

The matrix $\mathfrak{Q}_{\text{true}}$ generated by Algorithm 1.1 is globally valid and hence not dependent on the feasible region $\mathscr{R}$. The correction computed using the above updating formula has the property that it does not restrict the choice of the direction or magnitude of δ. The denominator needs to be safeguarded from becoming zero, and this can be accomplished with small changes to δ and/or γ within limits acceptable to the decision maker. Provided this is done, it can be shown that $\mathfrak{Q}_n$ remains positive-definite if $\mathfrak{Q}_c$ is positive-definite. As far as the present study is concerned, the interesting aspect of the algorithm is that, *after a finite number of steps, the sequence of $\mathfrak{Q}_c$ converges to the true weighting matrix of the decision maker*. This is stated in the following theorem.

Theorem 1.1 (Fiacco and McCormick, 1968; Theorem 41)
*Assume that $q_{\text{true}}(x)$ is a positive-definite quadratic form. Let the pair $(\tau\delta, \gamma)$ specified in **Step 1** be such that n* linearly independent *$\tau\delta$ vectors, and corresponding γ, are specified by the decision maker. Let each δ be a descent direction (1.15)*

at the corresponding x_c, $\langle \nabla q_{\mathrm{true}}(x_c), \delta \rangle < 0$, *and let each* $(\tau\delta, \gamma)$ *pair satisfy* $Q_{\mathrm{true}}(\tau\delta) = \gamma$. *Let the denominators the updating formula be nonzero. Then, starting with any positive-definite weighting matrix,* $\mathfrak{Q}_c$, *the sequence of matrices generated by Algorithm 1.1 converge (elementwise) to* $\mathfrak{Q}_{\mathrm{true}}$ *after at most* n *iterations.*

This result requires the pair $(\tau\delta, \gamma)$ to be consistent with the true weighting matrix and hence with q_{true} (x). In addition, there is no indication that there exists in reality a matrix $\mathfrak{Q}_{\mathrm{true}}$ that is valid globally on $\mathscr{R}$, for the decision maker. Thus, $\mathfrak{Q}_c$ in (1.2) can at best be seen as *a local approximation to the weighting matrix of the decision maker*. Furthermore, the above theorem does not apply for the approximation $\gamma \simeq \phi \mathfrak{Q}_c \delta$. In subsequent sections, and Chapters 4 and 5, the use of this approximation is explored for linearly constrained or, in general, convex $\mathscr{R}$. The justification for the resulting method and the properties of the updating formulae are discussed in Section 3.

2 THE INTERACTIVE METHOD FOR DETERMINING THE WEIGHTING MATRIX

Let $\Omega \subset \mathbb{R}^n$ denote the set of admissible values of x from the point of the decision maker. Given the desired value x^d and the region $\mathscr{R}$, the 'nearness' of the solution of (1.1) to x^d is only affected by the weighting matrix $\mathfrak{Q}_c$. Initially, we require $\mathfrak{Q}_c$ to be positive-definite and subsequently relax this to positive-semidefinite $\mathfrak{Q}_c$. In the latter case, the target values are allowed to be a subset of the vector x. In the former case, x and the targets are identical.

Different values of $\mathfrak{Q}_c$ define different points on $\mathscr{R}$ as the nearest point to x^d. Thus, given x^d, the only way of producing a solution of (1.1) that also satisfies Ω is to re-specify $\mathfrak{Q}_c$. It should be noted that, in contrast to the algebraic equalities and inequalities describing $\mathscr{R}$, the set Ω is assumed to exist only in the mind of the rational decision maker. It is further assumed that

$$\Omega \cap \mathscr{R} \neq \varnothing \tag{2.1}$$

and that this intersection is convex. Clearly, this assumption is satisfied for convex $\mathscr{R}$ and Ω. Thus, problem (1.1) has to be *solved a number of times, by re-specifying* $\mathfrak{Q}_c$, *until a solution is found such that* $x_c \in \Omega \cap \mathscr{R}$.

Let $\mathfrak{Q}_c$ denote the *current weighting matrix* of (1.2). The solution of (1.1) using this matrix will be called the *current optimal solution* x_c. This solution is presented to the decision maker, who may decide that $x_c \notin \Omega$ since some of the elements of x_c are not quite what she wants them to be. Consequently, an alternative solution of (1.1) has to be obtained by altering $\mathfrak{Q}_c$. In order to do this, the decision maker is required to specify a value for x which she would prefer, instead of x_c. This *preferred value* is denoted by x_p and obviously satisfies

$$x_p \in \Omega \tag{2.2}$$

but *not necessarily*

$$x_{\mathrm{p}} \in \mathscr{R}. \tag{2.3}$$

It should be noted that if $x_{\mathrm{p}} \in \mathscr{R}$ then the decision maker can achieve this value, and the search, or algorithm, stops. The vector x_{p} incorporates all the alterations to x_{c} such that x_{p} is preferred, by the decision maker, to x_{c}. Given x_{c} and x_{p}, the displacement or *correction vector* is defined as

$$\delta = x_{\mathrm{p}} - x_{\mathrm{c}}. \tag{2.4}$$

Using $\delta, \mathfrak{Q}_{\mathrm{c}}$ is altered to obtain the *new weighting matrix* $\mathfrak{Q}_{\mathrm{n}}$, with the rank-one correction

$$\mathfrak{Q}_{\mathrm{n}} = \mathfrak{Q}_{\mathrm{c}} + \eta \frac{\mathfrak{Q}_{\mathrm{c}}\delta\,\delta^{\mathrm{T}}\mathfrak{Q}_{\mathrm{c}}}{\langle \delta, \mathfrak{Q}_{\mathrm{c}}\delta \rangle}, \tag{2.5}$$

where $\eta \geqslant 0$ is a scalar, chosen to reflect the emphasis to be given to the update.

The new matrix $\mathfrak{Q}_{\mathrm{n}}$ is used in (1.1) to compute a *new optimal solution* x_{n}. As illustrated in Figure 2.1, the method starts by generating a current optimal solution with a given $\mathfrak{Q}_{\mathrm{c}}$ and x^{d}. The basic idea is to alter $\mathfrak{Q}_{\mathrm{c}}$ to $\mathfrak{Q}_{\mathrm{n}}$ so that x_{n} obtained by solving (1.1) with $\mathfrak{Q}_{\mathrm{n}}$, instead of $\mathfrak{Q}_{\mathrm{c}}$, is also in Ω. The role of η in determining x_{n} and the desirable characteristics of x_{n}, including the fact that it is an improvement on x_{c}, are discussed in Section 3.

The simple conjecture that a single update of the form (2.5) may not necessarily yield a completely satisfactory new optimal solution (i.e. $x_{\mathrm{n}} \notin \Omega$) leads to an iterative procedure. Thus, the process of adjusting the weights

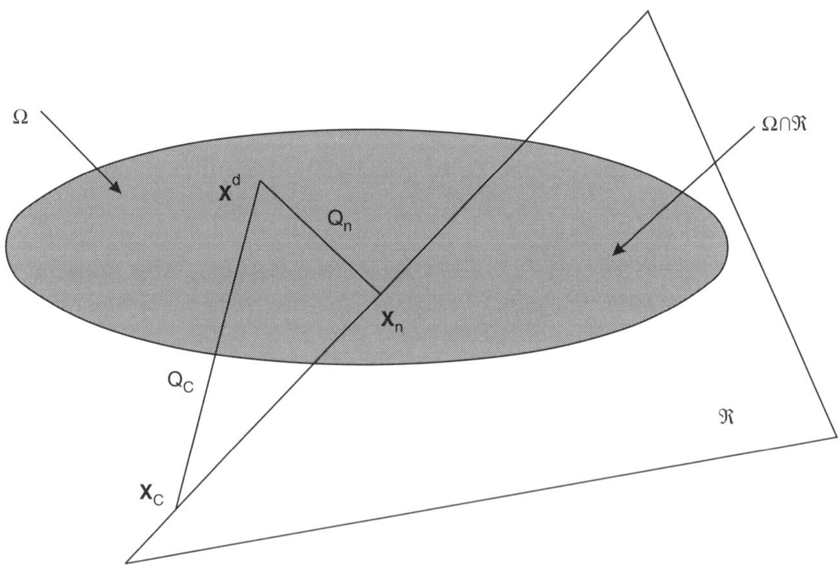

Figure 2.1 The influence of the weighting matrix in altering the optimal solution.

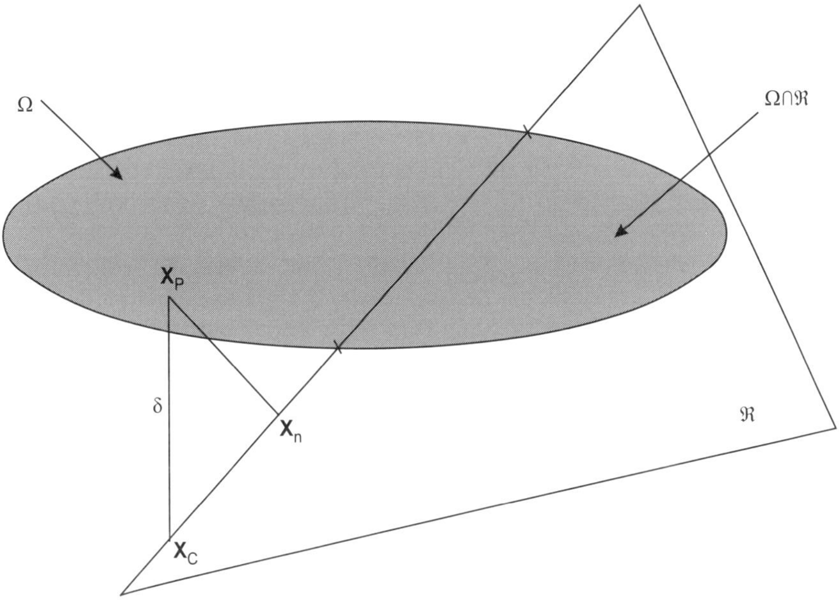

Figure 2.1 The influence of the weighting matrix in altering the optimal solution.

using (2.5) may need to be repeated until an acceptable optimal solution is attained. Figure 2.2 illustrates this iterative aspect: if $x_n \notin \Omega$, the adjustment (2.5) has to be repeated. The resulting algorithm is an interactive procedure between the decision maker and the model. It represents a cognitive process in which the decision maker is able to analyse the relationship between feasible and desirable decisions. In Chapter 5, the interactive updating of the weighting matrix is also characterized as the specification of constraints on the problem, as these become evident to the decision maker. Thus, (2.5) is utilized iteratively until $x_n \in \Omega$, yielding the following algorithm.

Algorithm 2.1 (Respecification of the quadratic objective)

Step 0: Given the desired value x^d and the feasible region $\mathscr{R}$, assume some initial symmetric, positive-semidefinite weighting matrix $\mathfrak{Q}_c$.

Step 1: Using $\mathfrak{Q}_c$, solve (1.1) to obtain x_c.

Step 2: (*The interaction between the decision maker and the method.*) If, according to the decision maker $x_c \in \Omega$, and hence $x_c \in \Omega \cap \mathscr{R}$, **stop**. Otherwise ask the decision maker to specify the changes δ required in the current optimal solution x_c to make it acceptable. The preferred value x_p is thus specified as

$$x_p = x_c + \delta.$$

Alternatively, the decision maker might choose to specify x_p directly.

Step 3: Given $\mathfrak{Q}_c$ and δ, choose an $\eta \geqslant 0$ and compute $\mathfrak{Q}_n$ using (2.5).

Step 4: Set $\mathfrak{Q}_c = \mathfrak{Q}_n$ and **return to Step 1**.

The choice of η is bounded from above, and this is discussed in Section 3. Also, the denominator of (2.5) has to be protected from becoming zero. This can be accomplished with small changes to δ, within limits acceptable to the decision maker. Provided this is done, $\mathfrak{Q}_n$ remains symmetric positive-(semi)definite if $\mathfrak{Q}_c$ is symmetric positive-(semi)definite.

An obvious choice of x_p is $x_p = x^d$. However, x^d defines the position of the unconstrained minimum of the decision maker's objective function (1.2). On the other hand, the purpose of Algorithm 2.1 is to define the *curvature* of this function, given x^d. Thus, $x_p = x^d$ does not add any new information for the determination of $\mathfrak{Q}$. This point is further discussed in Section 4, where it is shown that the choice $x_p = x^d$ indeed yields $x_n = x_c$.

3 PROPERTIES OF THE METHOD: LINEAR EQUALITY CONSTRAINTS

In this section, we discuss the method when $\mathscr{R}$ is given by the linear equalities in (1.4). Convex and general nonlinear constraints are considered in Chapter 5. However, the basic results for linear equality constraints in this section are also used in these generalizations. The properties of the iterative procedure of the previous section can be analysed by inspecting the effect of updating $\mathfrak{Q}_c$ on the optimal solution of (1.1). The next two theorems characterize the effect of (2.5) by considering x_n in terms of an update on the current optimal solution x_c and the corresponding update for the shadow price λ_c.

Theorem 3.1
Let

(i) $\mathfrak{Q}_c$ be positive-definite;

(ii) the feasible region $\mathscr{R}$ be given by (1.4).

Then, for any $\delta \neq 0$ and $\eta \geqslant 0$, $\mathfrak{Q}_n$ given by (2.5) is positive-definite and x_n obtained by using $\mathfrak{Q}_n$ instead of $\mathfrak{Q}_c$ in (1.1) is expressed in terms of x_c as

$$x_n = x_c + \tau P \delta, \tag{3.1}$$

where P is given by (1.12) and

$$\tau = -\frac{\eta\langle \delta, \mathfrak{Q}_c(x_c - x^d)\rangle}{\langle \delta, \mathfrak{Q}_c \delta\rangle + \eta\langle \mathfrak{Q}_c\delta, P\mathfrak{Q}_c^{-1}(\mathfrak{Q}_c\delta)\rangle}. \tag{3.2}$$

The corresponding shadow price vector at x_n is

$$\lambda_n = \lambda_c - \tau(\mathcal{G}^T \mathfrak{Q}_c^{-1} \mathcal{G})^{-1} \mathcal{G}^T \delta. \tag{3.3}$$

Furthermore, we have $\tau \geqslant 0$ *for*

$$\langle \delta, \mathfrak{Q}_c(x_c - x^d) \rangle \leqslant 0. \tag{3.4}$$

Proof

The fact that $\mathfrak{Q}_n$ is positive-definite if $\mathfrak{Q}_c$ is positive-definite follows trivially from (2.5). Expressions (3.1) and (1.12) can be derived, as discussed in Rustem, Velupillai and Westcott (1978, Lemmas 1 and 2), by using (2.5) instead of $\mathfrak{Q}_c$ in (1.5) and by applying the Sherman–Morrison formula (see CN.4). Alternatively, it follows from (1.5) and (1.6) that

$$x_n = x^d + \mathfrak{Q}_n^{-1} \mathcal{G} \lambda_n, \tag{3.5}$$

and so (3.1) and (1.12) can be obtained from the derivation of λ_n given below. To derive the update (3.3), consider

$$\begin{aligned}
\lambda_n &= -(\mathcal{G}^T \mathfrak{Q}_n^{-1} \mathcal{G})^{-1} (\mathcal{G}^T x^d - g) \qquad (3.6)\\
&= -\left[\mathcal{G}^T \left(\mathfrak{Q}_c + \eta \frac{\mathfrak{Q}_c \delta\, \delta^T \mathfrak{Q}_c}{\langle \delta, \mathfrak{Q}_c \delta \rangle}\right)^{-1} \mathcal{G} \right]^{-1} (\mathcal{G}^T x^d - g)\\
&= -\left(\mathcal{G}^T \mathfrak{Q}_c^{-1} \mathcal{G} - \frac{\eta}{1+\eta} \frac{\mathcal{G}^T \delta\, \delta^T \mathcal{G}}{\langle \delta, \mathfrak{Q}_c \delta \rangle} \right)^{-1} (\mathcal{G}^T x^d - g)\\
&= -(\mathcal{G}^T \mathfrak{Q}_c^{-1} \mathcal{G})^{-1} (\mathcal{G}^T x^d - g)
\end{aligned}$$

$$- \frac{\eta (\mathcal{G}^T \mathfrak{Q}_c^{-1} \mathcal{G})^{-1} \mathcal{G}^T \delta\, \delta^T \mathcal{G} (\mathcal{G}^T \mathfrak{Q}_c^{-1} \mathcal{G})^{-1}}{(\eta + 1)\langle \delta, \mathfrak{Q}_c \delta \rangle - \eta \langle \delta, \mathcal{G}(\mathcal{G}^T \mathfrak{Q}_c^{-1} \mathcal{G})^{-1} \mathcal{G}^T \delta \rangle} (\mathcal{G}^T x^d - g) \tag{3.7a}$$

$$= \lambda_c - \frac{\eta \langle \delta, \mathcal{G}(\mathcal{G}^T \mathfrak{Q}_c^{-1} \mathcal{G})^{-1} (\mathcal{G}^T x^d - b) \rangle}{\eta \langle \mathfrak{Q}_c \delta, P \mathfrak{Q}_c^{-1} (\mathfrak{Q}_c \delta) \rangle + \langle \delta, \mathfrak{Q}_c \delta \rangle} (\mathcal{G}^T \mathfrak{Q}_c^{-1} \mathcal{G})^{-1} \mathcal{G}^T \delta. \tag{3.7b}$$

Expressions (1.6) and (1.9) may be used to simplify (3.7), since

$$\begin{aligned}
-\langle \delta, \mathcal{G}(\mathcal{G}^T \mathfrak{Q}_c^{-1} \mathcal{G})^{-1} (\mathcal{G}^T x^d - g) \rangle &= \langle \delta, \mathcal{G}(\mathcal{G}^T \mathfrak{Q}_c^{-1} \mathcal{G})^{-1} \mathcal{G}^T \mathfrak{Q}_c^{-1} \mathfrak{Q}_c (x_c - x^d) \rangle\\
&= \langle \delta, \mathcal{G} \lambda_c \rangle \qquad (3.8)\\
&= \langle \delta, \mathfrak{Q}_c (x_c - x^d) \rangle,
\end{aligned}$$

and (3.7) with (3.8) yields the required result (3.3) with τ given by (3.2). Expression (3.7a) can also be used along with (3.5) to derive (3.1). Thus,

$$\begin{aligned}x_{\mathrm{n}} &= x^{\mathrm{d}} + \left(\mathfrak{Q}_{\mathrm{c}}^{-1} - \frac{\eta}{1+\eta}\frac{\delta\,\delta^{\mathrm{T}}}{\langle\delta, \mathfrak{Q}_{\mathrm{c}}\delta\rangle}\right)\mathscr{G}[\lambda_{\mathrm{c}} - \tau(\mathscr{G}^{\mathrm{T}}\mathfrak{Q}_{\mathrm{c}}^{-1}\mathscr{G})^{-1}\mathscr{G}^{\mathrm{T}}\delta] \\ &= x^{\mathrm{d}} + \mathfrak{Q}_{\mathrm{c}}^{-1}\mathscr{G}\lambda_{\mathrm{c}} - \tau\mathfrak{Q}_{\mathrm{c}}^{-1}\mathscr{G}(\mathscr{G}^{\mathrm{T}}\mathfrak{Q}_{\mathrm{c}}^{-1}\mathscr{G})^{-1}\mathscr{G}^{\mathrm{T}}\delta \\ &\quad - \frac{\eta\langle\delta, \mathfrak{Q}_{\mathrm{c}}(x_{\mathrm{c}} - x^{\mathrm{d}})\rangle}{1+\eta}\frac{\delta}{\langle\delta, \mathfrak{Q}_{\mathrm{c}}\delta\rangle} \\ &\quad \times \left[1 + \frac{\eta\langle\delta, \mathscr{G}(\mathscr{G}^{\mathrm{T}}\mathfrak{Q}_{\mathrm{c}}^{-1}\mathscr{G})^{-1}\mathscr{G}^{\mathrm{T}}\delta\rangle}{(1+\eta)\langle\delta, \mathfrak{Q}_{\mathrm{c}}\delta\rangle - \eta\langle\delta, \mathscr{G}(\mathscr{G}^{\mathrm{T}}\mathfrak{Q}_{\mathrm{c}}^{-1}\mathscr{G})^{-1}\mathscr{G}^{\mathrm{T}}\delta\rangle}\right] \\ &= x_{\mathrm{c}} + \tau P\delta.\end{aligned}$$

Finally, it can be seen from (3.2) by inspection that if inequality (3.4) is satisfied then $\tau \geqslant 0$, since $P\mathfrak{Q}_{\mathrm{c}}^{-1}$ is symmetric positive-semidefinite. □

The matrix P given by (1.12) ensures that the correction term $\tau P\delta$ in (3.1) is along the projection of $\tau\delta$ onto the subspace $\mathscr{R}_0$. Following the discussion in Section 1, this implies that x_{n} is as 'near' to the preferred solution x_{p} as allowed by the feasible region $\mathscr{R}$. The magnitude of the stepsize τ can be controlled using η. The inequality (3.4) may be interpreted as a 'rationality condition' on the choice of δ. The reason for this lies in the form of (3.1). When (3.4) holds, $\tau \geqslant 0$, and thus the modification $\tau P\delta$ to the current optimal solution in (3.1) lies in the same direction as δ. This is the 'best' feasible alternative to δ, in the sense of the norm (1.3), when δ is infeasible and lies outside $\mathscr{R}$. Also, δ is, by definition, a descent direction for $q_{\mathrm{n}}(x)$ at x_{c}, since it is given by (2.4). Thus, the information that x_{p} is actually preferred to x_{c} is incorporated in $q_{\mathrm{n}}(x)$. Using (2.5) with (3.4) establishes this result

$$\begin{aligned}\langle\delta, \nabla q_{\mathrm{n}}(x_{\mathrm{c}})\rangle &= \langle\delta, \mathfrak{Q}_{\mathrm{n}}(x_{\mathrm{c}} - x^{\mathrm{d}})\rangle \\ &= \left\langle\delta, \left[\mathfrak{Q}_{\mathrm{c}} + \eta\frac{\mathfrak{Q}_{\mathrm{c}}\delta\,\delta^{\mathrm{T}}\mathfrak{Q}_{\mathrm{c}}}{\langle\delta, \mathfrak{Q}_{\mathrm{c}}\delta\rangle}\right](x_{\mathrm{c}} - x^{\mathrm{d}})\right\rangle \\ &= (1+\eta)\langle\delta, \mathfrak{Q}_{\mathrm{c}}(x_{\mathrm{c}} - x^{\mathrm{d}})\rangle \\ &= (1+\eta)\langle\delta, \nabla q_{\mathrm{c}}(x_{\mathrm{c}})\rangle.\end{aligned}$$

Thus, δ is a decent direction for $q_{\mathrm{n}}(x_{\mathrm{c}})$ if it is a descent direction for $q_{\mathrm{c}}(x_{\mathrm{c}})$, and condition (3.4) ensures the latter.

Further properties of the method, discussed in Chapter 5, are dependent on (3.1). In particular, the latter is used to show that η can control, and thus bound, the size of τ. Hence, the distance between x_{n} and x_{p} is ensured to be less than the distance between x_{c} and x_{p} (see Lemma 5.3.3).

The basic limitation of Theorem 3.1 is that $\mathfrak{Q}_{\mathrm{c}}$, and hence $\mathfrak{Q}_{\mathrm{n}}$, are assumed to be strictly positive-definite. This is rather restrictive, since the decision maker need not attach an objective or have a target for every element of x: some elements may have an objective, whereas others may be free variables in the region $\mathcal{R}$. It is worth noting that when $\mathfrak{Q}_c$ is positive-semidefinite, the update (2.5) exists when

$$\langle \delta, \mathfrak{Q}_c \delta \rangle > 0.$$

As mentioned in Section 2, this inequality can be maintained with small alterations to δ, within limits acceptable to the decision maker. Also, if δ is always specified to express corrections only in those variables that have objectives (i.e. those corresponding to a submatrix $\mathfrak{Q}$ that is positive-definite) then it can be shown that δ lies in the range of $\mathfrak{Q}_c$ and thereby satisfies $\langle \delta, \mathfrak{Q}_c \delta \rangle > 0$. The following result establishes analogous expressions to (3.1)–(3.3) when $\mathfrak{Q}_c$ is positive-semidefinite in general but its projection, on the intersection of the constraints, is positive-definite.

Theorem 3.2

Let

(i) $\mathfrak{Q}_c$ *be positive semidefinite;*

(ii) $\mathcal{R}$ *be given by (1.4);*

(iii) $Z \in \mathbb{R}^{n \times (n-e)}$ *be a matrix with linearly independent columns orthogonal to the matrix* $\mathcal{G} \in \mathbb{R}^{n \times e}$ *with linearly independent columns in (1.4), and* $Z^{\mathrm{T}} \mathfrak{Q}_c Z$ *positive-definite.*

Then, for any δ *satisfying* $\langle \delta, \mathfrak{Q}_c \delta \rangle > 0$, and $\eta \geqslant 0$, $\mathfrak{Q}_n$ *given by (2.5) is positive-semidefinite, and the new optimal solution is given by*

$$x_n = x_c + \tau_z P_z \mathfrak{Q}_c \delta, \tag{3.9}$$

where

$$P_z = Z(Z^{\mathrm{T}} \mathfrak{Q}_c Z)^{-1} Z^{\mathrm{T}}, \tag{3.10}$$

$$\tau_z = -\frac{\eta \langle \delta, \mathfrak{Q}_c (x_c - x^d) \rangle}{\langle \delta, \mathfrak{Q}_c \delta \rangle + \eta \langle \mathfrak{Q}_c \delta, P_z \mathfrak{Q}_c \delta \rangle}. \tag{3.11}$$

The corresponding shadow price vector at x_n *is given by*

$$\lambda_n = \lambda_c - \tau_z (\mathcal{G}^{\mathrm{T}} \mathcal{G})^{-1} \mathcal{G}^{\mathrm{T}} \mathfrak{Q}_c (I - P_z \mathfrak{Q}_c) \delta. \tag{3.12}$$

Furthermore, for δ *satisfying (3.4), we have* $\tau_z \geqslant 0$.

Proof

The positive-semidefiniteness of $\mathfrak{Q}_n$ follows directly from (2.5) for positive-semidefinite $\mathfrak{Q}_c$. To establish (3.9), consider the optimality condition (1.9) with $\mathfrak{Q}_n$, given by (2.5), replacing $\mathfrak{Q}_c$:

$$\left(\mathfrak{Q}_c + \eta \frac{\mathfrak{Q}_c \delta \, \delta^T \mathfrak{Q}_c}{\langle \delta, \mathfrak{Q}_c \delta \rangle}\right)(x_n - x_c + x_c - x^d) - \mathcal{G}\lambda_n = 0. \tag{3.13}$$

Any vector v satisfying $\mathcal{R}_0$ given by (1.11) can be written in the form $v = Z\omega$, where $\omega \in \mathbb{R}^{(n-e)}$, since $Z^T\mathcal{G} = 0$ (see Chapter 2, and also Gill and Murray, 1978). As both x_n and x_c satisfy $\mathcal{R}$, we have $x_n - x_c \in \mathcal{R}_0$, and thus

$$x_n - x_c = Z\omega \tag{3.14}$$

for some ω. Using (3.14) and premultiplying (3.13) by Z^T yields

$$Z^T\left(\mathfrak{Q}_c + \eta \frac{\mathfrak{Q}_c \delta \, \delta^T \mathfrak{Q}_c}{\langle \delta, \mathfrak{Q}_c \delta \rangle}\right)(Z\omega + x_c - x^d) = 0,$$

$$Z\omega = -Z\left[Z^T\left(\mathfrak{Q}_c + \eta \frac{\mathfrak{Q}_c \delta \, \delta^T \mathfrak{Q}_c}{\langle \delta, \mathfrak{Q}_c \delta \rangle}\right)Z\right]^{-1} Z^T\left(\mathfrak{Q}_c + \eta \frac{\mathfrak{Q}_c \delta \, \delta^T \mathfrak{Q}_c}{\langle \delta, \mathfrak{Q}_c \delta \rangle}\right)(x_c - x^d). \tag{3.15}$$

Expression (1.9) can be used to establish the equality

$$Z^T\mathfrak{Q}_c(x_c - x^d) = Z^T\mathcal{G}\lambda_c = 0.$$

Thus,

$$\begin{aligned} x_n - x_c &= Z\omega \\ &= -Z\left[Z^T\left(\mathfrak{Q}_c + \eta \frac{\mathfrak{Q}_c \delta \, \delta^T \mathfrak{Q}_c}{\langle \delta, \mathfrak{Q}_c \delta \rangle}\right)Z\right]^{-1} Z^T\left[\eta \frac{\mathfrak{Q}_c \delta \, \delta^T \mathfrak{Q}_c}{\langle \delta, \mathfrak{Q}_c \delta \rangle}\right](x_c - x^d), \end{aligned}$$

and application of the Sherman–Morrison formula yields

$$\begin{aligned} x_n = x_c - Z&\left[(Z^T\mathfrak{Q}_c Z)^{-1} - \frac{\eta (Z^T\mathfrak{Q}_c Z)^{-1} Z^T \mathfrak{Q}_c \delta \, \delta^T \mathfrak{Q}_c Z^T (Z^T\mathfrak{Q}_c Z)^{-1}}{\langle \delta, \mathfrak{Q}_c \delta \rangle + \eta \langle \mathfrak{Q}_c \delta, P_z \mathfrak{Q}_c \delta \rangle}\right] \\ &\times Z^T\left[\eta \frac{\mathfrak{Q}_c \delta \, \delta^T \mathfrak{Q}_c}{\langle \delta, \mathfrak{Q}_c \delta \rangle}\right](x_c - x^d), \end{aligned}$$

from which, after some rearrangement, the required result (3.9) follows.

An alternative expression to (1.6) for λ_c is given by the least squares solution of λ_c using (1.9):

$$\lambda_c = \left(\mathcal{G}^T\mathcal{G}\right)^{-1}\mathcal{G}^T\mathfrak{Q}_c(x_c - x^d).$$

For $\mathfrak{Q}_n$, this becomes

$$\lambda_n = \left(\mathcal{G}^T\mathcal{G}\right)^{-1}\mathcal{G}^T\left[\mathfrak{Q}_c + \eta \frac{\mathfrak{Q}_c \delta \, \delta^T \mathfrak{Q}_c}{\langle \delta, \mathfrak{Q}_c \delta \rangle}\right](x_n - x_c + x_c - x^d),$$

which can be simplified using (1.9) and (3.9) to yield

$$\lambda_n = \lambda_c + (\mathscr{G}^T\mathscr{G})^{-1}\mathscr{G}^T\left[\mathfrak{Q}_c(x_n - x_c) + \eta\frac{\mathfrak{Q}_c\delta\,\delta^T\mathfrak{Q}_c}{\langle\delta, \mathfrak{Q}_c\delta\rangle}(x_n - x_c + x_c - x^d)\right]$$
$$= \lambda_c + (\mathscr{G}^T\mathscr{G})^{-1}\mathscr{G}^T\mathfrak{Q}_c\left[\tau_z P_z\mathfrak{Q}_c\delta + \tau_z\eta\mathfrak{Q}_c\delta\frac{\langle\mathfrak{Q}_c\delta, P_z\mathfrak{Q}_c\delta\rangle}{\langle\delta, \mathfrak{Q}_c\delta\rangle}\right.$$
$$\left.+ \eta\mathfrak{Q}_c\delta\frac{\langle\delta, \mathfrak{Q}_c(x_c - x^d)\rangle}{\langle\delta, \mathfrak{Q}_c\delta\rangle}\right].$$

After some rearrangement involving (3.11), this can be expressed as (3.12). Finally, the fact that $\tau_z \geqslant 0$ when (3.4) is satisfied follows from the positive-semidefiniteness of $\mathfrak{Q}_c P_z \mathfrak{Q}_c$. □

Although Theorem 3.2 establishes a more general result, the projection property of the method is equivalently characterized both in Theorem 3.1 and in Theorem 3.2 when $\mathfrak{Q}_c$ is positive-definite. The correspondence between x_n and λ_n computed using the expressions derived in either of these Theorems is discussed in Lemma 3.1 below. Following Theorem 3.2, the positive-semidefinite case is expressed as an equivalent positive-definite problem of reduced dimension. This can also be achieved by assuming that the feasible region $\mathscr{R}$ bounds $q_c(x)$ from below and formulating an optimization problem in which variables with zero weighting in $q_c(x)$ are substituted out using the constraints. The approach in Theorem 3.2 is in more general terms, requiring only the positive-definiteness of $Z^T\mathfrak{Q}_cZ$ instead of $\mathfrak{Q}_c$. The matrix $Z^T\mathfrak{Q}_cZ$ can be regarded as the projection of $\mathfrak{Q}_c$ onto the constraints.

Lemma 3.1
Let $\mathfrak{Q}_c$ be positive-definite. Then x_n and λ_n computed in Theorems 3.1 and 3.2 are identical. Hence,

$$P\delta = P_z\mathfrak{Q}_c\delta, \tag{3.16}$$
$$\tau = \tau_z, \tag{3.17}$$
$$-(\mathscr{G}^T\mathfrak{Q}_c^{-1}\mathscr{G})^{-1}(\mathscr{G}^Tx^d - g) = (\mathscr{G}^T\mathscr{G})^{-1}\mathscr{G}^T\mathfrak{Q}_c(x_c - x^d), \tag{3.18}$$
$$(\mathscr{G}^T\mathfrak{Q}_c^{-1}\mathscr{G})^{-1}\mathscr{G}^T = (\mathscr{G}^T\mathscr{G})^{-1}\mathscr{G}^T\mathfrak{Q}_c(I - P_z\mathfrak{Q}_c)\delta. \tag{3.19}$$

Proof
To establish that x_n computed using (3.1) and (3.9) are equivalent, (3.16) and (3.17) must be shown to hold. Since $P\delta \in \mathscr{R}_0$, we have

$$P\delta = Z\omega$$

for some $\omega \in \mathbb{R}^{(n-m)}$. Pre-multiplying by $Z^T\mathfrak{Q}_c$ yields

$$Z^T\mathfrak{Q}_cP\delta = Z^T\mathfrak{Q}_c\left[I - \mathfrak{Q}_c^{-1}\mathscr{G}\left(\mathscr{G}^T\mathfrak{Q}_c^{-1}\mathscr{G}\right)^{-1}\mathscr{G}^T\right]\delta$$
$$= Z^T\mathfrak{Q}_cZ\omega,$$
$$Z^T\mathfrak{Q}_c\delta = Z^T\mathfrak{Q}_cZ\omega,$$

since $Z^{\mathrm{T}}\mathcal{G} = 0$. Thus, (3.1) and (3.9) are equivalent. It follows immediately that (3.2) and (3.11) are also equivalent, and hence (3.17) holds.

To show that (3.18) holds, consider the optimality condition (1.9). The left-hand side of (3.18) may be expressed as

$$\begin{aligned} -\left(\mathcal{G}^{\mathrm{T}}\mathfrak{Q}_{\mathrm{c}}^{-1}\mathcal{G}\right)^{-1}\left(\mathcal{G}^{\mathrm{T}}x^{\mathrm{d}} - g\right) &= \left(\mathcal{G}^{\mathrm{T}}\mathfrak{Q}_{\mathrm{c}}^{-1}\mathcal{G}\right)^{-1}\mathcal{G}^{\mathrm{T}}\left(x_{\mathrm{c}} - x^{\mathrm{d}}\right) \\ &= \left(\mathcal{G}^{\mathrm{T}}\mathfrak{Q}_{\mathrm{c}}^{-1}\mathcal{G}\right)^{-1}\mathcal{G}^{\mathrm{T}}\mathfrak{Q}_{\mathrm{c}}^{-1}\mathcal{G}\lambda_{\mathrm{c}} \\ &= \lambda_{\mathrm{c}}, \end{aligned}$$

where λ_{c} satisfies (1.9) for $\mathcal{G}$ with full column rank. The right-hand side of (3.18) can be expressed as

$$\begin{aligned} \left(\mathcal{G}^{\mathrm{T}}\mathcal{G}\right)^{-1}\mathcal{G}^{\mathrm{T}}\mathfrak{Q}_{\mathrm{c}}(x_{\mathrm{c}} - x^{\mathrm{d}}) &= \left(\mathcal{G}^{\mathrm{T}}\mathcal{G}\right)^{-1}\mathcal{G}^{\mathrm{T}}\mathcal{G}\lambda_{\mathrm{c}} \\ &= \lambda_{\mathrm{c}}. \end{aligned}$$

Thus, (3.18) is satisfied. Finally, using (3.16), the required result is established for (3.19):

$$\begin{aligned} \left(\mathcal{G}^{\mathrm{T}}\mathcal{G}\right)^{-1}\mathcal{G}^{\mathrm{T}}\mathfrak{Q}_{\mathrm{c}}(I - P_z\mathfrak{Q}_{\mathrm{c}})\delta &= \left(\mathcal{G}^{\mathrm{T}}\mathcal{G}\right)^{-1}\mathcal{G}^{\mathrm{T}}\mathfrak{Q}_{\mathrm{c}}\mathfrak{Q}_{\mathrm{c}}^{-1}\mathcal{G}\left(\mathcal{G}^{\mathrm{T}}\mathfrak{Q}_{\mathrm{c}}^{-1}\mathcal{G}\right)^{-1}\mathcal{G}^{\mathrm{T}}\delta \\ &= \left(\mathcal{G}^{\mathrm{T}}\mathfrak{Q}_{\mathrm{c}}^{-1}\mathcal{G}\right)^{-1}\mathcal{G}^{\mathrm{T}}\delta. \qquad \square \end{aligned}$$

Theorems 3.1 and 3.2 show that the new optimal solution x_{n} lies along the projection, on $\mathcal{R}$, of the direction $\delta(= x_{\mathrm{p}} - x_{\mathrm{c}})$ specified by the decision maker. Expressions (3.1) and (3.9) can thus be interpreted as yielding a new optimal solution such that the change $x_{\mathrm{n}} - x_{\mathrm{c}}$ is as close as is feasibly possible to the required change δ. Thus, the method achieves the best that is possible. Further results, such as

$$\begin{aligned} \|x_{\mathrm{n}} - x_{\mathrm{p}}\|_{\mathfrak{Q}_{\mathrm{c}}} &< \|x_{\mathrm{c}} - x_{\mathrm{p}}\|_{\mathfrak{Q}_{\mathrm{c}}}, \\ \|x_{\mathrm{n}} - x_{\mathrm{p}}\|_{\mathfrak{Q}_{\mathrm{n}}} &< \|x_{\mathrm{c}} - x_{\mathrm{p}}\|_{\mathfrak{Q}_{\mathrm{n}}}, \end{aligned}$$

are established in Chapter 5, along with extensions of these results to convex and general nonlinear constraints. The overall convergence of the method to a solution $x_{\mathrm{n}} \in \mathcal{R} \cap \Omega$ is shown by invoking Khachian's (1979) algorithm.

4 REVEALED PREFERENCES, CHOICE OF PREFERRED VALUE AND SENSITIVITY

4.1 Inverse optimal control and revealed preferences

Given that the method is able to construct a weighting matrix that yields an optimal solution acceptable to the decision maker, we need to address its potential use as a solution to the problem of revealing historical preferences, or inverse optimal control. In this section, we establish that such an interpretation of the method would be incorrect. The idea of 'uncovering the objective function', once the decision has been made by others, is an old and illusive one. It has arisen independently in optimal control theory, in economics and finance. The interest from these areas is due to the desire to reveal the objective function, or the past preferences of the decision maker, that has given rise to the past optimal strategy.

The iterative method of Sections 1 and 2 may, for the purpose of this section, be formulated so that the desired value x^{d} is set according to the desirable historical conditions, and the feasible region $\mathcal{R}$ is fixed as the model of the system. The preferred value x_{p} is fixed at the actual historical value throughout the procedure, and is not changed, as would normally be done in *Step 2* of Algorithm 2.1.

In this setting, the historical value, which, in this case, is also the preferred value, has to be a feasible point, since, by definition, the model of the system must explain the historical event. Thus, for a linear model such as (1.4), the preferred value and all current optimal solutions are feasible, and hence δ satisfies

$$\mathcal{G}^{\mathrm{T}}\delta = 0.$$

Using this and (1.9), we can express the stepsize τ in (3.1) (or τ_z in (3.11)) as

$$\begin{aligned}\tau &= -\frac{\eta\langle\delta, \mathfrak{Q}_{\mathrm{c}}(x_{\mathrm{c}} - x^{\mathrm{d}})\rangle}{\langle\delta, \mathfrak{Q}_{\mathrm{c}}\delta\rangle + \eta\langle\mathfrak{Q}_{\mathrm{c}}\delta, P\mathfrak{Q}_{\mathrm{c}}^{-1}(\mathfrak{Q}_{\mathrm{c}}\delta_{\mathrm{c}})\rangle} \\ &= -\frac{\eta\langle\delta, \mathcal{G}\lambda_{\mathrm{c}}\rangle}{\langle\delta, \mathfrak{Q}_{\mathrm{c}}\delta\rangle + \eta\langle\mathfrak{Q}_{\mathrm{c}}\delta, P\mathfrak{Q}_{\mathrm{c}}^{-1}(\mathfrak{Q}_{\mathrm{c}}\delta_{\mathrm{c}})\rangle} \\ &= 0.\end{aligned}$$

Hence, if $\delta \subset \mathcal{R}_0$ then the stepsize $\tau = \tau_z = 0$. This is not a limitation of the method itself, since, if x_{p} is feasible (i.e. $x_{\mathrm{p}} \in \mathcal{R}$), the decision maker can have exactly what she wants (i.e., since $x_{\mathrm{p}} \in \Omega \cap \mathcal{R}$, the method stops, setting $x_{\mathrm{n}} = x_{\mathrm{p}}$). However, if the problem is to reveal the weighting matrix corresponding to a past decision, $\tau = 0$ shows that the method is not suitable for this purpose.

4.2 Choosing the desired, or bliss, value as the preferred solution

Consider another obvious choice for x_{p}:

$$x_{\mathrm{p}} = x^{\mathrm{d}}.$$

This choice does not add any further information about the decision maker's local preferences. It is already known, from the specification of the original decision problem, that the decision maker prefers x^{d} to any $x \in \mathcal{R}$. Hence, x^{d}

is preferred to x_c ($\in \mathscr{R}$). The fact that $x_p, x_p \neq x^d$, is preferred to x_c constitutes additional information about the local preferences of the decision maker. The uselessness of setting $x_p = x^d$ can be verified by considering its effect on the method. Consider (3.1) and (1.9):

$$\begin{aligned} x_n - x_c &= -\tau\left[I - \mathfrak{Q}_c^{-1}\mathscr{G}\left(\mathscr{G}^T\mathfrak{Q}_c^{-1}\mathscr{G}\right)^{-1}\mathscr{G}^T\right](x_c - x^d) \\ &= -\tau\left[I - \mathfrak{Q}_c^{-1}\mathscr{G}\left(\mathscr{G}^T\mathfrak{Q}_c^{-1}\mathscr{G}\right)^{-1}\mathscr{G}^T\right]\mathfrak{Q}_c^{-1}\mathfrak{Q}_c(x_c - x^d) \\ &= -\tau\left[I - \mathfrak{Q}_c^{-1}\mathscr{G}\left(\mathscr{G}^T\mathfrak{Q}_c^{-1}\mathscr{G}\right)^{-1}\mathscr{G}^T\right]\mathfrak{Q}_c^{-1}\mathscr{G}\lambda_c \\ &= 0, \end{aligned}$$

which illustrates that the choice $x_p = x^d$ does not add new information about the decision maker's preferences. As mentioned in Section 2, x^d defines the position of the unconstrained minimum of the objective function (1.2). Algorithm 2.1, on the other hand, aims to determine the curvature, given this point. Thus, the choice $x_p = x^d$ cannot contribute further information.

4.3 The issue of sensitivity

We consider the issue of sensitivity given that $\mathfrak{Q}_c$, determined using Algorithm 2.1, can only be regarded as valid locally. Specifically, having obtained a $\mathfrak{Q}_c$ that, when used in (1.1), yields an optimum x_c, acceptable to the decision maker (i.e. $x_c \in \Omega \cap \mathscr{R}$), we are interested in establishing a region of validity of $\mathfrak{Q}_c$ when the feasible region $\mathscr{R}$ changes slightly. This may happen if, for example, having computed x_c using a model of the system (i.e. $\mathscr{R}$) that incorporates the values of exogenous assumptions, it is discovered that some of these assumptions are in error, to some extent. Exogenous assumptions, as well as other values, assumed fixed during the optimization (1.1) can be taken as part of the vector g defining the feasible region $\mathscr{R}$ in (1.4). Consider, therefore, a small change in g given by

$$\mathscr{G}^T x = \bar{g}, \quad \bar{g} = g + \Delta g.$$

We wish to investigate the validity of $\mathfrak{Q}_c$ in reflecting the decision maker's objectives when the conditions under which the original optimal decision was taken are changed. The more general case, which allows $\mathscr{G}$ to be altered, can be interpreted as changes, or perturbations, in the structure of restrictions in the optimal decision framework. If $\mathscr{G}^T x = g$ denotes the model of the system then these general changes, both in $\mathscr{G}$ and in g, can represent alterations of the model equations. The general changed system can thus be written as

$$\bar{\mathscr{G}} = \mathscr{G} + \Delta\mathscr{G}, \quad \bar{g} = g + \Delta g, \tag{4.1}$$

$$\mathscr{R}(\Delta g,\ \Delta\mathscr{G}) = \{x \in \mathbb{R}^n \mid \bar{\mathscr{G}}^T x = \bar{g}\}, \tag{4.2}$$

and the same question about the validity of $\mathfrak{Q}_c$ exists when $\mathscr{R}(\Delta g, \Delta\mathscr{G})$ replaces $\mathscr{R}$ in the optimal decision problem.

In this chapter, we assume that the desired, or bliss, values x^d remain fixed. One reason for this is that, as the position of the unconstrained minimum of the objective function, x^d should be independent of the constraints (i.e. $\mathscr{R}$ or $\mathscr{R}(\Delta g, \Delta\mathscr{G})$). Indeed, viewed as the bliss point in decision making problems with conflicting objectives, x^d is an objective choice. It is not generally dependent on the constraints. Another reason is that the choice of $\mathfrak{Q}$ is dependent on x^d, $\mathscr{R}$ and Ω. Thus, any change in $\mathfrak{Q}$ is conditional on x^d. In this respect, we consider changing x^d and keeping $\mathfrak{Q}$ fixed in Chapter 4. Within the context of the present chapter, the only case for changing x^d arises when the set Ω alters, because of changing circumstances, and x^d, assumed to be in Ω, is left outside the altered Ω. In this case, a new desired value can be specified by the decision maker, and a new weighting matrix can be tailored to meet the requirements of the altered Ω. We note that neither Theorem 3.1 nor Theorem 3.2, nor the related convergence results in Chapter 5, assume $x^d \in \Omega$. It is, nevertheless, conceptually desirable for x^d to be chosen such that $x^d \in \Omega$. However, x^d still need not be changed when Ω is altered, provided x_c remains in Ω. Whatever the potential change in Ω, $\mathfrak{Q}_c$ can be tailored such that the resulting optimal decision x_c lies within a ball S, of radius r such that S is contained in Ω. Thus, for some $r > 0$, x_c satisfies

$$\Omega \supset S, \tag{4.3}$$

$$S = \{x \subset \mathbb{R}^n \mid \|x - x_c\|_{\mathfrak{Q}_c} \leqslant r\}. \tag{4.4}$$

In this case, x_c is guaranteed to remain in any altered Ω as long as the alteration is correspondingly bounded. Thus, there will not be any need to alter x^d or $\mathfrak{Q}_c$ for changes in Ω.

We return to the issue of sensitivity of $\mathfrak{Q}_c$ to general changes in $\mathscr{G}$ and g. The changes are characterized as a series of additions of new columns to $\mathscr{G}$ and removals of existing columns from $\mathscr{G}$. Similarly, new elements may be added to and old ones removed from the vector g.

Theorem 4.1 (Sensitivity)
Let $\mathfrak{Q}_c$ be positive-definite and $\mathscr{G}, \bar{\mathscr{G}}$ have linearly independent columns.
(i) Let

$$P[\bar{\mathscr{G}}] \stackrel{\Delta}{=} I - \mathfrak{Q}_c^{-1}\bar{\mathscr{G}}\left(\bar{\mathscr{G}}^T\mathfrak{Q}_c^{-1}\bar{\mathscr{G}}\right)^{-1}\bar{\mathscr{G}}^T, \quad \bar{\mathscr{G}}^* \stackrel{\Delta}{=} \mathfrak{Q}_c^{-1}\bar{\mathscr{G}}(\bar{\mathscr{G}}^T\mathfrak{Q}_c^{-1}\bar{\mathscr{G}})^{-1},$$

$$\bar{x} = \arg\min\{\langle x - x^d, \mathfrak{Q}_c(x - x^d)\rangle \mid x \in \mathscr{R}(\Delta g, \Delta\mathscr{G})\}. \tag{4.5}$$

with associated shadow price vector $\bar{\lambda}$. Then, we have

$$\bar{x} - x_c = \bar{\mathscr{G}}^{*T}(\bar{g} - \bar{\mathscr{G}}^T x_c) - P[\bar{\mathscr{G}}]\mathfrak{Q}_c^{-1}\mathscr{G}\lambda_c. \tag{4.6}$$

(ii) Let $\Delta\mathscr{G} = 0$ (i.e. $\bar{\mathscr{G}} = \mathscr{G}$ and $x \in \mathscr{R}(\Delta g, 0)$). Then, we have

$$\bar{x} = \arg\min\{\langle x - x_c, \mathfrak{Q}_c(x - x_c)\rangle \mid \mathscr{G}^T x = \bar{g}\}. \tag{4.7}$$

(iii) *Let $\bar{\mathscr{G}}$ be obtained by the addition of a new column to $\mathscr{G}$ and $\bar{g}$ be obtained by the addition of a corresponding element to g (i.e.). $\mathscr{R}(\Delta g, \Delta\mathscr{G})$ is obtained by the addition of a further constraint to $\mathscr{R}$). Then, we have*

$$\bar{x} = \arg\min\{\langle x - x_c, \mathfrak{Q}_c(x - x_c)\rangle \mid \bar{\mathscr{G}}^T x = \bar{g}\}. \tag{4.8}$$

(iv) *Let $\bar{\mathscr{G}}$ be obtained by deleting an existing (e.g. last) column of $\mathscr{G}$, and $\bar{g}$ be obtained by deleting the corresponding element of g (i.e. $\mathscr{R}(\Delta g, \Delta\mathscr{G})$ is obtained by deleting an existing constraint of $\mathscr{R}$). Then, we have*

$$\bar{x} - x_c = \frac{n^* \lambda_c^\ell}{\langle n^*, \mathfrak{Q}_c n^*\rangle} \tag{4.9}$$

where n^*T is the (last) row of $\left(\mathscr{G}^T \mathfrak{Q}_c^{-1}\mathscr{G}\right)^{-1}\mathscr{G}^T\mathfrak{Q}_c^{-1}$ corresponding to the removed constraint, and λ_c^ℓ is the (last) element of the vector λ_c, and is thus the shadow price of the removed constraint. Furthermore, in this case we have

$$\|\bar{x} - x^d\|^2_{\mathfrak{Q}_c} \leqslant \|x_c - x^d\|^2_{\mathfrak{Q}_c}. \tag{4.10}$$

Proof

We first derive the general expression for $\bar{x} - x_c$ from the basic optimality conditions for (4.5) and (1.1). These are

$$\mathfrak{Q}_c(\bar{x} - x^d) + (\mathscr{G} + \Delta\mathscr{G})\bar{\lambda} = 0, \qquad \mathfrak{Q}_c(x_c - x^d) + \mathscr{G}\lambda_c = 0, \tag{4.11}$$

$$(\mathscr{G} + \Delta\mathscr{G})^T\bar{x} = g + \Delta g, \qquad \mathscr{G}^T x_c = g, \tag{4.12}$$

where $\bar{\lambda}$ is the shadow price vectorof (4.5). From (4.12), we and write

$$\bar{\mathscr{G}}^T(\bar{x} - x_c) = \bar{g} - \bar{\mathscr{G}}^T x_c, \tag{4.13}$$

and, from (4.11), we can write

$$\begin{aligned} 0 &= \mathfrak{Q}_c(\bar{x} - x_c + x_c - x^d) + (\mathscr{G} + \Delta\mathscr{G})\bar{\lambda} \\ &= \mathfrak{Q}_c(\bar{x} - x_c) + (\mathscr{G} + \Delta\mathscr{G})(\bar{\lambda} - \lambda_c) + \Delta\mathscr{G}\lambda_c. \end{aligned}$$

Thus,

$$\bar{x} - x_c = -\mathfrak{Q}_c^{-1}\bar{\mathscr{G}}(\bar{\lambda} - \lambda_c) - \mathfrak{Q}_c^{-1}\Delta\mathscr{G}\lambda_c, \tag{4.14}$$

and, using (4.13), we have

$$\bar{\lambda} - \lambda_c = -(\bar{\mathscr{G}}^T\mathfrak{Q}_c^{-1}\bar{\mathscr{G}})^{-1}(\bar{g} - \bar{\mathscr{G}}^T x_c) - (\bar{\mathscr{G}}^T\mathfrak{Q}_c^{-1}\bar{\mathscr{G}})^{-1}\bar{\mathscr{G}}^T\mathfrak{Q}_c^{-1}\Delta\mathscr{G}\lambda_c,$$

which, combined with (4.14), yields

$$\bar{x} - x_c = \mathfrak{Q}_c^{-1}\bar{\mathscr{G}}(\bar{\mathscr{G}}^T\mathfrak{Q}_c^{-1}\bar{\mathscr{G}})^{-1}(\bar{g} - \bar{\mathscr{G}}^T x_c) - P[\bar{\mathscr{G}}]\mathfrak{Q}_c^{-1}\Delta\mathscr{G}\lambda_c. \tag{4.15}$$

Using (4.11), the last term in (4.15) can be expressed as

$$\begin{aligned} P[\bar{\mathscr{G}}]\mathfrak{Q}_c^{-1}\Delta\mathscr{G}\lambda_c &= P[\bar{\mathscr{G}}]\mathfrak{Q}_c^{-1}\left[\Delta\mathscr{G}\lambda_c + \mathscr{G}\lambda_c + \mathfrak{Q}_c(x_c - x^d)\right] \\ &= P[\bar{\mathscr{G}}](x_c - x^d), \end{aligned} \tag{4.16}$$

since $P[\bar{\mathscr{G}}]\mathfrak{Q}_c^{-1}\bar{\mathscr{G}} = 0$. Alternatively, we can express (4.16) as

$$P[\bar{\mathcal{G}}](x_c - x^d) = -P[\bar{\mathcal{G}}]\mathfrak{Q}_c^{-1}\mathcal{G}\lambda_c, \tag{4.17}$$

which yields the required result (4.6).

When $\bar{\mathcal{G}} = \mathcal{G}$, we have $P[\mathcal{G}]\mathfrak{Q}_c^{-1}\mathcal{G} = 0$, and hence (4.6) reduces to the solution of (4.7).

When a constraint is added to the original system, this is characterized by the addition of a column to $\mathcal{G}$ and a further row to g. For simplicity, let these be the last column of $\bar{\mathcal{G}}$ and the last row of $\bar{g}$. The addition of a constraint, with normal denoted by n, affects $P[\mathcal{G}]\mathfrak{Q}_c^{-1}$ with a rank-one change, and $P[\bar{\mathcal{G}}]\mathfrak{Q}_c^{-1}$ is given by

$$P[\bar{\mathcal{G}}]\mathfrak{Q}_c^{-1} = P[\mathcal{G}]\mathfrak{Q}_c^{-1} - \frac{P[\mathcal{G}]\mathfrak{Q}_c^{-1}nn^T\mathfrak{Q}_c^{-1}P^T[\mathcal{G}]}{\langle n, P[\mathcal{G}]\mathfrak{Q}_c^{-1}n\rangle}$$

(see Rustem, 1981, Chap. 3). Noting that $P[\mathcal{G}]\mathfrak{Q}_c^{-1}$ is symmetric, the last term in (4.6) vanishes, once again, since $P[\mathcal{G}]\mathfrak{Q}_c^{-1}\mathcal{G} = 0$. Thus, (4.6) becomes the solution of (4.8).

Finally, consider the case when the constraint corresponding to the last column of $\mathcal{G}$, and the last row of g, is removed. Since the ordering of the constraints is not important, any constraint may be expressed as the last constraint. As the remaining constraints are already satisfied by the current optimum, x_c, we have $\bar{g} - \bar{\mathcal{G}}^T x_c = 0$, and hence the first term on the right of (4.6) vanishes. When a constraint is removed, $P[\bar{\mathcal{G}}]\mathfrak{Q}_c^{-1}$ can be expressed in terms of a rank-one change to $P[\mathcal{G}]\mathfrak{Q}_c^{-1}$. Hence,

$$P[\bar{\mathcal{G}}]\mathfrak{Q}_c^{-1} = P[\mathcal{G}]\mathfrak{Q}_c^{-1} + \frac{n^* n^{*^T}}{\langle n^*, \mathfrak{Q}_c n^*\rangle}, \tag{4.18}$$

when n^{*^T} is the last row of $(\mathcal{G}^T\mathfrak{Q}_c^{-1}\mathcal{G})^{-1}\,\mathcal{G}^T\mathfrak{Q}_c^{-1}$, corresponding to the removed constraints (see Rustem, 1981, Chap. 3). It should be noted that n^* is orthogonal to all columns of $\mathcal{G}$ except the last column, corresponding to the removed constraint. Using (4.18) in (4.6) and invoking, once again, $P[\mathcal{G}]\mathfrak{Q}_c^{-1}\mathcal{G} = 0$, we have

$$\begin{aligned}\bar{x} - x_c &= -\frac{n^* n^{*^T}}{\langle n^*, \mathfrak{Q}_c n^*\rangle}\mathcal{G}\lambda_c \\ &= -\frac{n^* n_c^{\ell}}{\langle n^*, \mathfrak{Q}_c n^*\rangle},\end{aligned}$$

where λ_c^{ℓ} is the last element of λ_c and is thus the shadow price of the removed constraint. Expression (4.10) follows immediately from the fact that when a constraint is removed, $\mathcal{R}(\Delta g, \Delta\mathcal{G}) \supset \mathcal{R}$. □

The implication of Theorem 4.1 is that, provided the method in Section 2 is used to generate an optimal decision that satisfies (4.3) and (4.4), the sensitivity issues can be resolved in a satisfactory manner. Cases (ii) and (iii) discussed in the theorem show that the solutions of the perturbed problem (4.5) lie as close to the current optimal solution as is feasibly possible. This is due to the fact that both (4.7) and (4.8) are identical to the constrained minimization of the norm

$$\|x - x_c\|^2_{\mathfrak{Q}_c},$$

which also appears in (4.4).

Case (iv), corresponding to the removal of a constraint, ensures through (4.10) that the solution to the perturbed problem, $\bar{x}$, actually improves the value of the objective function. Thus, the distance between $\bar{x}$ and x^d is less than x_c and x^d. Expression (4.9) also provides the information concerning the direction $\bar{x} - x_c$. By using (4.11), we can evaluate the angle between $\mathfrak{Q}_c(x^d - x_{c)}$ and $\bar{x} - x_c$:

$$\begin{aligned}\langle \bar{x} - x_c, \ \mathfrak{Q}_c(x^d - x_c)\rangle &= \frac{\langle \lambda_c, \ \mathcal{G}^T n^* \rangle}{\langle n^*, \ \mathfrak{Q}_c n^* \rangle} \lambda_c^\ell \\ &= \frac{(\lambda_c^\ell)^2}{\langle n^*, \ \mathfrak{Q}_c n^* \rangle}\end{aligned}$$

since all the columns of $\mathcal{G}$, except the last one, are orthogonal to n^*. Thus, $\bar{x} - x_c$ makes an acute angle with $\mathfrak{Q}_c(x^d - x_c)$, and if $x^d \in \Omega$ then this also suggests that the perturbed solution will be closer to x^d than x_c.

5 CONCLUDING REMARKS

The method discussed in this chapter provides an interactive approach for the decision maker to search for an acceptable solution of the multi-objective optimization problem. It is clear from the discussion in Sections 1 and 2 that, given only preferred directions for key target variables and an element of consistency on the part of the decision maker (Sen, 1970, p. 63), it is possible to translate this 'qualitative' information into quantitative modification of the weights.

The quadratic function used by the method allows the decision maker to express her trade-offs in achieving each objective versus the others. Using a rank-one modification, the weighting matrix is tailored to meet the decision maker's requirements.

The effect of changing the weights are expressed in terms of optimal solution updates to the shadow prices. This is discussed further in Chapter 4 to demonstrate the possibility of altering shadow prices arbitrarily, while keeping the optimal solution fixed. The convergence properties of the method are discussed in Chapter 5 by invoking the results associated with Khachian's ellipsoid algorithm.

6 COMMENTS AND NOTES

CN.1 Quadratic functions

For any positive-semidefinite quadratic function, we have

$$\arg\min\{\langle a, x\rangle + \frac{1}{2}\langle x, \mathfrak{Q}_c x\rangle \mid x \in \mathscr{R}\} = \arg\min\{q_c(x) \mid x \in \mathscr{R}\}$$

for $\mathfrak{Q}_c x^d = -a$. This can be verified by inspecting the first-order optimality conditions (see Chapter 1) and noting that

$$a + \mathfrak{Q}_c x = \mathfrak{Q}_c(x - x^d).$$

Thus, provided this restriction on x^d is satisfied, arguments concerning the constrained minimum of $q_c(x)$ can also be extended to $\langle a, x\rangle + \frac{1}{2}\langle x, \mathfrak{Q}_c x\rangle$.

CN.2 Symmetric weighting matrices

As the weighting matrix is symmetric, the quadratic functions (1.2) are said to penalize a deviation of the optimal solution of (1.2) from the desired value in a desirable direction as much as a deviation in an undesirable direction. However, the definition of the desired values implies that each element of x^d is assigned the best possible value—implying Pareto optimality—for that element. Hence, in this case, a deviation in a desirable direction of the actual solution from x^d can only imply a mis-specification of x^d. Furthermore, nonsymmetric preferences can be modelled using the approach described in section 1.

CN.3 Orthogonal decomposition of $\mathscr{G}$

This orthogonal decomposition is given by $\mathscr{G}^T = [L \;\; 0]\,[Q]$, where L is an $m \times m$ lower-triangular matrix (for $g \in \mathbb{R}^e$) and Q an orthogonal matrix ($QQ^T = QQ^T = I$). Furthermore, Q is partitioned such that its first e rows are set to the $e \times e$ submatrix Q_1, the last $n - e$ rows are denoted by Q_2, and Z is chosen to be Q_2^T. Further details of the derivation of such orthogonal matrices are discussed in Chapter 2 (see CN.4 there).

CN.4 The Sherman–Morrison formula for matrix updates

Let $B \in \mathbb{R}^{n\times n}$ be a nonsingular matrix, $u, v \in \mathbb{R}^n$, and $\theta \in \mathbb{R}^1$. Then the inverse of

$$B - \theta uv^T$$

is given by the Sherman–Morrison formula

$$(B - \theta uv^T)^{-1} = B^{-1} - \xi B^{-1}uv^T B^{-1}$$
$$\theta^{-1} + \xi^{-1} = \langle v, B^{-1}u\rangle$$

(see e.g. Householder, 1964). The result can be verified by evaluating

$$(B - \theta uv^T)(B^{-1} - \xi B^{-1}uv^T B^{-1}).$$

REFERENCES

Bray, J. (1972). Some considerations in the practice of economic management. *PREM Discussion Paper*, No. 2.

Bray, J. (1975). Optimal control of a noisy economy with the U.K. as an example. *J.R. Statist. Soc., Ser.* A, **183**, Part 3.

Broyden, C. (1967). Quasi-Newton methods and their applications to functional minimization. *Math. Comput.* **21**, 368–381.

Davidon, W.C. (1968). Variance algorithm for minimization. *Comput. J.* **10**, 406–410.

Dixon, L.C.W. (1981). Multicriteria optimisation: some inherent difficulties. *Inst. Math. Applic. Bull.* **May/June**, 106–109.

Fiacco, A. and G. McCormick (1968). *Nonlinear Programming Sequential Unconstrained Minimization Technique*. Wiley, New York.

French, S., R. Hartley, L.C. Thomas, and D.J. White (1983). *Multi-Objective Decision Making*. Academic Press, London.

Gill, P.E. and W. Murray (1974). Newton-type methods for linearly constrained optimization. In *Numerical Methods for Constrained Optimization* (P.E. Gill and W. Murray, eds). Academic Press, London.

Gill, P.E. and W. Murray (1978). Numerically stable methods for quadratic programming. *Math. Programming* **14**, 349–372.

Goldfarb, D. (1969). Extensions of Davidon's variable metric method to maximization under linear inequality and equality constraints. *SIAM J. Appl. Math.* **17**, 739–764.

Goldfarb, D. and M.J. Todd (1982). Modifications and implementation of the ellipsoid algorithm for linear programming. *Math. Programming* **23**, 1–19.

Hillier, F.S. and G.J. Lieberman (1996). *Introduction to Operations Research*. McGraw-Hill, New York.

Householder, A.S. (1964). *The Theory of Matrices in Numerical Analysis*. Blaisdell, New York.

Khachian, L.G. (1979). A polynomial algorithm in linear programming. *Soviet Math. Dokl.* **20**, 191–194.

Khachian, L.G. (1980). Polynomial algorithms in linear programming. *USSR Comput. Math. Math. Phys.* **20**, 53–72.

Pinto, R.L.V. (1995). A logic-based modelling language and integer-programming framework for multicriteria optimization, Ph.D. Thesis, Department of Computing, Imperial College, University of London.

Pinto, R.L.V. and B. Rustem (1996a). Building and solving multi-criteria models involving logical conditions, In *Advances in Computational Economics*, M. Gilli ed. Kluwer, Dordrecht.

Pinto, R.L.V. and B. Rustem (1996b). Solving a mixed-integer multiobjective bond portfolio model involving logical conditions, Department of Computing, Imperial College.

Rustem, B. (1981). *Projection Methods in Constrained Optimization and Applications to Optimal Policy Decisions*. Springer-Verlag, Berlin.

Rustem, B. (1983). Rival models: min–max problems and algorithms. In *Dynamic Modelling and Control of National Economies* (T. Basar and L. Pau, eds). Pergamon Press, Oxford.

Rustem, B., Velupillai, K. and Westcott, J.H. (1978). Respecifying the weighting matrix of a quadratic objective function. *Automatica* **14**, 567–582.

Rustem, B. and M.B. Zarrop (1981). Newton-type algorithm for a class of N-player dynamic games using nonlinear econometric models. In *Dynamic Modelling and Control of National Economies* (J.M.L. Janssen, L. Pau and A. Straszak, eds). Pergamon Press, Oxford.

Sen, A. (1970). *Collective Choice and Social Welfare*. Holden-Day, San Francisco.

UNIDO (1972). *Guidelines for Project Evaluation*. United Nations Publications, UNIDO, Vienna.

White, D.J. (1983). A selection of multi-objective interactive programming methods. In *Multi-Objective Decision Making* (S. French, R. Hartley, L.C. Thomas and D.J. White, eds). Academic Press, London.

CHAPTER 4

Multiple-Objective Optimization 2 Interactive Search for Acceptable Decisions: Updating Bliss Points and Arbitrariness of Shadow Prices

The interactive method in Chapter 3 generates a full symmetric weighting matrix. In this chapter, we discuss an alternative method that only requires a diagonal weighting matrix. The diagonal matrix is computationally desirable and its interpretation is more straightforward in terms of the weighting of each objective.

The method updates the desired, or bliss, values using similar concepts as in Chapter 3. It is shown that the methods of that chapter 3 and the present one are equivalent. Extensions for updating the desired, or bliss, values to nonlinear constraints are also considered.

Shadow prices provide an important measure of the sensitivity of the optimal solution to the constraints. Therefore, they are used in decision analysis to indicate the value of scarce resources, represented by the active constraints. We use the tools developed in this chapter to establish that shadow prices are arbitrary for multiple objective decision problems represented by the quadratic objective and equality-constrained optimization problems. Starting initially with linear constraints, either Algorithm 3.2.1, or the algorithm in the present chapter, can be used to arbitrarily alter the shadow prices in one step without changing the optimal solution. This is also extended to nonlinear problems.

1 INTRODUCTION: DIAGONAL QUADRATIC OBJECTIVE FUNCTIONS

In this chapter, we discuss two issues related to the quadratic objective function in constrained optimization problems. The first is the specification of a diagonalized quadratic objective function and a study of its non-diagonal equivalent. By a diagonal quadratic function, we mean one with a diagonal Hessian. The specification of diagonal and non-diagonal Hessians is subsequently utilized to discuss the second problem, which is the arbitrariness of the shadow prices, or Lagrange multipliers. The criterion used for the specification of the objective function is the acceptability of the optimal solution to the decision maker. The possibility of formulating optimization problems conveniently using an appropriate Lagrangian is not new (see Geoffrion, 1971; Ponstein 1983). We consider the converse question and discuss the possibility of altering the Lagrange multiplier, or shadow price, values without changing the decision (Rustem, 1990; 1991).

An interesting application of the method in Chapter 3 is in the area of shadow price determination. Since there is a clear dual relationship between the weights of a preference function and the constraints, and thereby the Lagrange multipliers, a direct application to the *desirable* specification of the shadow prices is possible. This is achieved by reformulating the methods we have discussed. In the literature on the determination of shadow prices, particularly with respect to the problem of economic development, there seems to be an imputation of a halo of objectivity to such prices. The fact that they are largely dependent on the weights of the objective function is not always recognized. A 'truer' appreciation of the interlocking nature of weights, desired, or bliss, values and the decision is evident due to the iterative nature of the method in Chapter 3. An awareness of the constraints of the system is also required by the method. Together with the weights, the constraints are the principal determinants of shadow prices. In Chapter 3, we established the way in which changes in the weighting matrix determines shadow prices. Hence, any objectivity associated with these values would not be justified. Though this has been explicitly recognized in the UNIDO (1972, Chap. 18) exposition, no formal solution is provided. In this chapter, we show that, for an important class of problems, the shadow prices can be assigned arbitrarily, without altering the optimal solution.

Consider the linear–quadratic optimization problem

$$\min \{\langle x - x^{\mathrm{d}}, D(x - x^{\mathrm{d}})\rangle \mid \mathscr{G}^{\mathrm{T}} x = g\}. \tag{1.1}$$

where, as in Chapter 3, $x \in \mathbb{R}^n, x^{\mathrm{d}}$ is the desired or bliss value of x or the unconstrained optimum of the quadratic objective function, $g \in \mathbb{R}^e$ is a fixed vector and D is a diagonal matrix with nonnegative elements. The columns of the matrix $\mathscr{G} \in \mathbb{R}^{n \times e}$ are assumed, without loss of generality, to be linearly independent. The diagonal weighting matrix in (1.1) occurs in decision making and in general in linear–quadratic optimal control problems. In the latter case, (1.1) can be regarded as the transcription of the dynamic problem into static form (e.g. Chapter 1 and Polak, 1971).

A diagonal D is desirable both for computational reasons and for the interpretation of the problem and the optimal solution. Evidently, a general, symmetric matrix, which we denote by $\mathfrak{Q}$ and assume that it is positive-definite, can be factorized such that $\mathfrak{Q} = LDL^{\mathrm{T}}$ (see CN.1). The matrix L is a lower-diagonal matrix. The transformation $y = L^{\mathrm{T}}x, y^{\mathrm{d}} = L^{\mathrm{T}}x^{\mathrm{d}}$ yields the diagonal quadratic objective function

$$\langle y - y^{\mathrm{d}}, D(y - y^{\mathrm{d}})\rangle.$$

However, the transformed constraints

$$\mathscr{G}^{\mathrm{T}} L^{-\mathrm{T}} y = b$$

may pose computational problems. This is true in particular when the constraints are nonlinear. We shall not dwell on this aspect, since we have introduced the problem with only linear constraints. The same difficulty also occurs in the orthogonal decomposition of $\mathfrak{Q}$.

A more important difficulty is the interpretation of the off-diagonal elements of a general symmetric matrix. These elements are usually understood to represent trade-offs between achieving alternative objectives. However, it is difficult to assign an interpretation as distinct from the diagonal elements, which represent the relative importance of each objective. In other words, the diagonal elements are perceived to serve almost the same purpose. Another reason for preferring (1.1) is that in dynamic problems solved via dynamic programming, maintaining the block-diagonality of D, in terms of the time structure, is desirable at least for the physical interpretation of the resultant optimal linear feedback laws (e.g. Kwakernaak and Sivan, 1972).

In most decision making problems under conflicting objectives, the precise value of the weighting matrix is not known. Even a judicious choice of the weighting matrix need not necessarily yield a solution of problem (1.1) that is acceptable to the decision maker. In the discussion below, we discuss an iterative algorithm, involving interactions with the decision maker, to specify a diagonal quadratic objective function that generates a solution of (1.1) that is *acceptable* to the decision maker. The algorithm involves the modification of the desired, or bliss, values of the objective function. For dynamic policy optimization problems, this objective function permits the generation of linear optimal feedback laws inherent in the solution, in terms of the original, rather than transformed variables. This contrasts with the case when nondiagonal weighting matrices have to be factorized and the original variables have to be transformed in order to obtain feedback laws that will usually apply only to the transformed variables.

It is shown below that the method is equivalent to the respecification of the general nondiagonal symmetric matrix in Chapter 3, where the desired values remain fixed but the weighting matrix generated does not maintain a diagonal structure. This correspondence is helpful computationally, as diagonal matrices are simpler to compute. It also yields the nondiagonal weighting matrix that would generate the same optimal solution, with original desired values, as the solution with diagonal weighting matrix and modified desired values. Furthermore, the correspondence provides the means to establish the complexity of the algorithm. Using the results in Chapter 5, this leads to its polynomial-time termination property.

2 SPECIFYING DIAGONAL QUADRATIC OBJECTIVE FUNCTIONS

As a departure from the notation in Chapter 3, where the subscript 'c' is used to indicate the *current* solution, in this chapter, and Chapter 5, the subscript 'k' used. Similarly, subscript $k+1$ replaces 'n', the *new* solution, in Chapter 3. The reason for the change is that in Chapter 3 we are more concerned with the single-step properties such as the relation of the current, or kth-solution to the new, or $(k+1)$th, solution. In this Chapter, and Chapter 5, one of the important issues considered is the overall convergence of the algorithm. The use of k as an iteration count assists the discussion on convergence.

Consider the solution of (1.1), for given D and given initial desired values x_0^d,

$$x_0 = \arg\min\{\langle x - x_0^{\mathrm{d}}, D(x - x_0^{\mathrm{d}})\rangle \mid \mathscr{G}^{\mathrm{T}} x = g\}.$$

The solution is presented to the decision maker to decide if $x_0 \in \Omega$, where Ω is the set of acceptable policies. If $x_0 \notin \Omega$, the decision maker is required to specify an alternative to x_0 that is in Ω. As in Chapter 3, the set Ω is the decision maker's set of acceptable policies. It is supposed to exist only in the mind of the policy maker. Following Chapter 3, we assume that there does not exist any analytical characterization of this set. If such a characterization were possible, it could be used to augment the constraints of the decision problem, and the resultant problem could be easily solved. The characterization is inherently difficult or impossible, and sometimes also undesirable, from the decision maker's point of view.

As in Chapter 3, the decision problem we consider can be formulated as the computation of the policy, optimally determined via (1.1) *and* also acceptable to the policy maker, hence in the set Ω. We assume that, for $\mathscr{R}$ given by (3.1.4),

$$\mathscr{R} \cap \Omega \neq \varnothing.$$

The decision maker's preferred alternative to x_0 is denoted by x_{p}. By definition, we have $x_{\mathrm{p}} \in \Omega$ *but not necessarily*

$$x_{\mathrm{p}} \in \mathscr{R} \cap \Omega.$$

Indeed, if this is satisfied, such a preferred alternative would conceptually solve the decision problem. Let

$$\delta_0 = x_{\mathrm{p}} - x_0,$$

where δ_0 is the *correction* vector that needs to be added to x_0 in order to ensure that $x_0 + \delta_0 \in \Omega$. Using δ_0, we revise the desired, or bliss, value as

$$x_1^{\mathrm{d}} = x_0^{\mathrm{d}} + \tau_0 \delta_0,$$

where $\tau_0 \geqslant 0$ is a scalar. Using this new desired value, problem (1.1) is solved once again to yield a new optimal solution, x_1. This solution is shown below to have desirable characteristics. However, since there is no guarantee that $x_1 \in \Omega$, the above procedure may need to be repeated. The resulting algorithm is summarized below. We discuss the complexity and termination properties of the algorithm in Section 3 and Chapter 5.

Algorithm 2.1 (Updates of desired values with fixed diagonal weighting matrix)

Step 0: Given D, x_0^{d}, the sequence $\{\tau_k\}$ and the constraints, set $k = 0$.

Step 1: Compute the solution of the optimization problem

$$x_k = \arg\min\{\langle x - x_k^{\mathrm{d}}, D(x - x_k^{\mathrm{d}})\rangle \mid \mathcal{G}^{\mathrm{T}} x = g\}. \tag{2.1}$$

Step 2: Interact with the decision maker. If $x_k \in \Omega$, **stop**. Otherwise, the decision maker is required to specify the preferred value x_{p}, and hence,

$$\delta_k = x_{\mathrm{p}} - x_k. \tag{2.2}$$

Step 3: Update the desired values

$$x_{k+1}^{\mathrm{d}} = x_k^{\mathrm{d}} + \tau_k \delta_k, \tag{2.3}$$

set $k = k + 1$ and **go to Step 1**.

The effect of changing the desired values as described above alters the optimal solution systematically. The relationship of x_{k+1}, x_k and τ_k, δ_k is summarized in the following results. The possibility of altering the desired values in order to influence the solution is reasonably common in the practice of optimal control. It is also the case that the choice of the objective function may be based on criteria other than $x_k \in \Omega$. For example, in the linear stochastic dynamical systems, the weighting matrices might be chosen to yield a stable minimum variance controller (see Engwerda and Otter, 1989). However, in this chapter, we consider the equivalence properties of the approach adopted in this section to respecification, discussed in Chapter 3 and Section 3 below, of nondiagonal weighting matrices while keeping x^{d} fixed. This equivalence also provides the key to the complexity and convergence of the policy design process. In addition, since it is possible to decompose a symmetric matrix into a sequence of rank one updates (see Theorem 3.1.1), the method in this section and the equivalence result allow the possibility of expressing non diagonal weighting matrices in terms of diagonalized objective functions.

Proposition 2.1
Let the columns of $\mathcal{G}$ be linearly independent.

(i) Let D be positive-definite. Then the two successive solutions of Algorithm 2.1 are related by

$$x_{k+1} = x_k + \tau_k P \delta_k, \tag{2.4}$$

where

$$P = I - D^{-1}\mathcal{G}\left(\mathcal{G}^{\mathrm{T}} D^{-1}\mathcal{G}\right)^{-1}\mathcal{G}^{\mathrm{T}} \tag{2.5}$$

and the corresponding change in the Lagrange multipliers or shadow prices is given by

$$\lambda_{k+1} = \lambda_k - \tau_k(\mathcal{G}^{\mathrm{T}} D^{-1}\mathcal{G})^{-1}\mathcal{G}^{\mathrm{T}}\delta_k. \tag{2.6}$$

(ii) Let D be positive-semidefinite. Then the optimal solution and Lagrange multipliers can be written as

$$x_{k+1} = x_k + \tau_k Z(Z^{\mathrm{T}} D Z)^{-1} Z^{\mathrm{T}} D \delta_k \tag{2.7}$$

where $Z \in \mathbb{R}^{n \times (n-m)}$ is an orthogonal matrix such that $Z^{\mathrm{T}}\mathcal{G} = 0$, and

$$\lambda_{k+1} = \lambda_k - \tau_k (\mathcal{G}^{\mathrm{T}}\mathcal{G})^{-1} \mathcal{G}^{\mathrm{T}} D[I - Z(Z^{\mathrm{T}} D Z)^{-1} Z^{\mathrm{T}} D]\delta_k. \tag{2.8}$$

Remark

The choice of the orthogonal matrix Z has been discussed in detail in Chapter 2. The numerically stable way of generating Z is by considering the QR decomposition of $\mathcal{G}$. The matrix Z is given by the last $n - m$ columns of the matrix Q of this decomposition (see Chapter 2, and Gill, Murray and Wright, 1981).

As discussed in Chapter 3, the matrix P is a projection operator that projects vectors onto the subspace $\mathcal{R}_0$ given by (3.1.11). Thus, $P\delta_k$ is the *nearest feasible alternative of* δ_k. When D is positive-definite, the operators $Z(Z^{\mathrm{T}}DZ)^{-1}Z^{\mathrm{T}}D$ and P are, in effect, identical (see Lemma 3.3.1). This discussion can be extended to demonstrate the equivalence of (2.4), (2.7) and (2.6), (2.8).

Proof of Proposition 2.1

In order to establish (2.4), we note that at $k+1$, the solution of (2.1) is given by

$$x_{k+1} = x_{k+1}^{\mathrm{d}} - D^{-1}\mathcal{G}(\mathcal{G}^{\mathrm{T}} D^{-1} \mathcal{G})^{-1} (\mathcal{G}^{\mathrm{T}} x_{k+1}^{\mathrm{d}} - g). \tag{2.9}$$

Using (2.3) and the equivalent of (2.9) for x_k, we have (2.4). Similarly, for (2.6), λ_{k+1} is given by

$$\lambda_{k+1} = -(\mathcal{G}^{\mathrm{T}} D^{-1} \mathcal{G})^{-1} (\mathcal{G}^{\mathrm{T}} x_{k+1}^{\mathrm{d}} - g), \tag{2.10}$$

and, using (2.3), we obtain (2.6).

To establish (2.7), we note that $x_{k+1} - x_k \in \mathcal{R}_0$, and, since $\mathcal{G}^{\mathrm{T}} Z = 0$, any such vector can be written as a linear combination of the columns of Z. Thus, we have $x_{k+1} - x_k = Z\omega$, for some vector $\omega \in \mathbb{R}^{n-m}$, and, from the first order optimality condition of (2.1), we can write

$$Z^{\mathrm{T}} D(Z\omega + x_k - x_k^{\mathrm{d}} - \tau_k \delta_k) = Z^{\mathrm{T}} \mathcal{G} \lambda_{k+1} = 0,$$

and thus

$$x_{k+1} - x_k = Z\omega = -Z(Z^{\mathrm{T}} D Z)^{-1} Z^{\mathrm{T}} D(x_k - x_k^{\mathrm{d}} - \tau_k \delta_k). \tag{2.11}$$

From the optimality condition at iteration k, we have

$$Z^{\mathrm{T}} D(x_k - x_k^{\mathrm{d}}) = Z^{\mathrm{T}} \mathcal{G} \lambda_k = 0.$$

Thus, expression (2.7) follows from (2.11).

For (2.8), we invoke the optimality condition at $k+1$ to yield

$$\lambda_{k+1} = (\mathscr{G}^{\mathrm{T}}\mathscr{G})^{-1}\mathscr{G}^{\mathrm{T}}D(x_{k+1} - x_k + x_k - x_k^{\mathrm{d}} - \tau_k\delta_k). \tag{2.12}$$

Using (2.7) and the optimality condition at k, $D\ (x_k - x_k^{\mathrm{d}}) = \mathscr{G}\lambda_k$, leads to expression (2.8). □

3 DIAGONAL VERSION OF NONDIAGONAL QUADRATICS AND COMPLEXITY OF THE ALGORITHM

3.1 Equivalence of diagonal and nondiagonal approaches

In this section, we consider the diagonal equivalent of the Algorithm 3.2.1 discussed in Chapter 3. The equivalence is subsequently exploited in the complexity of Algorithm 2.1. Algorithm 3.1 below is basically Algorithm 3.2.1, expressed using the iterative framework in this chapter and Chapter 5. It uses the same δ_k as in Algorithm 2.1, and keeps the desired values fixed but updates the weighting matrix of the quadratic optimization problem.

Algorithm 3.1 (Fixed desired values and nondiagonal weighting matrix)

Step 0: Given a positive-semidefinite weighting matrix $\mathfrak{Q}_0$, the sequence η_k, the desired values x^{d} and the constraints, set $k = 0$.

Step 1: Compute the solution of the optimization problem

$$x_k = \arg\min\{\langle x - x^{\mathrm{d}}, \mathfrak{Q}_k(x - x^{\mathrm{d}})\rangle \mid \mathscr{G}^{\mathrm{T}}x = g\}. \tag{3.1}$$

Step 2: Interact with the decision maker. If $x_k \in \Omega$, **stop**. Otherwise, the decision maker is required to specify the preferred value x_{p}, and, hence,

$$\delta_k = x_{\mathrm{p}} - x_k. \tag{3.2}$$

Step 3: Update the weighting matrix:

$$\mathfrak{Q}_{k+1} = \mathfrak{Q}_k + \eta_k \frac{\mathfrak{Q}_k\delta_k\delta_k^{\mathrm{T}}\mathfrak{Q}_k}{\langle\delta_k, \mathfrak{Q}_k\delta_k\rangle}. \tag{3.3}$$

Set $k = k+1$ and **go to Step 1**.

As in Chapter 3, we assume that the denominator of (3.3) can be computed, at least approximately, for positive-semidefinite $\mathfrak{Q}_k$. The matrix $\mathfrak{Q}_k$ computed in Algorithm 3.1 is in general nondiagonal. It is shown below that, starting with an initial diagonal matrix, Algorithms 2.1 and 3.1 are equivalent. At each stage, the nondiagonal version above has a constant diagonal equivalent in Algorithm 2.1. The equivalent results to Proposition 2.1 related to Algorithm 3.1 are discussed in Chapter 3. We summarize these results using the iterative framework of Algorithm 2.1. When $\mathfrak{Q}_k$ is positive-definite, each subsequent iterate of Algorithm 3.1 is given by

$$x_{k+1} = x_k + \hat{\tau}_k P_k \delta_k, \tag{3.4a}$$

$$\lambda_{k+1} = \lambda_k - \hat{\tau}_k \left(\mathcal{G}^{\mathrm{T}} \mathfrak{Q}_k^{-1} \mathcal{G}\right)^{-1} \mathcal{G}^{\mathrm{T}} \delta_k, \tag{3.4b}$$

$$P_k = I - \mathfrak{Q}_k^{-1} \mathcal{G} \left(\mathcal{G}^{\mathrm{T}} \mathfrak{Q}_k^{-1} \mathcal{G}\right)^{-1} \mathcal{G}^{\mathrm{T}}, \tag{3.5a}$$

$$\hat{\tau}_k = -\frac{\eta_k \langle \delta_k, \mathfrak{Q}_k (x_k - x^{\mathrm{d}}) \rangle}{\langle \delta_k, \mathfrak{Q}_k \delta_k \rangle + \eta_k \langle \mathfrak{Q}_k \delta_k, P_k \mathfrak{Q}_k^{-1} (\mathfrak{Q}_k \delta_k) \rangle}, \tag{3.5b}$$

and when $\mathfrak{Q}_k$ is positive-semidefinite, (3.4) becomes

$$x_{k+1} = x_k + \hat{\tau}_k Z \left(Z^{\mathrm{T}} \mathfrak{Q}_k Z\right)^{-1} Z^{\mathrm{T}} \mathfrak{Q}_k \delta_k, \tag{3.5c}$$

$$\lambda_{k+1} = \lambda_k - \hat{\tau}_k (\mathcal{G}^{\mathrm{T}} \mathcal{G})^{-1} \mathcal{G}^{\mathrm{T}} \mathfrak{Q}_k [I - Z (Z^{\mathrm{T}} \mathfrak{Q}_k Z)^{-1} Z^{\mathrm{T}} \mathfrak{Q}_k] \delta_k, \tag{3.5d}$$

$$\hat{\tau}_k = -\frac{\eta_k \langle \delta_k, \mathfrak{Q}_k (x_k - x^{\mathrm{d}}) \rangle}{\langle \delta_k, \mathfrak{Q}_k \delta_k \rangle + \eta_k \langle \mathfrak{Q}_k \delta_k, Z (Z^{\mathrm{T}} \mathfrak{Q}_k Z)^{-1} Z^{\mathrm{T}} \mathfrak{Q}_k \delta_k \rangle} \tag{3.5e}$$

(see Theorems 3.3.1 and 3.3.2 and Lemma 3.3.1). The value of $\hat{\tau}_k$ in (3.5e) can be shown to be identical to that given by (3.5b) when $\mathfrak{Q}_k$ is positive-definite. Similarly, (3.4a,b) and (3.5a,b) are identical when $\mathfrak{Q}_k$ is positive-definite.

We can now show the equivalence of the solution of a diagonal quadratic optimization problem with only the desired or bliss values modified, and the quadratic optimization in which only the diagonal matrix has been modified to a nondiagonal form.

Proposition 3.1
Let $\mathfrak{Q}_k$ be positive-definite, and let x_k^{d} and δ_k be given.

(i) Then, there exist η_k, τ_k and $\hat{\tau}_k$, given by (3.5b), such that for

$$x_{k+1}^{\mathrm{d}} = x_k^{\mathrm{d}} + \tau_k \delta_k, \qquad \mathfrak{Q}_{k+1} = \mathfrak{Q}_k + \eta_k \frac{\mathfrak{Q}_k \delta_k \delta_k^{\mathrm{T}} \mathfrak{Q}_k}{\langle \delta_k, \mathfrak{Q}_k \delta_k \rangle}, \tag{3.6}$$

we have

$$\begin{aligned} x_{k+1} &= \arg\min \{ \langle x - x_{k+1}^{\mathrm{d}}, \mathfrak{Q}_k (x - x_{k+1}^{\mathrm{d}}) \rangle \mid \mathcal{G}^{\mathrm{T}} x = g \} \\ &= \arg\min \{ \langle x - x_k^{\mathrm{d}}, \mathfrak{Q}_{k+1} (x - x_k^{\mathrm{d}}) \rangle \mid \mathcal{G}^{\mathrm{T}} x = g \}, \end{aligned} \tag{3.7}$$

and for

$$\tau_k = -\eta_k \frac{\langle \delta_k, \mathfrak{Q}_k (x_{k+1} - x_k^{\mathrm{d}}) \rangle}{\langle \delta_k, \mathfrak{Q}_k \delta_k \rangle} \tag{3.8}$$

we have $\hat{\tau}_k = \tau_k$.

(ii) Furthermore, *let τ_k, η_k be restricted such that*

$$\tau_k,\ \eta_k \geqslant 0.$$

Then, the choice of δ_k *is restricted by*

$$\langle \delta_k, \mathfrak{Q}_k(x_k - x_k^{\mathrm{d}}) \rangle \leqslant 0.$$

Remark

The above proposition clearly holds for $\mathfrak{Q}_k = D$, and $\mathfrak{Q}_{k+1}$, as given above, is D with a rank-one update and hence is no longer, in general, diagonal.

When $\mathfrak{Q}_k$ is *positive semi-definite*, (3.6) can be written as

$$\mathfrak{Q}_k x_{k+1}^{\mathrm{d}} = \mathfrak{Q}_k(x_k^{\mathrm{d}} + \tau_k \delta_k),$$

from which the relevant parts of x_{k+1}^{d} can be recovered. For example, when $\mathfrak{Q}_k = D$, a diagonal matrix, clearly only those elements of x_{k+1}^{d} corresponding to nonzero diagonal elements of D can be recovered. The correspondence of $\hat{\tau}_k$ given by (3.5e) and τ_k can be established in the same way as in the following proof, except that (3.5c) is used for x_{k+1}.

Proof of Proposition 3.1

Consider the optimality conditions of the minimization problems on both sides of (3.7). With the solution and the Lagrange multipliers denoted respectively by x_{k+1} and λ_{k+1}, the left side yields

$$\mathfrak{Q}_k(x_{k+1} - x_{k+1}^{\mathrm{d}}) - \mathscr{G}\lambda_{k+1} = 0. \tag{3.9}$$

The right side yields

$$\left(\mathfrak{Q}_k + \eta_k \frac{\mathfrak{Q}_k \delta_k \, \delta_k^{\mathrm{T}} \mathfrak{Q}_k}{\langle \delta_k, \mathfrak{Q}_k \delta_k \rangle} \right)(x_{k+1} - x_k^{\mathrm{d}}) - \mathscr{G}\lambda_{k+1} = 0. \tag{3.10}$$

Equating both optimality conditions yields

$$\mathfrak{Q}_k x_{k+1}^{\mathrm{d}} = \mathfrak{Q}_k x_k^{\mathrm{d}} - \left(\eta_k \frac{\mathfrak{Q}_k \delta_k \, \delta_k^{\mathrm{T}} \mathfrak{Q}_k}{\langle \delta_k, \mathfrak{Q}_k \delta_k \rangle} \right)(x_{k+1} - x_k^{\mathrm{d}}), \tag{3.11}$$

and hence $x_{k+1}^{\mathrm{d}} = x_k^{\mathrm{d}} + \tau_k \delta_k$, where

We now show that τ_k given in (3.8) is equivalent to $\hat{\tau}_k$ in (3.5b). We first show that *if* τ_k *in (2.4) is given by* τ_k *in (3.8) then this value and* $\hat{\tau}_k$ *in (3.5b) are equivalent*. Conversely, *if* τ_k *in (2.4) is given by* $\hat{\tau}_k$ *in (3.5b) then* τ_k *in (3.8) and* $\hat{\tau}_k$ *in (3.5b) are equivalent*. For the first part, therefore, we have

$$\begin{aligned} \tau_k &= -\,\eta_k \frac{\langle \delta_k, \mathfrak{Q}_k(x_{k+1} - x_k^{\mathrm{d}}) \rangle}{\langle \delta_k, \mathfrak{Q}_k \delta_k \rangle} \\ &= -\,\eta_k \frac{\langle \delta_k, \mathfrak{Q}_k(x_k + \tau_k P_k \delta_k - x_k^{\mathrm{d}}) \rangle}{\langle \delta_k, \mathfrak{Q}_k \delta_k \rangle} \end{aligned}$$

Solving for τ_k yields $\tau_k = \hat{\tau}_k$, where $\hat{\tau}_k$ is given by (3.5b). To show the converse, suppose that in (2.4), $\tau_k = \hat{\tau}_k$,

$$
\begin{aligned}
-\eta_k & \frac{\langle \delta_k, \mathfrak{Q}_k(x_{k+1} - x_k^{\mathrm{d}})\rangle}{\langle \delta_k, \mathfrak{Q}_k \delta_k\rangle} \\
&= -\eta_k \left[\frac{\langle \delta_k, \mathfrak{Q}_k(x_k - x_k^{\mathrm{d}})\rangle}{\langle \delta_k, \mathfrak{Q}_k \delta_k\rangle} - \frac{\eta_k \langle \delta_k, \mathfrak{Q}_k(x_k - x_k^{\mathrm{d}})\rangle}{\langle \delta_k, \mathfrak{Q}_k \delta_k\rangle + \eta_k \langle \delta_k, \mathfrak{Q}_k P_k \delta_k\rangle} \frac{\langle \delta_k, \mathfrak{Q}_k P_k \delta_k\rangle}{\langle \delta_k, \mathfrak{Q}_k \delta_k\rangle} \right] \\
&= -\eta_k \left[\frac{\langle \delta_k, \mathfrak{Q}_k(x_k - x_k^{\mathrm{d}})\rangle}{\langle \delta_k, \mathfrak{Q}_k \delta_k\rangle} \left(1 - \frac{\eta_k \langle \delta_k, \mathfrak{Q}_k P_k \delta_k\rangle}{\langle \delta_k, \mathfrak{Q}_k \delta_k\rangle + \eta_k \langle \delta_k, \mathfrak{Q}_k P_k \delta_k\rangle} \right) \right] \\
&= -\eta_k \left[\frac{\langle \delta_k, \mathfrak{Q}_k(x_k - x_k^{\mathrm{d}})\rangle}{\langle \delta_k, \mathfrak{Q}_k \delta_k\rangle} \left(\frac{\langle \delta_k, \mathfrak{Q}_k \delta_k\rangle}{\langle \delta_k, \mathfrak{Q}_k \delta_k\rangle + \eta_k \langle \delta_k, \mathfrak{Q}_k P_k \delta_k\rangle} \right) \right] \\
&= -\eta_k \frac{\langle \delta_k, \mathfrak{Q}_k(x_k - x_k^{\mathrm{d}})\rangle}{\langle \delta_k, \mathfrak{Q}_k \delta_k\rangle + \eta_k \langle \delta_k, \mathfrak{Q}_k P_k \delta_k)\rangle} \\
&= \hat{\tau}_k
\end{aligned}
$$

Hence, τ_k in (3.8) and $\hat{\tau}_k$ in (3.5b) are equivalent.

The inequality $\langle \delta_k, \mathfrak{Q}_k(x_k - x_k^{\mathrm{d}})\rangle \leqslant 0$ ensures that $\tau_k, \eta_k \geqslant 0$. To demonstrate this when $\mathfrak{Q}_{k+1}$ is given as above, we see that τ_k given by (3.8) is equivalent to $\hat{\tau}_k$, and this is nonnegative if the above inequality and $\eta_k \geqslant 0$ are satisfied. When the desired value is being updated and an equivalent update to $\mathfrak{Q}_k$ is being computed, for $\tau_k \geqslant 0$,

$$
\eta_k = -\tau_k \frac{\langle \delta_k, \mathfrak{Q}_k \delta_k\rangle}{\langle \delta_k, \mathfrak{Q}_k(x_{k+1} - x_k^{\mathrm{d}})\rangle}. \tag{3.12}
$$

We now show that

$$
\langle \delta_k, \mathfrak{Q}_k(x_k - x_k^{\mathrm{d}})\rangle \leqslant 0 \Rightarrow \langle \delta_k, \mathfrak{Q}_k(x_{k+1} - x_k^{\mathrm{d}})\rangle \leqslant 0.
$$

By Lemma 3.1.2, the inequality

$$
\langle x_{k+1} - x_k, \mathfrak{Q}_{k+1}(x_{k+1} - x_k^{\mathrm{d}})\rangle \leqslant 0
$$

follows from the optimality of x_{k+1} for the quadratic optimization with $\mathfrak{Q}_{k+1}$. Using the expression for $\mathfrak{Q}_{k+1}$, we have

$$
\begin{aligned}
0 &\geqslant \langle x_{k+1} - x_k, \mathfrak{Q}_{k+1}(x_{k+1} - x_k^{\mathrm{d}})\rangle \\
&= \langle x_{k+1} - x_k, \mathfrak{Q}_k(x_{k+1} - x_k^{\mathrm{d}})\rangle \\
&\qquad + \frac{\eta_k}{\langle \delta_k, \mathfrak{Q}_k \delta_k\rangle} \langle x_{k+1} - x_k, \mathfrak{Q}_k \delta_k\rangle \langle \delta_k, \mathfrak{Q}_k(x_{k+1} - x_k^{\mathrm{d}})\rangle.
\end{aligned}
$$

By Lemma 3.1.2, the inequality

$$
\langle x_{k+1} - x_k, \mathfrak{Q}_k(x_k - x_k^{\mathrm{d}})\rangle \geqslant 0
$$

follows from the optimality of x_k with $\mathfrak{Q}_k$. Also, the inequality

$$\langle \delta_k, \mathfrak{Q}_k(x_{k+1} - x_k) \rangle \geqslant 0$$

holds if $\langle \delta_k, \mathfrak{Q}_k(x_k - x_k^{\mathrm{d}}) \rangle \leqslant 0$ (see Lemma 5.3.2). Then we have

$$\langle \delta_k, \mathfrak{Q}_k(x_{k+1} - x_k^{\mathrm{d}}) \rangle \leqslant 0 \Rightarrow \eta_k \geqslant 0,$$

and the corresponding η_k is given by (3.12). □

The extension of the above result to nonlinear constraints is straightforward. The method can thus be easily extended to the nonlinear constrained case when the diagonal equivalent of a nondiagonal quadratic function is desired. The useful analytical equivalence of τ and $\hat{\tau}$ cannot be established exactly in the nonlinear case. Clearly, if the departure of x_{k+1} from x_k is sufficiently small then this equivalence can also be established by invoking a mean value theorem and thereby using a local linear representation of the constraints (see e.g. Theorem 5.3.4). The following corollary summarizes the straightforward extension, which can be refined, if required, as described above.

Corollary 3.1 (Extension to nonlinear constraints)
Let x_{k+1}^{d} and $\mathfrak{Q}_{k+1}$ be defined by (3.6) and let the constraints be given by

$$\mathbb{G} = \{x \in \mathbb{R}^n \mid g(x) = 0\}, \tag{3.13}$$

where g is twice differentiable and g: $\mathbb{R}^n \to \mathbb{R}^e$. Then the equivalence

$$\arg\min\{\langle x - x_{k+1}^{\mathrm{d}}, \mathfrak{Q}_k(x - x_{k+1}^{\mathrm{d}})\rangle \mid x \in \mathbb{G}\}$$
$$= \arg\min\{\langle x - x_k^{\mathrm{d}}, \mathfrak{Q}_{k+1}(x - x_k^{\mathrm{d}})\rangle \mid x \in \mathbb{G}\}$$

holds for τ_k given by (3.8).

Proof
The proof follows from the equivalence of the first-order optimality conditions of both problems. □

The extension to nonlinear constraints is thus implementable, since the basic ingredients that enter τ_k are δ_k and x_{k+1}. Both these vectors are known when any one of the two quadratic problems have already been solved.

The above result relates the effect of a single update of $\mathfrak{Q}_k$ that yields $\mathfrak{Q}_{k+1}$ or a single update of x_k^{d} that yields x_{k+1}^{d}. We now consider the sequence $\{\mathfrak{Q}_k\}$ generated by Algorithm 3.1 and the corresponding sequence $\{x_k^{\mathrm{d}}\}$ generated by Algorithm 2.1. The diagonalizibility result stated below is essentially a corollary of the above discussion.

Theorem 3.1 (Diagonalizibility of quadratic forms)
Let

(*i*) *the sequence* $\{\mathfrak{Q}_k\}$ *be generated by (3.3);*

(*ii*) $\{x_k^{\mathrm{d}}\}$ *be generated by (2.3);*

(*iii*) $\mathfrak{Q}_0 = D$.

Then, the equivalence between the diagonal and nondiagonal quadratic optimizations

$$\arg\min\{\langle x - x_k^{\mathrm{d}}, D(x - x_k^{\mathrm{d}})\rangle \mid \mathscr{G}^{\mathrm{T}} x = g\}$$
$$= \arg\min\{\langle x - x_0^{\mathrm{d}}, \mathfrak{Q}_k(x - x_0^{\mathrm{d}})\rangle \mid \mathscr{G}^{\mathrm{T}} x = g\} \quad (3.14\mathrm{a})$$

holds if

$$Dx_k^{\mathrm{d}} = Dx_0^{\mathrm{d}} - \sum_{i=0}^{k-1} \eta_i \frac{\langle \delta_i, \mathfrak{Q}_i(x_k - x_0^{\mathrm{d}})\rangle}{\langle \delta_i, \mathfrak{Q}_i \delta_i\rangle} \mathfrak{Q}_i \delta_i. \quad (3.14\mathrm{b})$$

Remark
If $\mathfrak{Q}_0$ is positive-definite then clearly the entire vector x_k^{d} can be recovered from (3.14b). Otherwise, only the part of x_k^{d} that is relevant to computing the optimal solution can be recovered.

Using the matrix update formula (3.3), expression (3.14b) can be expressed in terms of $\mathfrak{Q}_0$ only. However, in computational terms, this does not appear to be a significant advantage, since from each iteration only the vector $\mathfrak{Q}_i\delta_i$ needs to be stored. It can be verified that this vector, appropriately normalized by the scalar terms, is sufficient to construct x_k^{d}, given the current solution x_k and the initial desired value x_0^{d}.

Proof of Theorem 3.1
The equivalence of the optimality conditions of both problems yields

$$\begin{aligned} D(x_k - x_k^{\mathrm{d}}) &= \mathfrak{Q}_k(x_k - x_0^{\mathrm{d}}) \\ &= \left(\mathfrak{Q}_{k-1} + \eta_{k-1}\frac{\mathfrak{Q}_{k-1}\delta_{k-1}\delta_{k-1}^{\mathrm{T}}\mathfrak{Q}_{k-1}}{\langle \delta_{k-1}, \mathfrak{Q}_{k-1}\delta_{k-1}\rangle}\right)(x_k - x_0^{\mathrm{d}}) \\ &= \left(D + \sum_{i=0}^{k-1}\eta_i \frac{\mathfrak{Q}_i\delta_i\delta_i^{\mathrm{T}}\mathfrak{Q}_i}{\langle \delta_i, \mathfrak{Q}_i\delta_i\rangle}\right)(x_k - x_0^{\mathrm{d}}), \end{aligned}$$

from which (3.14b) follows. □

3.2 Complexity of the algorithm

The equivalence of Algorithms 2.1 and 3.1 is the basic vehicle for studying the complexity of Algorithm 2.1. It is shown in Chapter 5 that Algorithm 3.1 is closely related to Khachian's (1979, 1980) polynomial-time algorithm for linear programming. We invoke the equivalence to establish the same convergence property for Algorithm 2.1. To that end, we prepare the framework for relating these algorithms.

It is shown in Chapter 5 (Theorem 5.2.1) that Algorithm 3.1 is equivalent to Khachian's algorithm *provided* that x_0^{d} on the right of (3.14a) is shifted, for every k, to affect only the stepsize τ_k or $\hat{\tau}_k$. The shift is defined by

$$\hat{x}_k^{\mathrm{d}} = x_k - \psi_k(x_k - x_k^{\mathrm{d}}),$$

where

$$\psi_k = -\frac{\langle \delta_k, \mathfrak{Q}_k\delta_k\rangle + \eta_k\langle \mathfrak{Q}_k\delta_k, P_k\mathfrak{Q}_k^{-1}(\mathfrak{Q}_k\delta_k)\rangle}{\eta_k(n+1)\langle \delta_k, \mathfrak{Q}_k(x_k - x_k^{\mathrm{d}})\rangle\langle \delta_k, \mathfrak{Q}_k\delta_k\rangle^{1/2}}$$

and

$$\eta_k = \frac{2}{n-1},$$

$$\hat{\tau}_k = -\frac{\psi_k\eta_k\langle \delta_k, \mathfrak{Q}_k(x_k - x_k^{\mathrm{d}})\rangle}{\langle \delta_k, \mathfrak{Q}_k\delta_k\rangle + \eta_k\langle \mathfrak{Q}_k\delta_k, P_k\mathfrak{Q}_k^{-1}(\mathfrak{Q}_k\delta_k)\rangle} = \frac{1}{(n+1)\langle \delta_k, \mathfrak{Q}_k\delta_k\rangle^{1/2}}.$$

Khatchian's algorithm is essentially given by the above choice of $\hat{\tau}$ and η_k and

$$\mathfrak{Q}_{k+1} = \frac{n^2-1}{n}\left(\mathfrak{Q}_k + \frac{2}{n-1}\frac{\mathfrak{Q}_k\delta_k\,\delta_k^{\mathrm{T}}\mathfrak{Q}_k}{\langle \delta_k, \mathfrak{Q}_k\delta_k\rangle}\right).$$

It can be verified that for above choices of $\hat{x}_k^{\mathrm{d}}, \eta_k$ and $\mathfrak{Q}_k$ do not affect the solution of

$$\min\{\langle x - \hat{x}_k^{\mathrm{d}}, \mathfrak{Q}_k(x - \hat{x}_k^{\mathrm{d}})\rangle \mid \mathscr{G}^{\mathrm{T}}x = g\},$$

which is still x_k (see Lemma 4.1). The solution of

$$\min\{\langle x - \hat{x}_k^{\mathrm{d}}, \mathfrak{Q}_{k+1}(x - \hat{x}_k^{\mathrm{d}})\rangle \mid \mathscr{G}^{\mathrm{T}}x = g\}$$

is changed, because the value of τ_k or $\hat{\tau}_k$ is affected. Consider the equivalence of the last problem with the diagonal case

$$\begin{aligned}&\arg\min\{\langle x - x_{k+1}^{\mathrm{d}}, D(x - x_{k+1}^{\mathrm{d}})\rangle \mid \mathscr{G}^{\mathrm{T}}x = g\}\\ &\qquad = \arg\min\{\langle x - \hat{x}_k^{\mathrm{d}}, \mathfrak{Q}_{k+1}(x - \hat{x}_k^{\mathrm{d}})\rangle \mid \mathscr{G}^{\mathrm{T}}x = g\}.\end{aligned} \tag{3.15}$$

We can establish this equivalence using the foregoing discussion and the corresponding equivalence of the optimality conditions

$$D(x_{k+1} - x_{k+1}^{\mathrm{d}}) = \left(D + \sum_{i=0}^{k}\eta_i\frac{\mathfrak{Q}_i\delta_i\,\delta_i^{\mathrm{T}}\mathfrak{Q}_i}{\langle \delta_i, \mathfrak{Q}_i\delta_i\rangle}\right)(x_{k+1} - \hat{x}_k^{\mathrm{d}}). \tag{3.16}$$

Using (3.16), we have

$$Dx^{\mathrm{d}}_{k+1} = Dx_{k+1} - \mathfrak{Q}_k(x_{k+1} - \hat{x}^{\mathrm{d}}_k) - \eta_k \frac{\langle \delta_k, \mathfrak{Q}_k(x_{k+1} - \hat{x}^{\mathrm{d}}_k)\rangle}{\langle \delta_k, \mathfrak{Q}_k \delta_k\rangle} \mathfrak{Q}_k \delta_k$$

$$Dx^{\mathrm{d}}_{k+1} = D\hat{x}^{\mathrm{d}}_k - \sum_{i=0}^{k} \eta_i \frac{\langle \delta_i, \mathfrak{Q}_i(x_{k+1} - \hat{x}^{\mathrm{d}}_k)\rangle}{\langle \delta_i, \mathfrak{Q}_i \delta_i\rangle} \mathfrak{Q}_i \delta_i.$$

x^{d}_{k+1} can be recovered from either of the last two expressions. Thus, since the equivalence of the diagonal and nondiagonal problems has been established, we can conclude that the convergence of Algorithm 2.1 is the same as Algorithm 3.1, which is shown to be polynomial-time in Chapter 5.

The above result is also independent of the restriction $\langle \delta_k, \mathfrak{Q}_k(x_k - x^{\mathrm{d}}_k)\rangle \leqslant 0$, since, for the choice of ψ_k given above, the sign of $\hat{\tau}_k$ is independent of the sign of $\langle \delta_k, \mathfrak{Q}_k(x_k - x^{\mathrm{d}}_k)\rangle$.

4 THE ARBITRARINESS OF SHADOW PRICES

In this section, we discuss a procedure for assigning arbitrary values to the shadow prices of a class of multiple-objective decision problems. Let there exist a quadratic objective that reflects the policy maker's preferences. Hence, the quadratic objective function is assumed to yield an optimal policy that is accepted by the policy maker. It is not necessary that this quadratic function should have been specified through one of Algorithms 2.1 and 3.1.

We show below that either one of these algorithms can be used, with such a given quadratic function, to alter the values of the shadow prices of the problem *without* altering the optimal solution. The arbitrariness of the shadow prices for diagonal and nondiagonal objective functions follows from this observation. This may clearly have significant consequences in problems where shadow prices are used to indicate the 'price' of scarce resources (see e.g. UNIDO, 1972). The extension of this result to general nonlinear objective functions and nonlinear constraints is also discussed below. We begin with the nondiagonal quadratic case, and thus assume that we have a matrix $\mathfrak{Q}_k$ and x^{d} such that

$$\arg\min\{\langle x - x^{\mathrm{d}}, \mathfrak{Q}_k(x - x^{\mathrm{d}})\rangle \mid \mathcal{G}^{\mathrm{T}}x = g\} = x_k \in \Omega. \tag{4.1}$$

We do not need to assume that the inequality $\langle \delta_k, \mathfrak{Q}_k(x_k - x^{\mathrm{d}})\rangle \leqslant 0$ is satisfied. This is due to the transformation introduced in the following lemma.

Lemma 4.1
Let $\mathfrak{Q}_k, x^{\mathrm{d}}$ and δ_k be given.

(i) Let the inequality

$$\langle \delta_k, \mathfrak{Q}_k(x_k - x^{\mathrm{d}})\rangle \geqslant 0,$$

hold. Then, there exists a sufficiently small $\kappa_k \in (-\infty, 0)$ *such that the transformation*

$$\tilde{x}_k^{\mathrm{d}} = x_k - \kappa_k (x_k - x^{\mathrm{d}}) \tag{4.2}$$

ensures that

$$x_k = \arg\min \{ \langle x - \tilde{x}_k^{\mathrm{d}}, \mathfrak{Q}_k (x - \tilde{x}_k^{\mathrm{d}}) \rangle \mid \mathscr{G}^{\mathrm{T}} x = g \}$$

and also

$$\hat{\tau}_k \geqslant 0.$$

(ii) Let the inequality

$$\langle \delta_k, \mathfrak{Q}_k (x_k - x^{\mathrm{d}}) \rangle \leqslant 0$$

hold. Then, we have

$$\hat{\tau}_k \geqslant 0,$$

with $\kappa_k = 1$, *and thence* $\hat{x}_k^{\mathrm{d}} = x^{\mathrm{d}}$.

(iii) Let the restriction

$$\langle \delta_k, \mathfrak{Q}_k (x_k - x^{\mathrm{d}}) \rangle \neq 0$$

hold. Then, for any sign of this expression, there exists a sufficiently small $\kappa_k \in (-\infty, +\infty)$ *such that* $\hat{\tau}_k = 1$ *and* $\eta_k \geqslant 0$.

Proof

To show that x_k is still the solution of the revised problem, consider the solution of the latter, denoted by $\tilde{x}_k$. Since

$$\mathscr{G}^{\mathrm{T}} (\tilde{x}_k - x_k) = 0,$$

we have

$$\tilde{x}_k \quad x_k = Z\omega,$$

for some vector ω. Furthermore, the optimality condition of the original problem yields

$$\mathfrak{Q}_k (x_k - x^{\mathrm{d}}) - \mathscr{G} \lambda_k = 0.$$

Hence, we can write the optimality condition of the transformed problem as

$$\mathfrak{Q}_k (\tilde{x}_k - x_k + x_k - \tilde{x}_k^{\mathrm{d}}) - \mathscr{G} \tilde{\lambda}_k = 0,$$

where $\tilde{\lambda}_k$ is the associated shadow price vector. Thus,

$$\begin{aligned}
\mathfrak{Q}_k Z \omega &= \mathfrak{Q}_k (\tilde{x}_k - x_k) \\
&= -\mathfrak{Q}_k \{ x_k - [x_k - \kappa_k (x_k - x^{\mathrm{d}})] \} + \mathscr{G} \tilde{\lambda}_k \\
&= \mathscr{G} (\tilde{\lambda}_k - \kappa_k \lambda_k).
\end{aligned}$$

Premultiplying by Z^{T} yields

$$Z^{\mathrm{T}}\mathfrak{Q}_k Z\omega = 0.$$

Since part of the sufficient conditions for optimality, $Z^{\mathrm{T}}\mathfrak{Q}_k Z$ is positive-definite (see e.g. Gill, Murray and Wright, 1981), we have $\omega = 0$, and thence

$$\tilde{x}_k - x_k = Z\omega = 0.$$

To show that $\hat{\tau}_k \geqslant 0$, consider

$$\begin{aligned}\hat{\tau}_k &= -\frac{\eta_k\langle\delta_k, \mathfrak{Q}_k(x_k - \tilde{x}_k^{\mathrm{d}})\rangle}{\langle\delta_k, \mathfrak{Q}_k\delta_k\rangle + \eta_k\langle\mathfrak{Q}_k\delta_k, Z(Z^{\mathrm{T}}\mathfrak{Q}_k Z)^{-1}Z^{\mathrm{T}}\mathfrak{Q}_k\delta_k\rangle}\\ &= -\frac{\eta_k\kappa_k\langle\delta_k, \mathfrak{Q}_k(x_k - x^{\mathrm{d}})\rangle}{\langle\delta_k, \mathfrak{Q}_k\delta_k\rangle + \eta_k\langle\mathfrak{Q}_k\delta_k, Z(Z^{\mathrm{T}}\mathfrak{Q}_k Z)^{-1}Z^{\mathrm{T}}\mathfrak{Q}_k\delta_k\rangle}.\end{aligned}$$

Thus, κ_k can be chosen to yield $\hat{\tau}_k \geqslant 0$.

To show the second part of the lemma, we note that for

$$\langle\delta_k, \mathfrak{Q}_k(x_k - x^{\mathrm{d}})\rangle \leqslant 0,$$

the above expression yields $\hat{\tau}_k \geqslant 0$ for any positive value of κ_k, including $\kappa_k = 1$.

To establish the last part of the lemma, we note that $\hat{\tau}_k = 1$ if η_k is given by

$$\eta_k = -\frac{\langle\delta_k, \mathfrak{Q}_k\delta_k\rangle}{\langle\mathfrak{Q}_k\delta_k, Z(Z^{\mathrm{T}}\mathfrak{Q}_k Z)^{-1}Z^{\mathrm{T}}\mathfrak{Q}_k\delta_k\rangle + \kappa_k\langle\delta_k, \mathfrak{Q}_k(x_k - x^{\mathrm{d}})\rangle},$$

and $\eta_k \geqslant 0$ if κ_k satisfies

$$\kappa_k < -\frac{\langle\mathfrak{Q}_k\delta_k, Z(Z^{\mathrm{T}}\mathfrak{Q}_k Z)^{-1}Z^{\mathrm{T}}\mathfrak{Q}_k\delta_k\rangle}{\langle\delta_k, \mathfrak{Q}_k(x_k - x^{\mathrm{d}})\rangle}. \qquad \square$$

Proposition 4.1 (Arbitrariness of shadow prices using Algorithm 3.1)
Let $\mathfrak{Q}_k$ be given, and

(i) λ_k be the vector of shadow prices of problem (4.1);

(ii) λ_k be unacceptable to the decision maker as a measure of the 'price' of the corresponding scarce resources represented by the constraints;

(iii) λ_{p} be the preferred shadow prices of the decision maker;

(iv) δ_k in Algorithm 3.1 be defined by

$$\mathfrak{Q}_k\delta_k = -\mathcal{G}(\lambda_{\mathrm{p}} - \lambda_k); \tag{4.3}$$

using this δ_k, let $\mathfrak{Q}_{k+1}$ be computed with (3.3). Given $\mathfrak{Q}_{k+1}$ and $\tilde{x}_k^{\mathrm{d}}$ in (4.2), let one iteration of Algorithm 3.1 be evaluated to compute

$$x_{k+1} = \arg\min \{\langle x - \tilde{x}_k^{\rm d}, \mathfrak{Q}_{k+1}(x - \tilde{x}_k^{\rm d})\rangle \mid \mathscr{G}^{\rm T} x = g\}.$$

Then, we have

$$x_{k+1} = x_k, \tag{4.4a}$$

$$\lambda_{k+1} = \lambda_k + \hat{\tau}_k(\lambda_{\rm p} - \lambda_k), \tag{4.4b}$$

where $\hat{\tau}_k$ *is given by*

$$\hat{\tau}_k = -\frac{\eta_k \kappa_k \langle \delta_k, \mathfrak{Q}_k(x_k - x^{\rm d})\rangle}{\langle \delta_k, \mathfrak{Q}_k \delta_k\rangle + \eta_k \langle \mathfrak{Q}_k \delta_k, Z(Z^{\rm T}\mathfrak{Q}_k Z)^{-1} Z^{\rm T}\mathfrak{Q}_k \delta_k\rangle}, \tag{4.4c}$$

and for η_k, κ_k, *given by the last part of Lemma 4.1, we have*

$$\lambda_{k+1} = \lambda_{\rm p}. \tag{4.4d}$$

Remark

In the above result, we have not assumed that $\mathfrak{Q}_k$ is positive-definite. All that is needed is that $\langle \delta_k, \mathfrak{Q}_k(x_k - x^{\rm d})\rangle \neq 0$. Provided this condition is satisfied, we can arbitrarily change the shadow prices.

Proof of Proposition 4.1

We consider the general case when $\mathfrak{Q}_k$ is positive- semidefinite. By Lemma 4.1, we know that x_k is unchanged by the transformation of the desired values. Since $Z^{\rm T}\mathscr{G} = 0$, expression (3.5c) yields

$$x_{k+1} = x_k + \hat{\tau}_k Z(Z^{\rm T}\mathfrak{Q}_k Z)^{-1} Z^{\rm T}[-\mathscr{G}(\lambda_{\rm p} - \lambda_k)] = x_k.$$

Similarly, expression (3.5d) yields

$$\begin{aligned}\lambda_{k+1} &= \lambda_k - \hat{\tau}_k(\mathscr{G}^{\rm T}\mathscr{G})^{-1}\mathscr{G}^{\rm T}[I - \mathfrak{Q}_k Z(Z^{\rm T}\mathfrak{Q}_k Z)^{-1} Z^T][-\mathscr{G}(\lambda_{\rm p} - \lambda_k)]\\ &= \lambda_k + \hat{\tau}_k(\mathscr{G}^{\rm T}\mathscr{G})^{-1}\mathscr{G}^{\rm T}[-\mathscr{G}(\lambda_{\rm p} - \lambda_k)],\end{aligned}$$

and thence we have (4.4b).

The last part of the proposition follows from the ability to set $\hat{\tau}_k = 1$ by choosing κ_k and η_k as established in Lemma 4.1. □

Thus, it is possible to select a correction vector δ_k such that it will not alter the value of the optimal solution x_k but will change the shadow prices in order to achieve the preferred shadow prices specified as $\lambda_{\rm p}$.

We now consider the diagonal quadratic forms and updates of the desired values that will achieve the same effect as in Proposition 4.1. We state the next result with slightly greater generality in that we do not assume that $\mathfrak{Q}_k$ is diagonal. Only, we do not use δ_k to update $\mathfrak{Q}_k$. Instead, keeping $\mathfrak{Q}_k$ constant, we revise the desired values as in Section 2.

Proposition 4.2 (Arbitrariness of shadow prices using Algorithm 2.1)
Let $\mathfrak{Q}_k$ and x^{d} be given and (i)–(iii) in Proposition 4.1 hold. Let δ_k in Algorithm 2.1 be defined by (4.3), using this δ_k, let x^{d}_{k+1} be computed by

$$x^{\mathrm{d}}_{k+1} = x^{\mathrm{d}} + \tau_k \delta_k,$$

for given τ_k, and let one iteration of Algorithm 2.1 be evaluated to compute

$$x_{k+1} = \arg\min \{\langle x - x^{\mathrm{d}}_{k+1}, \mathfrak{Q}_k(x - x^{\mathrm{d}}_{k+1})\rangle \mid \mathcal{G}^{\mathrm{T}} x = g\}. \tag{4.5}$$

Then, we have (4.4a),

$$\lambda_{k+1} = \lambda_k + \tau_k(\lambda_{\mathrm{p}} - \lambda_k), \tag{4.6}$$

and, with the choice $\tau_k = 1$, we have $\lambda_{k+1} = \lambda_{\mathrm{p}}$.

Proof
Let x^{d} be the desired value used in (4.1). Using the same arguments as in (2.11), we have

$$\begin{aligned} x_{k+1} - x_k &= Z\omega \\ &= -Z(Z^{\mathrm{T}}\mathfrak{Q}_k Z)^{-1} Z^{\mathrm{T}} \mathfrak{Q}_k (x_k - x^{\mathrm{d}}_k - \tau_k \delta_k). \end{aligned}$$

The optimality condition of (4.1) yields

$$\mathfrak{Q}_k(x_k - x^{\mathrm{d}}) - \mathcal{G}\lambda_k = 0, \tag{4.7}$$

and δ_k is given by (4.3), and, since $Z^{\mathrm{T}}\mathcal{G} = 0$, we have $x_{k+1} - x_k = 0$.
To establish (4.6), consider (2.12), which can be rewritten as

$$\lambda_{k+1} = (\mathcal{G}^{\mathrm{T}}\mathcal{G})^{-1}\mathcal{G}^{\mathrm{T}}\mathfrak{Q}_k(x_{k+1} - x_k + x_k - x^{\mathrm{d}}_k - \tau_k\delta_k).$$

Using (4.7), (4.3) and $x_{k+1} - x_k = 0$, this yields (4.6).
The choice of $\tau_k = 1$ immediately yields $\lambda_{k+1} = \lambda_{\mathrm{p}}$ using the same arguments as above. □

We now consider the extension of the above arbitrary nature of the shadow prices to nonlinear constraints. For nonlinear objective functions, it is sufficient to argue that the quadratic approximation of the function in the neighbourhood of x_k is an adequate approximation for sufficiently small values of η_k in Proposition 4.1 and τ_k in Proposition 4.2.

Theorem 4.1 (Arbitrariness of shadow prices for nonlinear problems when updating weighting matrices)
Let $\mathfrak{Q}_k$ be given, let x^{d}_{k+1} be computed as in Proposition 4.2, and

(i) the feasible set $\mathbb{G}$ be given by the nonlinear system in (3.13);

(ii) x_k *be the solution of the problem*

$$\min\{\langle x - x_k^{\mathrm{d}}, \mathfrak{Q}_k(x - x_k^{\mathrm{d}})\rangle \mid x \in \mathbb{G}\}, \tag{4.8}$$

with λ_k *denoting the associated shadow prices;*

(iii) λ_{p} *be the shadow price value that is preferred by the decision maker;*

(iv) $\nabla g(x_k)$ *denote the matrix whose columns are the constraint normals at* x_k*;*

(v) δ_k *be given by*

$$\mathfrak{Q}_k\delta_k = -\kappa_k \nabla g(x_k)(\lambda_{\mathrm{p}} - \lambda_k),$$
$$\langle \delta_k, \mathfrak{Q}_k(x_k - x_k^{\mathrm{d}})\rangle \neq 0$$

and used to compute $\mathfrak{Q}_{k+1}$ *in (3.3).*

Then

(a) $\exists \kappa_k \neq 0$, *sufficiently small, such that*

$$x_{k+1}^{\mathrm{d}} = x_k - \kappa_k(x_k - x_k^{\mathrm{d}}),$$

$$\kappa_k\langle \delta_k, \mathfrak{Q}_k(x_k - x_k^{\mathrm{d}})\rangle > 0,$$

and η_k *in (3.3) is given by*

$$\eta_k = \frac{\langle \delta_k, \mathfrak{Q}_k\delta_k\rangle}{\kappa_k\langle \delta_k, \mathfrak{Q}_k(x_k - x_k^{\mathrm{d}})\rangle} > 0,$$

such that $\hat{\tau}_k = 1$,

(b) we have

$$x_k = \arg\min\{\langle x - x_{k+1}^{\mathrm{d}}, \mathfrak{Q}_k(x - x_{k+1}^{\mathrm{d}})\rangle \mid x \in \mathbb{G}\}, \tag{4.9}$$

$$x_{k+1} = x_k = \arg\min\{\langle x - x_{k+1}^{\mathrm{d}}, \mathfrak{Q}_{k+1}(x - x_{k+1}^{\mathrm{d}})\rangle \mid x \in \mathbb{G}\}, \tag{4.10}$$

and

$$\lambda_{k+1} = \lambda_{\mathrm{p}}.$$

Proof

The first-order optimality conditions of (4.9) yield

$$\mathfrak{Q}_k(x_k - x_{k+1}^{\mathrm{d}}) + \nabla g(x_k)\lambda_k = 0. \tag{4.11}$$

Evaluating the desired values

$$\mathfrak{Q}_k\{x_k - [x_k - \kappa_k(x_k - x_k^{\mathrm{d}})]\} + \nabla g(x_k)\lambda_k = 0,$$

and, thus,

$$\mathfrak{Q}_k(x_k - x_k^{\mathrm{d}}) + \nabla g(x_k)\left(\frac{1}{\kappa_k}\right)\lambda_k = 0.$$

Hence, x_k still solves the transformed problem, but λ_k is normalized by the scalar κ_k. Consider the same condition for problem (4.10). We establish that, for the given values of κ_k, η_k and $\hat{\tau}_k = 1$, this condition is identical to (4.11). Thus,

$$\begin{aligned}
&\mathfrak{Q}_{k+1}\{x_k - [x_k - \kappa_k(x_k - x_k^{\mathrm{d}})]\} \\
&\quad = \left(\mathfrak{Q}_k + \eta_k \frac{\mathfrak{Q}_k \delta_k \delta_k^{\mathrm{T}} \mathfrak{Q}_k}{\langle \delta_k, \mathfrak{Q}_k \delta_k \rangle}\right)\{x_k - [x_k - \kappa_k(x_k - x_k^{\mathrm{d}})]\} \\
&\quad = \kappa_k \mathfrak{Q}_k(x_k - x_k^{\mathrm{d}}) + \left[\eta_k \kappa_k \frac{\langle \delta_k, \mathfrak{Q}_k(x_k - x_k^{\mathrm{d}})\rangle}{\langle \delta_k, \mathfrak{Q}_k \delta_k \rangle}\right]\mathfrak{Q}_k \delta_k \\
&\quad = -\kappa_k \nabla g(x_k)\lambda_k + \mathfrak{Q}_k \delta_k \\
&\quad = -\kappa_k \nabla g(x_k)\lambda_k - \kappa_k \nabla g(x_k)(\lambda_{\mathrm{p}} - \lambda_k).
\end{aligned}$$

From the foregoing discussion, we have the optimality condition for (4.10):

$$\kappa_k \mathfrak{Q}_{k+1}(x_k - x_k^{\mathrm{d}}) = -\kappa_k \nabla g(x_k)\lambda_{\mathrm{p}}.$$

Hence the optimality condition of (4.10) is solved by x_k and λ_{p}. □

We finally consider the case when the desired values are updated to alter the shadow prices of the nonlinear problem.

Theorem 4.2 (Arbitrariness of shadow prices for nonlinear problems when updating desired values)
Let $\mathfrak{Q}_k, \lambda_{\mathrm{p}}$ and x_k^{d} be given and δ_k be chosen to satisfy

$$\mathfrak{Q}_k \delta_k = -\nabla g(x_k)(\lambda_{\mathrm{p}} - \lambda_k).$$

Then, the adjustment

$$x_{k+1}^{\mathrm{d}} = x_k^{\mathrm{d}} + \delta_k$$

ensures that

$$x_k = \arg\min \{\langle x - x_k^{\mathrm{d}}, \mathfrak{Q}_k(x - x_k^{\mathrm{d}})\rangle \mid x \in \mathbb{G}\}$$

and

$$x_{k+1} = x_k = \arg\min \{\langle x - x_{k+1}^{\mathrm{d}}, \mathfrak{Q}_k(x - x_{k+1}^{\mathrm{d}})\rangle \mid x \in \mathbb{G}\},$$

and, furthermore, $\lambda_{k+1} = \lambda_{\mathrm{p}}$.

Proof
From the optimality conditions of the both problems, we have

$$\mathfrak{Q}_k(x_k - x_k^{\rm d} - \delta_k) = \mathfrak{Q}_k(x_k - x_k^{\rm d}) + \nabla g(x_k)(\lambda_{\rm p} - \lambda_k)$$
$$= -\nabla g(x_k)\lambda_k + \nabla g(x_k)(\lambda_{\rm p} - \lambda_k)$$
$$= -\nabla g(x_k)\lambda_{\rm p}.$$

Hence, the solution of the second problem is given by x_k and $\lambda_{\rm p}$. □

5 CONCLUDING REMARKS

The possibility of constructing quadratic objective functions with diagonal weighting matrices, or Hessians, is desirable both in terms of computational convenience and interpretability in defining the relative weights in decision making under conflicting objectives. The extension of these results to nonlinear constraints permits the wider applicability of the results.

The arbitrariness of shadow prices seems to be an intuitively obvious concept, since, given the constraints, there is usually more than one objective function that leads to an acceptable optimal solution. Yet there have been numerous attempts in the past to use shadow prices as the 'price' of scarce resources or activities. The two procedures discussed—one by the updating of the weighting matrix and the other by the updating of the desired values—actually construct arbitrary shadow price values without altering the optimal solution. Extensions of both methods to nonlinear problems illustrate the general case for the nonuniqueness of shadow prices.

6 COMMENTS AND NOTES

CN.1 The $LDL^{\rm T}$ factorization of $\mathfrak{Q}$

The computation of the factorisation $LDL^{\rm T}$ is discussed in Golub and van Loan (1983). We note that the positive-definiteness assumption on $\mathfrak{Q}$, which requires all elements of D to be strictly positive, is more restrictive than (1.1), since in that positive-semidefinite formulation, D is allowed to have zero diagonal elements. This allows the zero-weighting of certain variables, if required.

REFERENCES

Engwerda, J.C. and P.W. Otter (1989). On the choice of weighting matrices in the minimum variance controller. *Automatica* **25**, 279–285.

Fiacco, A.V. and G.P. McCormick (1968). *Nonlinear Programming: Sequential Unconstrained Minimization Techniques*. Wiley, New York.

Geoffrion, A.M. (1971). Duality in nonlinear programming: a simplified applications oriented development. *SIAM Rev.* **13**, 1–37.

Gill, P.E., W. Murray and M.H. Wright (1981). *Practical Optimization*. Academic Press, London.

Golub, G.H. and C.F. van Loan (1983). *Matrix Computations*. North Oxford Academic, Oxford.

Khachian, L.G. (1979). A polynomial algorithm in linear programming. *Soviet Math. Dokl.* **20**, 191–194.

Khachian, L.G. (1980). Polynomial algorithm in linear programming. *USSR Comput. Math. Math. Phys.* **20**, 53–72.

Kwakernaak, H. and R. Sivan (1972). *Linear Optimal Control Systems*. Wiley, New York.

Polak, E. (1971). *Computational Methods in Optimization*. Academic Press, New York.

Ponstein, J. (1983). Applying some modern developments to choosing your own lagrange multipliers. *SIAM Rev.* **25**, 183–199.

Rustem, B. (1990). On the diagonalizibility of quadratic forms and the arbitrariness of shadow prices. In *Dynamic Modelling and Control of National Economies 1989* (N. Christodoulakis, ed.), pp. 299–306. Pergamon Press, Oxford.

Rustem, B. (1991). The diagonalizibility of quadratic functions and the arbitrariness of shadow prices. *Automatica* **27**, 573–578.

Rustem, B., K. Velupillai and J. Westcott (1978). Respecifying the weighting matrix of a quadratic objective function. *Automatica* **14**, 567–582.

UNIDO (1972). *Guidelines for Project Evaluation*. United Nations Publications, UNIDO, Vienna.

CHAPTER 5

Multiple-Objective Optimization 3 Convergence and Complexity of Decision Processes

We consider the convergence properties of Algorithm 3.2.1, for multiple-objective decision problems, along with the extensions of this method to convex and general nonlinearly constrained decision problems.

A desirable property of the method is that, at every iteration, the decision computed is optimal with respect to the current objective function. The optimality at each stage allows the decision maker to easily assess the shortcomings of the decision at every stage. The method then translates these shortcomings into modifications of the objective function.

In order to explore the convergence of the method, a second algorithm is introduced for which polynomial-time convergence is ensured. The desirable convergence property of the latter is offset by the fact that at each iteration the decision generated is not optimal. This conceptual difficulty is overcome by introducing a simple equivalence between the two algorithms. Hence, *both optimality, at each iteration, and termination in polynomial time can be ensured.*

An interactive approach to the multiple objective problem can be viewed as a process of adding constraints to refine the feasible region until an acceptable decision is obtained. The complexity of the decision making process with these methods is studied within this context. Polynomial-time convergence is demonstrated by invoking Khachian's (1979, 1980) ellipsoid algorithm for computing a feasible point of a system of linear inequality constraints. The search for an appropriate weighting matrix is shown to be similar to the search of Khachian's algorithm for a feasible point of a linearly constrained region.

1 INTRODUCTION

In this chapter, we discuss the properties of the algorithms in Chapters 3 and 4. The two approaches in Chapter 3, specifying the quadratic function for multiple-objective decision models, involve interactions with decision makers. The first, Algorithm 3.1.1, demands excessive and precise information from the decision maker concerning the gradient values of the decision maker's true objective function. It is shown to be convergent in Theorem 3.1.1, in at most as many steps as the dimension of the vector of variables in the decision problem. The information requirements of this method make it impractical: it can only be seen as an idealised model of the approach adopted in Section 3.2.

The second method is Algorithm 3.2.1, or its reformulation, Algorithm 4.3.1. This is an attempt to relax the excessive information required by Algorithm

3.1.1. In addition to its relaxed information requirements, Algorithm 4.3.1 also has desirable properties when the underlying decision model is linear, as discussed in Chapter 3. In this chapter, we discuss the extensions of Algorithm 4.3.1 to convex and general nonlinear constraints (or models). This method generates optimal policies, with respect to the current objective function, at each iteration. The optimality property allows the decision maker to identify any deficiency in the decision at each stage. The correction of this deficiency is then translated by the method into a modification of the objective function.

We introduce Algorithm 2.1 below, which is similar to Algorithm 4.3.1, and, using Khachian's (1979, 1980) results, can be shown to be convergent in polynomial time. This invokes Khachian's algorithm for linear programming. The interactive method for multiple objective decisions is interpreted as a search for a feasible point. The decision maker augments the constraint set at each iteration with an explicit constraint as the current solution transgresses the set of acceptable decisions, Ω. The main deficiency of this method is that optimality, with respect to a known objective function, cannot necessarily be associated with the decision computed at each iteration. The decision maker might therefore find it difficult to specify the correction δ_k to the decision at each iteration. Since Algorithm 2.1 also proceeds from one iteration to the next using the correction information, this poses a difficulty.

The difficulty is resolved by exploiting the similarity of Algorithms 4.3.1 and 2.1. The global convergence of Algorithm 4.3.1 can be ensured by a modification that leads to Algorithm 2.2. The latter has the property that it incorporates optimality of the decision computed at each iteration as well as the polynomial-time termination of the algorithm.

It must be said at the outset, however, that the practical importance of the issue of convergence may not be compelling in the context of iterative methods that involve interactions with decision makers. It may be sufficient to establish the fact that a method ensures at each stage that the desires of the decision maker are properly incorporated. Results given in Chapter 3 and Section 3 below are of the latter category. Global convergence results are nevertheless discussed in Sections 4 and 5. These yield some theoretical insight into the complexity of the decision process (Rustem and Velupillai, 1987). They are also helpful in defining the circumstances under which polynomial convergence may *not* occur.

2 POLYNOMIAL-TIME ALGORITHMS FOR MULTIPLE OBJECTIVES

Following Chapter 4, we use the subscript k to indicate the iteration count of the interactive method. Algorithm 3.2.1 and its reformulation, Algorithm 4.3.1, are for tailoring the weighting matrix $\mathfrak{Q}_k$ of the quadratic objective function to meet the requirements of the decision maker. These are specified in view of optimal values presented at each stage of the method to the decision maker. No demand is made for any direct information on the actual weighting the decision maker may have in mind. Nevertheless, the information obtained from the decision maker is, in effect, translated by the method to the latter type of information.

As in Chapters 3 and 4, the multiple-objective optimization framework consists of the problem

$$\min \{q_k(x) \mid x \in \mathscr{R}\} \tag{2.1}$$

where $x \in \mathbb{R}^n$ is the vector of decision values and $\mathbb{R}^n$ is n-dimensional Euclidean vector space. $\mathscr{R}$ is the feasible region given by the decision model. In Chapter 3, $\mathscr{R}$ is taken to be a set of linear equality constraints given by

$$\mathscr{R} = \{x \in \mathbb{R}^n \mid \mathscr{G}^{\mathrm{T}} x = g\} \tag{2.2}$$

where the columns of the $(m \times n)$ matrix $\mathscr{G}$ can be assumed, without loss of generality, to be linearly independent. The objective function $q_k(x)$ is given by

$$q_k(x) = \frac{1}{2} \langle x - x^{\mathrm{d}}, \mathfrak{Q}_k(x - x^{\mathrm{d}}) \rangle, \tag{2.3}$$

where x^{d} are the desired (or bliss) values of the decision variables and $\mathfrak{Q}_k$ is the symmetric $n \times n$ weighting matrix, which is assumed to be positive-definite in this chapter.

As in Chapters 3 and 4, we denote by $\Omega \subset \mathbb{R}^n$ the set of *acceptable values* from the point of view of the decision maker. The vector of desired values x^{d} needs, by definition, to satisfy $x^{\mathrm{d}} \in \Omega$, although we do not invoke this property in our results. However, if an optimal solution of (2.1) is to be acceptable to the decision maker, it needs to be in the set Ω. Also, as in Chapters 3 and 4, this set is not specified in terms of analytical restrictions, but assumed to exist in the mind of the decision maker.

The main character of the approach in Algorithms 3.2.1 and 4.3.1 has been established in Theorem 3.3.1. Further implications of these results are discussed in Theorems 3.1 and 3.2 below, along with their extensions to a general feasible region $\mathscr{R}$ with convex and general nonlinear constraints.

Algorithm (2.1) below that *ensures* the global convergence of x_k to a feasible solution, such that

$$x_k \in \mathscr{R} \cap \Omega.$$

The main deficiency of Algorithm 2.1, however, is that the successive values of x_k are not optimal solutions of (2.1). The important link, provided by the optimization framework, between the achievement of the desired, or bliss, values and the generation of $x_k \in \Omega \cap \mathscr{R}$, is absent. Nevertheless, the method provides an illustration of the relationship between multiple-objective decision making and Khachian's (1979, 1980) ellipsoid algorithm for computing a point that satisfies specified inequality constraints. The link is concidered in Sections

4 and 5, along with the important property that the method converges in a finite (polynomial) number of steps to an $x_k \in \mathscr{R} \cap \Omega$. Furthermore, if its features are incorporated in Algorithm 4.3.1, global convergence of the latter is ensured

We let I be the $n \times n$ identity matrix and let $\mathscr{L}$ denote a constant specified in Section 4. We assume that the dimension of x is $n \geqslant 2$.

Algorithm 2.1 (Nonoptimal feasible policies in polynomial time)

Step 0: Given an initial estimate, $\tilde{x}_0$, and the equality constraints in (2.2), set $k = 0$, $\mathfrak{Q}_k = 2^{-2\mathscr{L}} I$, compute the projection of x_0 onto $\mathscr{R}$, given by

$$x_0 = \tilde{x}_0 - \mathfrak{Q}_k^{-1}\mathscr{G}(\mathscr{G}^{\mathrm{T}}\mathfrak{Q}_k^{-1}\mathscr{G})^{-1}(\mathscr{G}^{\mathrm{T}}\tilde{x}_0 - g).$$

Step 1: Same as **Step 2** in Algorithm 4.3.1.

Step 2: Compute

$$x_{k+1} = x_k + \frac{1}{n+1}\frac{P_k\delta_k}{\langle\delta_k, \mathfrak{Q}_k\delta_k\rangle^{1/2}}. \tag{2.4}$$

Step 3: Use the updating formula to compute

$$\mathfrak{Q}_{k+1} = \frac{n^2 - 1}{n}\left(\mathfrak{Q}_k + \frac{2}{n-1}\frac{\mathfrak{Q}_k\delta_k\,\delta_k^{\mathrm{T}}\mathfrak{Q}_k}{\langle\delta_k, \mathfrak{Q}_k\delta_k\rangle}\right). \tag{2.5}$$

Step 4: Set $k = k + 1$ and **return to Step 1**.

We recall from Section 3.1.1 that the point x_k, computed in **Step 0**, is the point on $\mathscr{R}$ nearest to x_0 with respect to the norm

$$\|\cdot\|_{\mathfrak{Q}_k} = \langle\cdot, \mathfrak{Q}_k(\cdot)\rangle^{1/2}.$$

As mentioned above, x_k in Algorithm 2.1 is not, in general, the solution of the optimization problem (2.1). However, x_{k+1} still has the same property as (4.3.4a), since $x_{k+1} - x_k$ lies along the projection of the direction δ_k, specified by the decision maker. Thus, the results in Theorems 3.1 and 3.2 below also apply to Algorithm 2.1. In addition, this algorithm is shown in Section 4 to terminate in polynomial time either at a solution $x_k \in \mathscr{R} \cap \Omega$, or to identify the case $\mathscr{R} \cap \Omega = \varnothing$.

We consider the modification of Algorithm 4.3.1 to ensure polynomial convergence while maintaining x_k as a solution of the optimization problem (2.1). The main tool used for this purpose is a modification to x^d. Once more, we assume the dimension of the vector x to be $n \geqslant 2$.

Algorithm 2.2 (Modified Algorithm 4.3.1 for optimal policies in polynomial time)

Step 0: Given $\mathscr{R}$ and x^{d}, assume an initial weighting matrix $\mathfrak{Q}_0$, set $k = 0$, and $x_0^{\mathrm{d}} = x^{\mathrm{d}}$.

Step 1: Using $\mathfrak{Q}_k$ and x_k^{d}, compute

$$x_k = \arg\min\{\langle x - x_k^{\mathrm{d}}, \mathfrak{Q}_k(x - x_k^{\mathrm{d}})\rangle \mid x \in \mathscr{R}\}. \tag{2.6}$$

Step 2: Same as in Algorithm 4.3.1. Also, compute

$$\psi_k = -\frac{\langle \delta_k, \mathfrak{Q}_k\delta_k\rangle + \eta_k\langle \mathfrak{Q}_k\delta_k, P_k\mathfrak{Q}_k^{-1}(\mathfrak{Q}_k\delta_k)\rangle}{\eta_k(n+1)\langle \delta_k, \mathfrak{Q}_k(x_k - x^{\mathrm{d}})\rangle\langle \delta_k, \mathfrak{Q}_k\delta_k\rangle^{1/2}}, \tag{2.7a}$$

$$\tau_k = -\frac{\psi_k\eta_k\langle \delta_k, \mathfrak{Q}_k(x_k - x^{\mathrm{d}})\rangle}{\langle \delta_k, \mathfrak{Q}_k\delta_k\rangle + \eta_k\langle \mathfrak{Q}_k\delta_k, P_k\mathfrak{Q}_k^{-1}(\mathfrak{Q}_k\delta_k)\rangle} = \frac{1}{(n+1)\langle \delta_k, \mathfrak{Q}_k\delta_k\rangle^{1/2}} \tag{2.7b}$$

$$\eta_k = \frac{2}{n-1}, \tag{2.7c}$$

$$x_k^{\mathrm{d}} = x_k - \psi_k(x_k - x^{\mathrm{d}}). \tag{2.7d}$$

Step 3: Same as in Algorithm 2.1.

Step 4: Set $k = k + 1$ and **return to Step 1**.

Algorithm 2.2 maintains the optimality of Algorithm 4.3.1 and the polynomial termination of Algorithm 2.1. Without the redefinition of the desired values in **Step 2** above, the choice for η_k and δ_k in the above algorithm would be more restricted than in Algorithm 4.3.1. With $\psi_k = 1$, the restriction implied by (2.7) is

$$\langle \delta_k, \mathfrak{Q}_k(x_k - x^{\mathrm{d}})\rangle = -\frac{\langle \delta_k, \mathfrak{Q}_k\delta_k\rangle + \eta_k\langle \mathfrak{Q}_k\delta_k, P_k\mathfrak{Q}_k^{-1}(\mathfrak{Q}_k\delta_k)\rangle}{\eta_k(n+1)\langle \delta_k, \mathfrak{Q}_k\delta_k\rangle^{1/2}} \leqslant 0. \tag{2.8}$$

This is simply a more restrictive requirement than $\langle \delta_k, \mathfrak{Q}_k(x_k - x^d)\rangle \leqslant 0$ in Algorithm 4.3.1. The justification for the latter is discussed in Section 3.3. It is shown in Theorem 2.1 and Sections 4 and 5 below that the redefinitions in (2.7) help to satisfy (2.8) on η_k and δ_k and ensure that the new optimal solutions for both algorithms, given by (4.3.4a)–(4.3.5a,b) and (2.4), are identical. Hence, the modified algorithm enjoys the properties that x_k is the solution of the optimization problem (2.1), with modified desired values, and that the process converges in polynomial time.

Condition (2.7b) ensures the optimality of x_{k+1} as well as the termination of the algorithm in polynomial time. The latter is established in Sections 4 and 5. To show that (2.7) can be satisfied for a given vector δ_k, we study whether x_{k+1} computed in Algorithm 2.2 can be made equivalent to that computed in Algorithms 4.3.1 and 2.1. The modification of x^d in (2.7d) is introduced for this purpose, and ensures that x_{k+1} computed by the algorithms is the same. The modified x^d only changes the stepsize τ_k taken in Algorithm 4.3.1 such that x_{k+1} is the same in Algorithms 2.1 and 2.2. Thus, the polynomial-time characteristic of Algorithm 2.1, discussed below, can be associated with the *step-by-step* optimality of Algorithm 4.3.1 by modifying the desired values.

Consider the solution of the linear–quadratic model objective function problem (2.1)–(2.3), given by

$$x_k = x^{\mathrm{d}} - \mathfrak{Q}_k^{-1}\mathscr{G}(\mathscr{G}^{\mathrm{T}}\mathfrak{Q}_k^{-1}\mathscr{G})^{-1}(\mathscr{G}^{\mathrm{T}}x^{\mathrm{d}} - g) \tag{2.9}$$

(see Lemma 3.1.1). Suppose that the vector δ_k satisfies $\langle \delta_k, \mathfrak{Q}_k(x_k - x^{\mathrm{d}})\rangle \leqslant 0$. $\mathfrak{Q}_{k+1}$ is computed using the update formula (2.5), which is used to re-solve the problem (2.1)–(2.3) and obtain the new optimal solution given by (4.3.4), with τ_k also required to satisfy (2.7b) above. We discuss the modification of x^d in (2.7d) that ensures optimality for each step of Algorithms 2.1 and 2.2. The desired value vector changes with k, and this does not alter the current optimal solution. The subscript on the desired value x_k^d is omitted for simplicity in later sections.

Theorem 2.1 (Consistency of 'polynomial' Algorithm 2.1 and 'optimization' Algorithm 4.3.1)
Consider the modified desired value in Algorithm 2.2:

$$x_k^{\mathrm{d}} = x_k - \psi_k(x_k - x^{\mathrm{d}}), \qquad \psi_k \in (0, \infty). \tag{2.10}$$

Then, the following hold.

(*i*) *the solution of*

$$\min\{q_k(x) \mid \mathscr{G}^{\mathrm{T}}x = g\},$$

with x_k^d replacing x^d in $q_k(x)$, remains unchanged, and x_k in both cases below is given by (2.9); *hence,*

$$\begin{aligned}&\arg\min\{\langle x - x^{\mathrm{d}}, \mathfrak{Q}_k(x - x^{\mathrm{d}})\rangle \mid x \in \mathscr{R}\}\\ &\qquad = \arg\min\{\langle x - x_k^{\mathrm{d}}, \mathfrak{Q}_k(x - x_k^{\mathrm{d}})\rangle \mid x \in \mathscr{R}\}.\end{aligned}$$

(*ii*) *δ_k satisfying*

$$\langle \delta_k, \mathfrak{Q}_k(x_k - x^{\mathrm{d}})\rangle \leqslant 0$$

yields

$$\langle \delta_k, \mathfrak{Q}_k(x_k - x_k^{\mathrm{d}})\rangle \leqslant 0.$$

(*iii*) *The solution*

$$\hat{x}_{k+1} = \arg\min\{\langle x - x_k^{\mathrm{d}}, \mathfrak{Q}_{k+1}(x - x_k^{\mathrm{d}})\rangle \mid \mathscr{G}_x^{\mathrm{T}} = g\} \tag{2.11}$$

is given by x_{k+1} in (4.3.4), with

$$\tau_k = -\frac{\psi_k\eta_k\langle \delta_k, \mathfrak{Q}_k(x_k - x^{\mathrm{d}})\rangle}{\langle \delta_k, \mathfrak{Q}_k\delta_k\rangle + \eta_k\langle \mathfrak{Q}_k\delta_k, P_k\mathfrak{Q}_k^{-1}(\mathfrak{Q}_k\delta_k)\rangle}.$$

(*iv*) *For* δ_k *such that*

$$\langle \delta_k, \mathfrak{Q}_k(x_k - x^{\mathrm{d}}) \rangle < 0, \tag{2.12}$$

ψ_k *is given by*

$$\psi_k = -\frac{\langle \delta_k, \mathfrak{Q}_k \delta_k \rangle + \eta_k \langle \mathfrak{Q}_k \delta_k, P_k \mathfrak{Q}_k^{-1}(\mathfrak{Q}_k \delta_k) \rangle}{(n+1)\eta_k \langle \delta_k, \mathfrak{Q}_k(x_k - x^{\mathrm{d}}) \rangle \langle \delta_k, \mathfrak{Q}_k \delta_k \rangle^{1/2}} > 0,$$

ψ_k *satisfies (2.7) and* $\hat{x}_{k+1}$ *is equivalent to (2.4) (see CN.1).*

Proof

In order to establish the equivalence of x_k and the solution of

$$\min \{ \langle x - x_k^{\mathrm{d}}, \mathfrak{Q}_k(x - x_k^{\mathrm{d}}) \rangle \mid \mathscr{G}^{\mathrm{T}} x = g \},$$

we use (2.9) to write

$$\hat{x}_k = x_k^{\mathrm{d}} - \mathfrak{Q}_k^{-1} \mathscr{G} (\mathscr{G}^{\mathrm{T}} \mathfrak{Q}_k^{-1} \mathscr{G})^{-1} (\mathscr{G}^{\mathrm{T}} x_k^{\mathrm{d}} - g),$$
$$\hat{x}_k = x_k - \psi_k(x_k - x^{\mathrm{d}}) - \mathfrak{Q}_k^{-1} \mathscr{G} (\mathscr{G}^{\mathrm{T}} \mathfrak{Q}_k^{-1} \mathscr{G})^{-1} \{ \mathscr{G}^{\mathrm{T}} [x_k - \psi_k(x_k - x^{\mathrm{d}})] - g \}.$$

The optimality condition of (2.1)–(2.3) yields $x_k - x^d = \mathfrak{Q}_k^{-1} \mathscr{G} \lambda$, where λ is the associated Lagrange multiplier. Using this condition and $x_k \in \mathscr{R}$ yields $\hat{x}_k = x_k$, and, hence (i).

To show (ii), it is sufficient to note that

$$\begin{aligned} \langle \delta_k, \mathfrak{Q}_k(x_k - x_k^{\mathrm{d}}) \rangle &= \langle \delta_k, \mathfrak{Q}_k \{ x_k - [x_k - \psi_k(x_k - x^{\mathrm{d}})] \} \rangle \\ &= \psi_k \langle \delta_k, \mathfrak{Q}_k(x_k - x^{\mathrm{d}}) \rangle. \end{aligned}$$

Let $\hat{x}_{k+1}$ be the solution of (2.11). In order to show that this solution is given by (4.3.4), consider the analogous expression to (4.3.4) when $\hat{x}_k$ is computed using x_k^d. Since $\hat{x}_k = x_k$, we have

$$\begin{aligned} \hat{x}_{k+1} &= \hat{x}_k + \hat{\tau}_k P_k \delta_k \\ &= x_k + \hat{\tau}_k P_k \delta_k, \end{aligned} \tag{2.13a}$$

$$\hat{\tau}_k = -\frac{\psi_k \eta_k \langle \delta_k, \mathfrak{Q}_k(x_k - x^{\mathrm{d}}) \rangle}{\langle \delta_k, \mathfrak{Q}_k \delta_k \rangle + \eta_k \langle \mathfrak{Q}_k \delta_k, P_k \mathfrak{Q}_k^{-1}(\mathfrak{Q}_k \delta_k) \rangle}. \tag{2.13b}$$

This establishes (iii). It is straightforward to verify that this solution is equivalent to (2.4) if (2.7) is satisfied with (2.12) and $\eta_k > 0$. Thus, we have (iv). □

3 PROPERTIES OF THE METHODS: GENERAL CONUEX AND NONLINEAR CONSTRAINTS

A desirable local property of Algorithms 4.3.1 and 2.2, which compute optimal policies at each iteration, is that every time $\mathfrak{Q}_k$ is updated, a more satisfactory new *optimal* solution is computed. The decision maker can be shown to be more satisfied with the new optimal solution x_{k+1} than with the current optimal solution x_k. Theorems 3.1 and 3.2 below establish results that support this view for the linear equality-constrained case. These results are extended to general convex and also nonlinear constraints in Theorems 3.3 and 3.4.

Theorem 3.1
For $0 \leqslant \tau_k \leqslant 2$ in (4.3.4a), the inequality

$$\|x_{k+1} - x_{\mathrm{p}}\|_{\mathfrak{Q}_k} \leqslant \|x_k - x_{\mathrm{p}}\|_{\mathfrak{Q}_k} \tag{3.1}$$

holds. Furthermore, for $0 < \tau_k < 2$, (3.1) is a strict inequality.

Proof
Consider (4.3.4) with $\tau_k = 1$:

$$\hat{x}_1 = x_k + P_k \delta_k. \tag{3.2}$$

Hence, $\hat{x}_1 - x_k$ is the projection of δ_k onto

$$\{x \in \mathbb{R}^n \mid \mathscr{G}^{\mathrm{T}} x = 0\},$$

or $\hat{x}_1$ is the projection of x_{p} onto $\mathscr{R}$ given by (2.2). The inequality

$$\langle x_{\mathrm{p}} - \hat{x}_1, \mathfrak{Q}_k(x_k - \hat{x}_1)\rangle \leqslant 0 \tag{3.3}$$

follows from the projection property of $\hat{x}_1$ (see Lemma 3.1.2). Defining $x(\tau_k)$ as

$$x(\tau_k) = x_k + \tau_k(\hat{x}_1 - x_k) \tag{3.4}$$

for $0 \leqslant \tau_k \leqslant 2$ and using (3.3), we have

$$\begin{aligned}
\|x(\tau_k) - x_{\mathrm{p}}\|^2_{\mathfrak{Q}_k} &= \|x_k - x_{\mathrm{p}} + \tau_k(\hat{x}_1 - x_k)\|^2_{\mathfrak{Q}_k} \\
&= \|x_k - x_{\mathrm{p}}\|^2_{\mathfrak{Q}_k} + 2\tau_k\langle x_k - \hat{x}_1 + \hat{x}_1 - x_{\mathrm{p}}, \mathfrak{Q}_k(\hat{x}_1 - x_k)\rangle \\
&\quad + \tau_k^2 < \hat{x}_1 - x_k, \mathfrak{Q}_k(\hat{x}_1 - x_k)\rangle \\
&\leqslant \|x_k - x_{\mathrm{p}}\|^2_{\mathfrak{Q}_k},
\end{aligned}$$

from which the result follows with $x_{k+1} = x(\tau_k)$. Furthermore, it should be noted that for

$$0 < \tau_k < 2,$$

(3.2) becomes a strict inequality. □

Lemma 3.1
For $\hat{x}_1$ given by (3.3), the inequality

$$\langle x_{\mathrm{p}} - \hat{x}_1, \mathfrak{Q}_{k+1}(\hat{x}_1 - x_k)\rangle \geqslant 0 \tag{3.5}$$

holds.

Proof

Using the update formulae (4.3.3) or (2.5) in (3.8), we have

$$
\begin{aligned}
\langle x_p - \hat{x}_1, \mathfrak{Q}_{k+1}(\hat{x}_1 - x_k)\rangle &= \langle x_p - \hat{x}_1, \mathfrak{Q}_k(\hat{x}_1 - x_k)\rangle \\
&\quad + \eta_k \frac{\langle x_p - \hat{x}_1, \mathfrak{Q}_k(x_p - x_k)\rangle\langle \hat{x}_1 - x_k, \mathfrak{Q}_k(x_p - x_k)\rangle}{\langle x_p - x_k, \mathfrak{Q}_k(x_p - x_k)\rangle} \\
&= \langle x_p - \hat{x}_1, \mathfrak{Q}_k(\hat{x}_1 - x_k)\rangle \\
&\quad + \eta_k \frac{\langle x_p - \hat{x}_1, \mathfrak{Q}_k(x_p - \hat{x}_1 + \hat{x}_1 - x_k)\rangle\langle \hat{x}_1 - x_k, \mathfrak{Q}_k(x_p - \hat{x}_1 + \hat{x}_1 - x_k)\rangle}{\langle x_p - x_k, \mathfrak{Q}_k(x_p - x_k)\rangle} \\
&\geqslant 0,
\end{aligned}
$$

where the last inequality follows from the projection property. □

Theorem 3.2

For $0 \leqslant \tau_k \leqslant 2$ in (4.3.4a) the inequality

$$\|x_{k+1} - x_p\|_{\mathfrak{Q}_{k+1}} \leqslant \|x_k - x_p\|_{\mathfrak{Q}_{k+1}} \tag{3.6}$$

holds. Furthermore, for $0 \leqslant \tau_k \leqslant 2$, (3.6) is a strict inequality.

Proof

Let $x(\tau_k)$ be defined by (3.4). For $0 \leqslant \tau_k \leqslant 2$ and using (3.5), we have

$$
\begin{aligned}
\|x_p - x(\tau_k)\|^2_{\mathfrak{Q}_{k+1}} &= \|x_p - x_k - \tau_k(\hat{x}_1 - x_k)\|^2_{\mathfrak{Q}_{k+1}} \\
&= \|x_p - x_k\|^2_{\mathfrak{Q}_{k+1}} \\
&\quad - 2\tau_k\langle x_p - \hat{x}_1 + \hat{x}_1 - x_k, \mathfrak{Q}_{k+1}(\hat{x}_1 - x_k)\rangle \\
&\quad + \tau_k^2\langle \hat{x}_1 - x_k, \mathfrak{Q}_{k+1}(\hat{x}_1 - x_k)\rangle \\
&\leqslant \|x_p - x_k\|^2_{\mathfrak{Q}_{k+1}},
\end{aligned}
$$

which establishes the required result with $x_{k+1} = x(\tau_k)$. □

Let the feasible region $\mathscr{R}$ be a general convex set and let x_{k+1} be given by

$$x_{k+1}(\eta) = \arg\min \left\{ \frac{1}{2}\langle x - x^{\mathrm{d}}, \mathfrak{Q}_{k+1}(\eta)(x - x^{\mathrm{d}})\rangle \,\middle|\, x \in \mathscr{R} \right\}, \tag{3.7a}$$

$$x_{k+1} \equiv x_{k+1}(\eta_k) \tag{3.7b}$$

for $\eta, \eta_k \geqslant 0$, defining $\mathfrak{Q}_{k+1}(\eta)$ with (4.3.3). The difficulty in establishing similar results in this case arises from the fact that there may not be a point between x_k and x_{k+1}, along $x_{k+1} - x_k$, that can be expressed as the projection of x_p onto $\mathscr{R}$, similar to (3.2). Equivalently, varying η does not necessarily produce $x_{k+1}(\eta)$ that lies between x_k and $x_{k+1} = \hat{x}_1$. The importance of this becomes clear when the linear equality-constrained case is considered. If, for given

$\eta, x_{k+1}(\eta)$ is such that $\tau_k > 2$ in (4.3.4a), in the linear equality case, reducing η clearly reduces τ_k given by (4.3.4a). Hence, if $\tau_k > 2, \eta$ can be reduced to define $x_{k+1}(\eta)$ for which the bound $0 \leqslant \tau_k \leqslant 2$ is satisfied.

The main concept necessary for extending the results for Theorems 3.1 and 3.2 to general convex $\mathscr{R}$ is the line passing through x_{k+1} and x_k. By considering the projection of x_p on this line, it is shown in Theorems 3.3 and 3.4 below that (3.1) and (3.6) hold for x_{k+1} close to x_k. It is also shown that if this is not the case then reducing η_k brings x_{k+1} close to x_k so that, for small η_k, (3.1) and (3.6) still hold.

The following two lemmas establish the technical results required by Theorem 3.3. It must be noted, however, that these are dependent on the ability to vary η_k, and hence only apply to Algorithm 4.3.1. For Algorithms 2.1 and 2.2, the desirable properties are the projective character in Theorem 2.1 and polynomial-time convergence, as shown in Theorem 4.1.

Lemma 3.2

For $\eta_k \geqslant 0$ in (4.3.3) and for x_k and x_{k+1} given by the solutions of (2.1) and (3.7) respectively, with convex $\mathscr{R}$, the inequality

$$\langle x_{\mathrm{p}} - x_k, \mathfrak{Q}_k(x_{k+1} - x_k)\rangle \geqslant 0 \tag{3.8}$$

holds if $\langle \delta_k, \mathfrak{Q}_k(x_k - x^{\mathrm{d}})\rangle \leqslant 0$.

Proof

The inequality

$$\langle x_{k+1} - x_k, \mathfrak{Q}_{k+1}(x_{k+1} - x^{\mathrm{d}})\rangle \leqslant 0 \tag{3.9}$$

follows from the optimality of x_{k+1} for (3.7). Using (4.3.3), we have

$$\begin{aligned} 0 \geqslant & \langle x_{k+1} - x_k, \mathfrak{Q}_{k+1}(x_{k+1} - x^{\mathrm{d}})\rangle \\ &= \langle x_{k+1} - x_k, \mathfrak{Q}_k(x_{k+1} - x^{\mathrm{d}})\rangle \\ &\quad + \frac{\eta_k}{\langle \delta_k, \mathfrak{Q}_k \delta_k\rangle}\langle x_{k+1} - x_k, \mathfrak{Q}_k \delta_k\rangle\langle \delta_k, \mathfrak{Q}_k(x_{k+1} - x^{\mathrm{d}})\rangle \\ &= \langle x_{k+1} - x_k, \mathfrak{Q}_k(x_{k+1} - x_k + x_k - x^{\mathrm{d}})\rangle \\ &\quad + \frac{\eta_k}{\langle \delta_k, \mathfrak{Q}_k \delta_k\rangle}[\langle x_{k+1} - x_k, \mathfrak{Q}_k(x_{\mathrm{p}} - x_k)\rangle^2 \\ &\quad + \langle x_k - x^{\mathrm{d}}, \mathfrak{Q}_k(x_{\mathrm{p}} - x_k)\rangle\langle x_{\mathrm{p}} - x_k, \mathfrak{Q}_k(x_{k+1} - x_k)\rangle]. \end{aligned} \tag{3.10}$$

Since

$$\langle x_{k+1} - x_k, \mathfrak{Q}_k(x_k - x^{\mathrm{d}})\rangle \geqslant 0 \tag{3.11}$$

follows from the optimality of x_k for (2.1), and the relationship

$$< x_k - x^{\mathrm{d}}, \mathfrak{Q}_k(x_{\mathrm{p}} - x_k) > \leqslant 0$$

follows by hypothesis, inequality (3.10) can only be satisfied if (3.8) holds. □

It should be noted that the condition $\langle \delta_k, \mathfrak{Q}_k(x_k - x^{\mathrm{d}})\rangle \leqslant 0$ also plays an important role in establishing (3.8).

Lemma 3.3

If the conditions of Lemma 3.2 hold then $\eta_k \to 0$ implies that $\|x_k - x_{k+1}\| \to 0$. Furthermore, for $\tau_k \geqslant 0$ sufficiently small, there exists $\eta_k \geqslant 0$ that satisfies (4.3.5b) (or (4.3.5e), when $\mathfrak{Q}_k$ is not strictly positive-definite).

Proof

In (3.10), the term multiplying η_k is negative because of (3.8) and (3.9). It has also been established in (3.10) and (3.11) that

$$\langle x_{k+1} - x_k, \mathfrak{Q}_k(x_{k+1} - x^{\mathrm{d}})\rangle \geqslant 0. \tag{3.12}$$

Thus, $\eta_k \to 0$ implies

$$\langle x_{k+1} - x_k, \mathfrak{Q}_k(x_{k+1} - x^{\mathrm{d}})\rangle \to 0.$$

Since $\|x_{k+1} - x^{\mathrm{d}}\| \neq 0$, we have $\|x_k - x_{k+1}\| \to 0$. If, on the other hand,

$$\langle x_{k+1} - x_k, \mathfrak{Q}_k(x_{k+1} - x^{\mathrm{d}})\rangle = 0$$

due to orthogonality, this implies that x_{k+1} solves (2.1), and hence $x_{k+1} = x_k$.

Finally, consider τ_k given by (4.3.5b) (or (4.3.5e): due to the equivalence with (4.3.5b), established in Lemma 3.3.1, only (4.3.5b) is considered). Given τ_k, we have

$$\eta_k = -\frac{\tau_k\langle \delta_k, \mathfrak{Q}_k\delta_k\rangle}{\tau_k\langle \delta_k, \mathfrak{Q}_k P_k\delta_k\rangle + \langle \delta_k, \mathfrak{Q}_k(x_k - x^{\mathrm{d}})\rangle}. \tag{3.13}$$

As $\langle \delta_k, \mathfrak{Q}_k(x_k - x^{\mathrm{d}})\rangle \leqslant 0$ by hypothesis, for sufficiently small $\tau_k \geqslant 0$, it can be verified that $\eta_k \geqslant 0$. □

Lemma 3.3 implies that the scalar η_k can be used by the decision maker to control the size of the norm $\|x_k - x_{k+1}\|$. This result is clearly only applicable to Algorithm 4.3.1, and is utilized in Theorem 3.3.

Theorem 3.3

Let

(i) the conditions of Lemma 3.2 be satisfied;

(ii) P_k be the operator projecting, under the norm $\|\cdot\|_{\mathfrak{Q}_k}$, vectors in $\mathbb{R}^n$ onto the line passing through x_k and x_{k+1};

(iii) $\hat{x}_1$ be the projection of x_p onto this line, given by

$$\hat{x}_1 = x_k + P_k\delta_k. \tag{3.14}$$

Then,

(a) for $x(\tau_k)$ given by $x(\tau_k) = x_k + \tau_k(\hat{x}_1 - x_k)$ and $0 \leqslant \tau_k \leqslant 2$,

$$\|x(\tau_k) - x_{\mathrm{p}}\|_{\mathfrak{Q}_k} \leqslant \|x_k - x_{\mathrm{p}}\|_{\mathfrak{Q}_k}, \tag{3.15}$$

$$\|x(\tau_k) - x_{\mathrm{p}}\|_{\mathfrak{Q}_{k+1}} \leqslant \|x_k - x_{\mathrm{p}}\|_{\mathfrak{Q}_{k+1}}; \tag{3.16}$$

(b) x_{k+1} in (3.7) can be expressed as

$$x_{k+1} - x_k = \xi(\hat{x}_1 - x_k), \tag{3.17}$$

with $\xi \geqslant 0$, and hence both $\hat{x}_1 - x_k$ and $x_{k+1} - x_k$ lie in the same direction;

(c) if, furthermore,

$$x_{k+1}(\eta) \notin \mathbb{X} = \{x \mid x = x_k + \tau_k(\hat{x}_1 - x_k); \tau_k \in [0, 2]\},$$

there exists an $\eta_k \geqslant 0$, small enough, so that the resulting solution of (3.7) satisfies $\mathbb{X}$ and thereby (3.15) and (3.16) with $x_{k+1} = x(\tau_k)$.

Proof

Inequalities (3.15) and (3.16) are established in exactly the same way as in Theorems 1 and 2, with the feasible region $\mathscr{R}$ replaced by the line passing through x_k and x_{k+1}.

To show that $\xi \geqslant 0$ in (3.17), consider (3.8), which yields

$$0 \leqslant \langle x_{\mathrm{p}} - x_k, \mathfrak{Q}_k(x_{k+1} - x_k)\rangle = \xi\langle x_{\mathrm{p}} - x_k, \mathfrak{Q}_k(\hat{x}_1 - x_k)\rangle. \tag{3.18}$$

Since the inequality

$$\langle x_{\mathrm{p}} - \hat{x}_1 + \hat{x}_1 - x_k, \mathfrak{Q}_k(\hat{x}_1 - x_k)\rangle \geqslant 0$$

follows from (3.3), in order to preserve the non-negativity of (3.18), the condition $\xi \geqslant 0$ must be satisfied. Thus, if x_{k+1} can be expressed as

$$x_{k+1} = x(\tau_k) = x_k + \tau_k(\hat{x}_1 - x_k) \tag{3.19}$$

for some $\tau_k \in [0, 2]$ then (3.15) and (3.6) hold. However, if x_{k+1} is further away from x_k so that $\tau_k > 2$ then η can be reduced. According to Lemma 3.3,

$$\eta \to 0 \Rightarrow \|x_{k+1}(\eta) - x_k\| \to 0.$$

In view of (3.19), $\|x_{k+1} - x_k\| \to 0$ implies that $\tau_k \to 0$, since $\|\hat{x}_1 - x_k\|$ is not related to η. □

The extension of these results to general nonlinear constraints is possible, since reducing η brings $\mathfrak{Q}_{k+1}$ closer to $\mathfrak{Q}_k$, independently of Lemmas 3.2 and 3.3. Thus, there exists on $\eta_k > 0$ for which inequalities (3.9) and (3.11) are satisfied, and thence the results of Theorem 3.3 hold for general nonlinear constraints.

The above results are particularly helpful if x_k is at a vertex of linear inequality constraints. If such vertices are excluded, Theorem 3.4 below provides an *alternative* characterization of the results of Theorem 3.3 in the presence of general nonlinear constraints.

Theorem 3.4

(*i*) *Let the feasible region be given by*

$$\mathscr{R} = \{x \in \mathbb{R}^n \mid h(x) \leqslant 0\},$$

where $h(x) \in \mathbb{C}^1$. *Then, there exist scalars* $\eta_k \geqslant 0 (\eta_k > 0$ *with vertices excluded) and* $\sigma(\eta_k) \geqslant 0 (\sigma(\eta_k) > 0$, *excluding vertices) such that*

$$\|x_{k+1} - x_k\| \leqslant \sigma(\eta_k),$$

and all constraints satisfied as equalities (i.e. active) at x_k *and* x_{k+1} *are the same.*

(*ii*) *Let these active constraints may be expressed as*

$$\bar{h}(x_k) - \bar{h}(x_{k+1}) = 0 = \mathscr{H}^{\mathrm{T}}(x_k, x_{k+1})(x_k - x_{k+1}),$$

where $\bar{h}$ *is the vector of* m_{a} *active constraints and*

$$\mathscr{H}(x_k, x_{k+1}) \equiv [\nabla \bar{h}^1(x_k - t^1(x_{k+1} - x_k)) \vdots \cdots \vdots \nabla \bar{h}^{m_{\mathrm{a}}}(x_k - t^{m_{\mathrm{a}}}(x_{k+1} - x_k))],$$

$t^1, \cdots, t^{m_{\mathrm{a}}} \in (0, 1)$. *Then, the line segment connecting* x_k *and* x_{k+1} *satisfies the above equality, and, for small* η_k, *the results of Theorem 3.3 hold for general nonlinear inequality constraints.*

Proof

The linear equalities for $\bar{h}(x)$ are obtained by a simple application of the mean value theorem (see **CN.6.5**, and Ortega and Rheinboldt, 1970, Theorem 3.2.2). In order to show that for small η_k, the active constraints at x_k are the same as those active at x_{k+1}, consider the first-order optimality conditions for (2.1) and (3.7), with h as the vector of nonlinear inequality constraints. For (2.1), these conditions are

$$\mathfrak{Q}_k(x_k - x^{\mathrm{d}}) + \mathscr{H}_k \mu_k = 0,$$
$$h(x_k) \leqslant 0, \quad \langle h(x_k), \mu_k \rangle = 0, \quad \mu_k \geqslant 0,$$

where $\mathscr{H}_k$ is the matrix of normals of all the constraints, $h(x)$, evaluated at x_k, and μ_k is the vector of associated multipliers. Using (3.7), we have

$$\left(\mathfrak{Q}_k + \eta \frac{\mathfrak{Q}_k \delta_k \delta_k^{\mathrm{T}} \mathfrak{Q}_k}{\langle \delta_k, \mathfrak{Q}_k \delta_k \rangle}\right)[x_{k+1}(\eta) - x^{\mathrm{d}}] + \mathscr{H}_{k+1} \mu_{k+1}(\eta) = 0,$$
$$h(x_{k+1}(\eta)) \leqslant 0, \langle h(x_{k+1}(\eta)), \mu_{k+1}(\eta) \rangle = 0, \quad \mu_{k+1}(\eta) \geqslant 0.$$

Thus, as $\eta \to 0, x_{k+1}(\eta) \to x_k$, and, with strict complementarity holding at both x_k and $x_{k+1}(\eta)$, the active constraints at x_k predict the active constraints at $x_{k+1}(\eta)$, and vice versa for some $\eta > 0$, since $\mu^i_{k+1}(\eta) > 0 \Rightarrow \mu^i_k > 0$ for active constraint i as $x_{k+1}(\eta) \to x_k$.

Having constructed the intersection of hyperplanes that characterize the active constraints between x_{k+1} and x_k and in which the line through x_{k+1} and x_k passes, the rest of the proof is identical to that of Theorem 3.3. □

4 KHACHIAN'S ELLIPSOID ALGORITHM: COMPLEXITY OF THE DECISION PROCESS

In this section, we discuss the termination property of Algorithms 2.1 and 2.2 for the decision process outlined in Section 1 and Chapter 3. In particular, we show that the decision process terminates after a finite number of iterations if the decision model and the inequality constraints bounding the region Ω are assumed to be linear. The region Ω and the inequality constraints bounding it are still assumed to exist only in the mind of the decision maker. The consideration of the convergence of decision process under such circumstances may not be an important issue in real applications. Indeed, as discussed below, the convergence of the method ultimately depends on the decision maker, who may arbitrarily extend or truncate this process. Nevertheless, the 'condition' under which the method converges, does provide an insight into the method from a different vantage point, and indicates the reason why the process is extended or truncated. Convergence is studied by establishing an equivalence of the algorithms in Section 2 with Khachian's (1979, 1980) ellipsoid algorithm for linear programming. The latter algorithm has been shown to terminate in polynomial time (i.e. the number of iterations required to arrive at a solution—or to establish the absence of one—is bounded by a polynomial in the original data of the problem (Khachian, 1980; Kozlov, Tarasov and Khachian, 1980; Aspvall and Stone, 1979). This result is summarized in Theorem 4.1 below.

In order to introduce the ellipsoid algorithm, consider first the problem of finding a feasible point satisfying the following system of inequalities:

$$\langle \Upsilon_j, x \rangle \leqslant \nu_j, \quad j = 1, \cdots, p, \tag{4.1}$$

where $\Upsilon_j \in \mathbb{R}^n$, $p \geqslant 2$, $n \geqslant 2$. Khachian's algorithm, summarized below, finds such a point, or establishes its nonexistence, in a finite number of iterations. Let $\mathscr{L}$ be the length of the binary encoding of the input data $\Upsilon_j, \nu_j, j = 1, \cdots, p$, i.e. the number of 0s and 1s needed to write these coefficients in binary form:

$$\mathscr{L} = \sum_{i,j=1}^{n,p} \log_2 (|\Upsilon_{ij}| + 1) + \sum_{j=1}^{p} \log_2 (|\nu_j| + 1) + \log_2 np + 2. \tag{4.2}$$

where Υ_{ij} is the ith element of the vector Υ_j. Khachian's algorithm assumes that coefficients Υ_{ij}, ν_j are integers. This can trivially be achieved, in general, by suitably scaling each inequality. Algorithm 4.1, discussed below, can also be used directly for noninteger Υ_{ij}, ν_j. In this case, the changes in the properties of the algorithm are discussed is Goldfarb and Todd (1982).

Algorithm 4.1 (Khachian's algorithm for computing feasible points of a system of linear inequalities)

Step 1: (Initialization) Set $x_0 = 1$ (*1* is the vector with all its elements unity), $\mathfrak{Q}_0^{-1} = 2^{2\mathscr{L}} I, k = 0$

Step 2: If x_k satisfies

$$\langle \Upsilon_i, x_k \rangle < \nu_i + 2^{-\mathscr{L}}, \quad \forall i = 1, \dots p, \tag{4.3}$$

then terminate the algorithm with x_k as a feasible solution. If

$$k < 4(n+1)^2 \mathscr{L}$$

then **go to Step 3.** Otherwise, terminate the algorithm, responding that no solution exists.

Step 3: Select any inequality for which

$$\langle \Upsilon_i, x_k \rangle \geqslant \nu_i + 2^{-\mathscr{L}}, \tag{4.4}$$

and set

$$x_{k+1} - x_k \quad \frac{1}{n+1} \frac{\mathfrak{Q}_k^{-1} \Upsilon_i}{\langle \Upsilon_i, \mathfrak{Q}_k^{-1} \Upsilon_i \rangle^{1/2}} \tag{4.5}$$

and

$$\mathfrak{Q}_{k+1}^{-1} = \frac{n^2}{n^2 - 1} \left(\mathfrak{Q}_k^{-1} - \frac{2}{n+1} \frac{\mathfrak{Q}_k^{-1} \Upsilon_i \Upsilon_i^{\mathrm{T}} \mathfrak{Q}_k^{-1}}{\langle \Upsilon_i, \mathfrak{Q}_k^{-1} \Upsilon_i \rangle} \right). \tag{4.6}$$

Set $k = k + 1$ and **go to Step 2.**

It can be verified that $\mathfrak{Q}_{k+1}^{-1}$ is symmetric positive-definite if $\mathfrak{Q}_k^{-1}$ has these properties (see Aspvall and Stone, 1979, Lemma 3). Thus, there is no danger of the denominator of the above expressions vanishing, provided $\Upsilon_i \neq 0$. Algorithm 4.1 actually finds a feasible solution for the system of inequalities (4.3). However, the following lemma ensures that this is compatible with the requirement of the system (4.1).

Lemma 4.1 (Aspvall and Stone, 1979, Lemma 6)
The system of inequalities (4.1) has a solution if, and only if, the system of strict inequalities (4.3) has a solution.

The following theorem implies that Algorithm 4.1 returns a feasible solution or establishes the non-existence of one in at most $4(n+1)^2 \mathscr{L}$ iterations.

Theorem 4.1 (Aspvall and Stone, 1979; Theorem 1)
Algorithm 4.1 returns a feasible solution if and only if (4.3) is satisfiable.

The algorithm ensures that a feasible solution is obtained in, at most, the number of iterations mentioned above.

Consider now the problem of finding the solution to the system of inequalities (4.1) in the presence of linear equalities

$$\mathscr{G}^{\mathrm{T}} x = g. \tag{4.7}$$

In this case, given a starting point $\tilde{x}_0$, the initial solution estimate is defined by

$$x_0 = \tilde{x}_0 - \mathfrak{Q}_k^{-1}\mathscr{G}(\mathscr{G}^{\mathrm{T}}\mathfrak{Q}_k^{-1}\mathscr{G})^{-1}(\mathscr{G}^{\mathrm{T}}\tilde{x}_0 - g).$$

It can be verified that $\mathscr{G}^{\mathrm{T}} x_0 = g$. In order that all x_k generated by the algorithms satisfy $\mathscr{G}^{\mathrm{T}} x_k = g$, we have to replace (4.3.4a) by

$$x_{k+1} = x_k - \frac{1}{n+1}\frac{P_k\mathfrak{Q}_k^{-1}\Upsilon_i}{\langle \Upsilon_i, \mathfrak{Q}_k^{-1}\Upsilon_i\rangle^{1/2}}, \tag{4.8}$$

where

$$P_k = I - \mathfrak{Q}_k^{-1}\mathscr{G}(\mathscr{G}^{\mathrm{T}}\mathfrak{Q}_k^{-1}\mathscr{G})^{-1}\mathscr{G}^{\mathrm{T}},$$

as in (4.3.5a) (or $P_k = Z(Z^{\mathrm{T}}\mathfrak{Q}_k Z)^{-1}Z_k^{\mathrm{T}}\mathfrak{Q}_k$ as in (4.3.5c)) (see CN.2). It can also be verified that

$$P_k(x_{k+1} - x_k) = x_{k+1} - x_k.$$

All other steps of the algorithm remain unchanged. Thus, any feasible point generated by the algorithm also satisfies the linear equalities given by (4.7).

Let us now return to Algorithm 4.3.1, and attempt to identify the reason why the decision maker may wish to specify a given direction δ_k as a direction of 'improvement'. We may effectively interpret x_{p} in (4.3.2), specified by the decision maker in Algorithm 4.3.1, as the point in Ω nearest to x_k. This 'nearness' is measured with respect to the current weighting $\mathfrak{Q}_k$. Thus, x_{p} is the solution of

$$\min\left\{\frac{1}{2}\|x - x_k\|^2_{\mathfrak{Q}_k}\,\middle|\,\langle \Upsilon, x\rangle \leqslant \nu\right\}, \tag{4.9}$$

where $\langle \Upsilon, x\rangle \leqslant \nu$ is one of the implicit constraints describing the region Ω and violated at x_k. *The decision maker may not know that such a constraint exists until she notices that x_k is violating it* (i.e. $\langle \Upsilon, x_k\rangle > \nu$).

Clearly, the decision maker also does not know Υ and ν, but can only specify x_{p}. It is argued below that this is sufficient to identify Υ to some degree, and thereby quantify Ω, if x_{p} is interpreted as the solution of (4.9). This interpretation is shown to allow the use of Algorithm 4.1 to solve for a feasible point of $\Omega \cap \mathscr{R}$ in polynomial time by updating $\mathfrak{Q}^{-1}$ and using (4.5) and (4.6).

The solution of (4.9) can be obtained by writing the first-order necessary conditions of optimality:

$$x_{\mathrm{p}} - x_k = -\mathfrak{Q}_k^{-1}\Upsilon\mu, \tag{4.10a}$$

$$\langle\Upsilon, x_{\mathrm{p}}\rangle \leqslant \nu, \quad \mu\{\langle\Upsilon, x_p\rangle - \nu\} = 0, \quad \mu \geqslant 0, \tag{4.10b}$$

where μ is the associated multiplier. Thus, we have

$$\delta_k = x_{\mathrm{p}} - x_k = -\mathfrak{Q}_k^{-1}\Upsilon\mu,$$

where the Lagrange multiplier (shadow price) μ is non- negative. It can be seen from (4.10) that

$$\mu = \frac{-\langle\Upsilon, x_{\mathrm{p}} - x_k\rangle}{\langle\Upsilon, \mathfrak{Q}_k^{-1}\Upsilon\rangle},$$

or

$$\mu = \frac{-\langle x_{\mathrm{p}} - x_k, \mathfrak{Q}_k(x_{\mathrm{p}} - x_k)\rangle}{\langle\Upsilon, x_{\mathrm{p}} - x_k\rangle}.$$

Since $\mu \geqslant 0$, these yield

$$\begin{aligned}\mu &= \frac{\langle x_{\mathrm{p}} - x_k, \mathfrak{Q}_k(x_{\mathrm{p}} - x_k)\rangle^{1/2}}{\langle\Upsilon, \mathfrak{Q}_k^{-1}\Upsilon\rangle^{1/2}} \\ &= \frac{\|\delta_k\|_{\mathfrak{Q}_k}}{\|\Upsilon\|_{\mathfrak{Q}_k^{-1}}}.\end{aligned}$$

Hence, we have

$$\frac{\delta_k}{\|\delta_k\|_{\mathfrak{Q}_k}} = -\frac{\mathfrak{Q}_k^{-1}\Upsilon}{\|\Upsilon\|_{\mathfrak{Q}_k^{-1}}}. \tag{4.11}$$

It may be that more than one constraint is violated at x_k. Assume that the system

$$\mathcal{N}^{\mathrm{T}}x \leqslant \mathcal{N}$$

is violated at x_k (i.e. $\mathcal{N}^{\mathrm{T}}x_k > \mathcal{N}$) for some appropriate-dimensional matrix $\mathcal{N}$ and vector $\mathcal{R}$. Then (4.9) can be written as

$$\min\left\{\frac{1}{2}\|x - x_k\|_{\mathfrak{Q}_k}^2 \,\middle|\, \mathcal{N}^{\mathrm{T}}x \leqslant \mathcal{N}\right\}, \tag{4.12}$$

for which the first-order optimality conditions are

$$x_{\mathrm{p}} - x_k = -\mathfrak{Q}_k^{-1}\mathcal{N}\mu, \tag{4.13a}$$

$$\langle\mu, \mathcal{N}^{\mathrm{T}}x_{\mathrm{p}} - \mathcal{R}\rangle = 0, \quad \mathcal{N}^{\mathrm{T}}x_{\mathrm{p}} \leqslant \mathcal{R}, \quad \mu \geqslant 0, \tag{4.13b}$$

for the appropriate-dimensional Lagrange multiplier μ. Thus, $\delta_k = -\mathfrak{Q}_k^{-1}\mathcal{N}\mu$, and μ is given by the solution of the quadratic programming problem (4.12). Algorithm 4.1 can use μ and the constraints $\mathcal{N}^{\mathrm{T}}x \leqslant \mathcal{N}$ to define

$$\Upsilon = \mathcal{N}\mu, \qquad \nu = \langle \mathcal{N}, \mu \rangle.$$

This is called a 'surrogate' cut, and is discussed in Goldfarb and Todd (1982) and Bland, Goldfarb and Todd (1981).

5 DISCUSSION OF THE MATRIX UPDATES

The previous section illustrates the correspondence of the correction vector δ_k used in Algorithms 4.3.1, 2.1 and 2.2, and the constraint Υ used in Algorithm 4.1. We now reformulate the rank-one update (4.3.3) as

$$\mathfrak{Q}_{k+1} = \frac{n^2-1}{n^2}\left(\mathfrak{Q}_k + \frac{2}{n-1}\frac{\mathfrak{Q}_k\delta_k\delta_k^{\mathrm{T}}\mathfrak{Q}_k}{\langle \delta_k, \mathfrak{Q}_k\delta_k\rangle}\right), \tag{5.1}$$

where η_k in (4.3.3) is given by (2.7c) and $n \geqslant 2$. In this case, the results given by (4.3.4a) and (4.3.5a, b) can be formulated such that (4.3.4a) remains unchanged, and τ_k is given by

$$\tau_k = -\frac{2\langle \delta_k, \mathfrak{Q}_k(x_k - x^{\mathrm{d}})\rangle}{(n-1)\langle \delta_k, \mathfrak{Q}_k\delta_k\rangle + 2\langle \mathfrak{Q}_k\delta_k, P_k\mathfrak{Q}_k^{-1}(\mathfrak{Q}_k\delta_k)\rangle}. \tag{5.2}$$

The value of η_k in (2.7c) eliminates the need for specifying this value explicitly. The vector of shadow prices associated with the equality constraints can also be written as

$$\lambda_{k+1} = \frac{n^2-1}{n}[\lambda_k - \tau_k(\mathcal{G}^{\mathrm{T}}\mathfrak{Q}_k^{-1}\mathcal{G})^{-1}\mathcal{G}^{\mathrm{T}}\delta_k]$$

(see Chapter 3). Using (5.1), the inverse of $\mathfrak{Q}_{k+1}$ can be written as

$$\mathfrak{Q}_{k+1}^{-1} = \frac{n^2}{n^2-1}\left(\mathfrak{Q}_k^{-1} - \frac{2}{n+1}\frac{\delta_k\delta_k^{\mathrm{T}}}{\langle \delta_k, \mathfrak{Q}_k\delta_k\rangle}\right). \tag{5.3}$$

Invoking the equivalence between δ_k and Υ given by (4.11), we can express (4.3.4a) as

$$x_{k+1} = x_k - \bar{\tau}_k P_k\mathfrak{Q}_k^{-1}\frac{\Upsilon}{\langle \Upsilon, \mathfrak{Q}_k^{-1}\Upsilon\rangle^{1/2}}, \tag{5.4}$$

where

$$\bar{\tau}_k = \frac{\eta_k\langle \Upsilon, \mathfrak{Q}_k^{-1}\Upsilon\rangle^{1/2}\langle \Upsilon, (x_k - x^{\mathrm{d}})\rangle}{\langle \Upsilon, \mathfrak{Q}_k^{-1}\Upsilon\rangle + \eta_k\langle \Upsilon, P_k\mathfrak{Q}_k^{-1}\Upsilon\rangle}, \tag{5.5}$$

and (5.3) becomes

$$\mathfrak{Q}_{k+1}^{-1} = \frac{n^2}{n^2-1}\left(\mathfrak{Q}_k^{-1} - \frac{2}{n+1}\frac{\mathfrak{Q}_k^{-1}\Upsilon\Upsilon^{\mathrm{T}}\mathfrak{Q}_k^{-1}}{\langle\Upsilon, \mathfrak{Q}_k^{-1}\Upsilon\rangle}\right). \tag{5.6}$$

We note that expressions (4.6) and (4.8) of Algorithm 4.1 are identical to (5.6) and (5.4) respectively, with $\bar{\tau}_k$ set to

$$\bar{\tau}_k = \frac{1}{n+1}. \tag{5.7}$$

The equivalence between the values of $\bar{\tau}_k$ in (5.5) and (5.7) can be achieved by suitably redefining x^{d} for $\langle\Upsilon, x_k - x^{\mathrm{d}}\rangle < 0$, following a similar approach to that of Theorem 2.1. Also, in the case when δ_k is given by (4.13), we have

$$x_{k+1} = x_k - \tau_k \frac{P_k\mathfrak{Q}_k^{-1}\mathscr{N}\mu}{\langle\mathscr{N}\mu, \mathfrak{Q}_k^{-1}\mathscr{N}\mu\rangle^{1/2}}$$

replacing (5.4). Similarly, (4.8) is replaced by

$$x_{k+1} = x_k - \frac{1}{n+1}\frac{P_k\mathfrak{Q}_k^{-1}\mathscr{N}\mu}{\langle\mathscr{N}\mu, \mathfrak{Q}_k^{-1}\mathscr{N}\mu\rangle^{1/2}}.$$

Thus, by exploring the similarities between Algorithm 4.1 in the presence of equality constraints, and Algorithm 4.3.1, we have constructed an algorithm that has specific values for η_k and τ_k, which, by Theorem 4.1, has a guaranteed termination property in a finite number of steps, or iterations. Assuming that $\mathscr{R} \cap \Omega \neq \emptyset$, we have Algorithm 2.1.

In order to modify Algorithm 4.3.1, to ensure convergence, we start with the same value for η_k as in (5.1). The τ_k that is generated by Algorithm 4.3.1 is given by (5.5) or, using (4.11), by

$$\bar{\tau}_k = -\frac{\eta_k\langle\delta_k, \mathfrak{Q}_k\delta_k\rangle^{1/2}\langle\delta_k, \mathfrak{Q}_k(x_k - x^{\mathrm{d}})\rangle}{\langle\delta_k, \mathfrak{Q}_k\delta_k\rangle + \eta_k\langle\mathfrak{Q}_k\delta_k, P_k\mathfrak{Q}_k^{-1}(\mathfrak{Q}_k\delta_k)\rangle},$$

with (4.3.4a) written as in (5.4). As discussed in Theorem 2.1, provided (2.12) is satisfied, we can ensure that this stepsize is identical to (5.7) by introducing the restriction (2.7), which yields Algorithm 2.2. Hence Algorithms 2.2 and 4.1 are identical, and the polynomial termination property of Algorithms 2.1 and 2.2 is given by the following result.

Theorem 5.1
Algorithms 2.1 and 2.2 have the same polynomial termination property as Algorithm 4.1.

Proof

The proof follows from the polynomial convergence of Algorithm 4.1 (see Lemma 4.1 and Theorem 4.1) and the equivalence of Algorithms 2.1 and 2.2 to Algorithm 4.1. □

Algorithms 2.1 and 2.2 and the associated concept of finite termination provide an insight into the decision process. Nevertheless, this finite termination property may easily be undermined by an indecisive decision maker who may choose to change the structure of Ω (e.g. by shrinking this region) during the decision process. Another difficulty arises when the assumption $\mathscr{R} \cap \Omega \neq \emptyset$ breaks down. Although Algorithm 4.1 can identify $\mathscr{R} \cap \Omega = \emptyset$ by not returning a feasible point after

$$4\mathscr{L}(n+1)^2$$

iterations, the value of $\mathscr{L}$ is difficult to estimate precisely for Algorithms 2.1 and 2.2. However, it may be possible to determine some upper limit to $\mathscr{L}$. Furthermore, changes in the structure of Ω seem to be in the nature of decision making. This would, at worst, increase p in (4.2), and hence $\mathscr{L}$. It may also be possible to determine an upper limit to p at the beginning of the algorithms. Thus, the discussion in this chapter has at least some analytical value in demonstrating the finiteness of the decision process with Algorithms 2.1 and 2.2.

6 CONCLUDING REMARKS

The desirable properties of the Algorithm 4.3.1 for constructing objective functions are described in Section 3. In effect, these imply that, at each stage of the algorithm, the decision maker specifies her requirements and is satisfied by the results generated by the method. The overall convergence issue need not necessarily be important practice. However, the 'conditions' under which Algorithm 2.2 converges provide an insight into these methods and indicate some of the reasons why the decision maker may wish to extend or truncate the decision process.

Finally, the similarities between Khachian's ellipsoid algorithm and Algorithms 2.1 and 2.2 highlight two points. The first is the complexity equivalence of the latter two algorithms. The second point is the implication that a search for an appropriate weighting matrix is similar to the search of Khachian's algorithm for computing a feasible point in $\Omega \cap \mathscr{R}$.

7 COMMENTS AND NOTES

CN.1 Positive-semidefinite version of Theorem 2.1

Theorem 2.1 and its proof can also be stated for positive-semidefinite $\mathfrak{Q}_k$ by using an orthogonal matrix Z such that $Z^{\mathrm{T}}\mathscr{G} = 0, \hat{x}_k - x_k = Zw$, and

$$\mathfrak{Q}_k(\hat{x}_k - x_k + x_k - x_k^{\mathrm{d}}) = \mathfrak{Q}_k Z w + \mathscr{G}\lambda_k = \mathscr{G}\hat{\lambda}_k.$$

Premultiplying the above expression by Z^{T}, we establish $\hat{x}_k = x_k$. To establish the result related to $\hat{\tau}_k$ is straightforward, and is accomplished using the expression for τ_k in Theorem 3.3.2. The reason for assuming $\mathfrak{Q}_k$ to be positive-definite is that the discussion of the complexity of the algorithm requires this property. Thus, as far as the discussion below is concerned, any rank deficiency in $\mathfrak{Q}_k$ is eliminated by considering $\mathfrak{Q}_k$ to be the projected Hessian in the intersection of a number of equality constraints.

CN.2 The equality-constrained case

In the presence of equality constraints, these constraints are used to eliminate some of the variables from the problem, hence resulting in a problem with reduced dimensions (see Goldfarb and Todd, 1982). This, in turn, naturally reduces $\mathscr{L}$ in (4.2), and hence the bound on the number of iterations.

REFERENCES

Aspvall, B. and R.E. Stone (1979). 'Khachian's Linear Programming Algorithm'. STAN—CS—79–776, Stanford University.

Bland, R.G., D. Goldfarb and M.J. Todd (1981). The ellipsoid method: a survey. *Oper. Res.* **29**, 1039–1091.

Goldfarb, D. and M.J. Todd (1982). Modifications and implementation of the ellipsoid algorithm for linear programming. *Math. Programming* **23**, 1–19.

Khachian, L.G. (1979). A polynomial algorithm in linear programming. *Soviet Math. Dokla.* **20**, 191–194.

Khachian, L.G. (1980). Polynomial algorithm in linear programming. *USSR Comput. Math. Math. Phys.* **20**, 53–72.

Kozlov, M.K., S.P. Tarasov and L.G. Khachian (1979). Polynomial solvability of convex quadratic programming. *Soviet Math. Dokla.* **20**, 1108–1111.

Ortega, J.M. and W.C. Rheinboldt (1970). *Iterative Solution of Nonlinear Equations in Several Variables*. Academic Press, London.

Rustem, B. and K. Velupillai (1987). Objective functions and the complexity of policy design. *J. Econ. Dynam. Control* **11**, 185–192.

PART III

Algorithms for Nonlinear Optimization and Equilibria

CHAPTER 6

Nonlinear Optimization with Convex Constraints The Goldstein–Levitin–Polyak Algorithm

The basic multi-objective decision problem over time can often be formulated as the static optimization of a single objective function. The latter is sometimes a weighted combination of the multiple objectives or sometimes a more complicated function of the individual objectives. In Chapters 3–5 we have discussed an approach to the problem of combining the objectives. In this chapter, we consider an algorithm for solving the optimization problem under convex constraints. Although the convexity of the constraints is treated in its generality, in practice this means linear equalities and inequalities. A special class of such constraints are upper and lower bounds on the decision variables. Apart from convexity, the main assumption on the constraints is that they are at least once-differentiable.

The method belongs to the powerful class of algorithms developed by Goldstein, Levitin and Polyak. It requires that the constraints be satisfied at every iteration. This is an inconvenience in general. Yet, for linear constraints and for special constraint structures, it is simple to maintain feasibility. The method is particularly attractive for problems with only bound constraints due to the simplicity of the projections that need to be performed to maintain feasibility. Two important advantages that are offered in return for this constraint satisfaction are the strong results concerning unit stepsize achievement and superlinear convergence. In Section 4, it is established that the *necessary and sufficient* condition for Q-superlinear rate of convergence is the two-sided projected Hessian condition, which, in other algorithms discussed in Chapters 7 and 8, can only ensure lesser Q-superlinear rates.

1 THE CONVEX OPTIMIZATION PROBLEM

We return to the original optimal decision problem of Chapter 1 and consider an algorithm for solving problem (1.1.4). The convexity of the constraints allows the development of a special class algorithm for this type of problem. In particular, we use the fact that if two points belong to the convex set described by the constraints then any point on the line segment joining the two points is also in the convex set. The point chosen by the algorithm on the line segment is the one that corresponds to an improved value of the objective. We recall the problem

$$\min \{\mathscr{F}(x) \mid h(x) \leqslant 0\}, \tag{1.1}$$

where $\mathscr{F} : \mathbb{R}^n \to \mathbb{R}^1$, and the elements of $h : \mathbb{R}^n \to \mathbb{R}^i$ are differentiable functions. The feasible region

$$\mathscr{R} = \{x \in \mathbb{R}^n \mid h(x) \leqslant 0\} \tag{1.2}$$

is assumed to be *convex* and $\mathscr{F}(x)$ is assumed to be bounded below on $\mathscr{R}$.

The method we describe for solving (1.1) under this convexity assumption is a generalization of the Goldstein–Levitin–Polyak (GLP) algorithm (Goldstein, 1964; Levitin and Polyak, 1966). The original GLP algorithm has two distinguishing features: it solves (1.1) by computing arcs on $\mathscr{R}$ using the steepest descent direction, and the sequence of points it generates to converge to the optimal solution is such that every member of the sequence remains feasible. The algorithm has been extensively used to solve large-scale optimization problems (Gafni and Bertsekas, 1984).

The basic GLP algorithm consists of the iterative scheme

$$x_{k+1} = P_{\mathscr{R}}(x_k + \alpha_k \mathrm{d}_k),$$

where d_k is a descent direction such as the steepest descent direction, $-\nabla\mathscr{F}(x_k) \cdot \alpha_k$ is the stepsize and $P_{\mathscr{R}}$ is the projection of $x_k + \alpha_k d_k$ onto the convex feasible region $\mathscr{R}$.

The version of the GLP algorithm considered in this chapter follows Rustem (1984). It is not restricted to convex $\mathscr{F}(\mathrm{x})$, and has been successfully applied to engineering process design problems (Darlington *et al.*, 1996). It involves projections, quasi-Newton descent directions and an additional stepsize strategy for which two alternatives are considered. The steplength strategies employed by earlier attempts are often difficult to implement. For example, Goldstein (1964) and Levitin and Polyak (1966) suggest choosing α_k from a range bounded by a Lipschitz constant. McCormick and Tapia (1972) require univariate minimization along the steepest descent arc. The algorithm below is closer to the work of Allwright (1980) and Psenichny and Danilin (1978), who suggest the use of Armijo-type stepsize strategies along projected directions. The global linear rate of convergence of the strategy (see CN.1) is established by Allwright (1980) for a general descent direction and a general projection. Psenichny and Danilin (1978) consider a Newton version of the algorithm, and establish its quadratic rate of convergence. Below, the determination of the stepsize is divided into two stages. The first stage consists of a fixed steplength that determines the length of the unconstrained step. This stepsize can be chosen to be a sequence in the range

$$\alpha \in [1, 2)$$

converging to unity, and determines the position of a possibly infeasible vector $\bar{x}$. The option of taking steps larger than unity can be useful in the initial iterations of the algorithm. The second stage is the determination of the stepsize from the range

$$\tau \in [0, 1],$$

which shortens the projected step if the projection of $\bar{x}$ on $\mathscr{R}$ does not reduce the objective functions sufficiently. Since $\mathscr{R}$ is convex, a step shorter than unity is ensured to remain in $\mathscr{R}$ (see Section 1.2). The attainment of unit steplengths and convergence rates are analysed for $\mathscr{F}(x)$, which need not have a positive-definite Hessian at the solution, while the algorithm utilizes a positive-definite Hessian approximation. Using a positive-definite approximate Hessian simplifies the quadratic subproblem (see (2.3) below), which, in the case of linear constraints, reduces to a positive-definite quadratic programming problem.

The algorithm is introduced in Section 2. In Section 3, its global convergence is considered and the stepsize strategies for α_k and τ_k are justified. In Section 4, it is shown that unit steplengths and subsequently Q-superlinear convergence are attained if the two-sided projection of the quasi-Newton approximation converges to the projected Hessian at the solution of (1.1).

The use of positive-definite Hessian approximations, even when the exact Hessian at the solution is not positive-definite, relaxes the positive-definite Hessian requirement in the GLP algorithm, while preserving the Q-superlinear convergence rate of property. This is analogous to the approach of Powell (1978a), in sequential quadratic programming algorithms for general constrained optimization that do not require feasibility at every iteration (see chapter 7). Powell (1978a) establishes two-step Q-superlinear and R-superlinear convergence. Positive-definite Hessian approximations are only required to be accurate on the *tangent manifold of the constraints*. The convergence rate result in Section 4, for the GLP algorithm, is the *necessary and sufficient* condition for its Q-superlinear rate. The reason for this stronger result is the feasibility of the sequence generated by the GLP algorithm. In Chapter 8, we study detailed Q-superlinear convergence results for algorithms that do not require feasibility at every iteration, and observe that the absence of feasibility, until convergence, results in lesser superlinear rates than that observed in Section 4 (see Corollary 8.3.1).

2 THE GLP ALGORITHM

Let the local quadratic approximation to $\mathscr{F}(x)$, at x_k, be given by

$$q_k(x) = \langle \nabla \mathscr{F}(x_k), x - x_k \rangle + \frac{1}{2} \langle x - x_k, \hat{\mathfrak{Q}}_k (x - x_k) \rangle, \tag{2.1}$$

where $\hat{\mathfrak{Q}}_k$ is a symmetric positive-definite approximation to the Hessian $\mathfrak{Q}_k$ of $\mathscr{F}(x)$ at x_k. The assumption that $\hat{\mathfrak{Q}}_k$ is positive-definite is not restrictive, since, as established in Sections 3 and 4, it does not impair the convergence properties of the algorithm, even when the true Hessian is not strictly positive-definite. In particular, it is shown in Section 4 that the convergence of the stepsize to unity and the Q-superlinear convergence rate of the algorithm depends on the accuracy of the projection of $\hat{\mathfrak{Q}}_k$ onto the tangent manifold of the constraints.

2.1 The algorithm

The algorithm is based on the projections of the unconstrained step

$$\bar{x} = x_k - \alpha_k \hat{\mathfrak{Q}}_k^{-1} \nabla \mathscr{F}(x_k), \qquad k = 0, 1, 2, \ldots, \tag{2.2}$$

with $\alpha_k \in [1, \bar{\alpha}], \bar{\alpha} \in [1, 2)$, onto the feasible region $\mathscr{R}$ using the subproblem

$$\min \left\{ \frac{1}{2} \|x - \bar{x}\|^2_{\hat{\mathfrak{Q}}_k} \middle| x \in \mathscr{R} \right\}, \tag{2.3}$$

where $\|v\|^2_Q = \langle v, \hat{\mathfrak{Q}} v \rangle$, $v \in \mathbb{R}^n$. Let x_k^{p} be the value of x that solves (2.3).

Remark
As $\mathscr{R}$ is convex, for $x_k, x_k^{\mathrm{p}} \in \mathscr{R}$, we have

$$x(\tau) = x_k + \tau(x_k^{\mathrm{p}} - x_k) \in \mathscr{R}, \quad \forall \tau \in [0, 1].$$

The next iterate, x_{k+1}, is computed using

$$x_{k+1} = x_k + \tau_k (x_k^{\mathrm{p}} - x_k), \tag{2.4}$$

with $\tau_k \in [0, 1]$ given by the smallest value of $j = 0, 1, 2, 3, \ldots, \tau_k = (\bar{\tau})^j$, $\bar{\tau} \in (0, 1)$, satisfying

$$\mathscr{F}(x_{k+1}) - \mathscr{F}(x_k) \leqslant \tau_k \rho_1 q_k(x_k^{\mathrm{p}}), \qquad \rho_1 \in (0, 1), \tag{2.5}$$

with ρ_1 an arbitrary number in the above range. An alternative to (2.5) is the Armijo-type stepsize strategy

$$\mathscr{F}(x_{k+1}) - \mathscr{F}(x_k) \leqslant \tau_k \rho_2 \langle \nabla \mathscr{F}(x_k), x_k^{\mathrm{p}} - x_k \rangle, \qquad \rho_2 \in \left(0, 1 - \frac{1}{2}\bar{\alpha}\right), \tag{2.6}$$

with ρ_2 an arbitrary number in the above range.

The matrix $\hat{\mathfrak{Q}}_k$ is approximated using Powell's (1978b) modification to a quasi-Newton formula due to Broyden (1969, 1970), Fletcher (1970), Goldfarb (1970) and Shanno (1970) for approximating the Hessian of a Lagrangian. This modified BFGS (Broyden–Fletcher–Goldfarb–Shanno) formula is given by

$$\hat{\mathfrak{Q}}_{k+1} = \hat{\mathfrak{Q}}_k - \frac{\hat{\mathfrak{Q}}_k \delta_x \delta_x^{\mathrm{T}} \hat{\mathfrak{Q}}_k}{\langle \delta_x, \hat{\mathfrak{Q}}_k \delta_x \rangle} + \frac{\nu \nu^{\mathrm{T}}}{\langle \delta_x, \nu \rangle}, \tag{2.8a}$$

where

$$\delta_x = x_{k+1} - x_k,$$

$$\gamma_k = f(x_{k+1}) - f(x_k), \tag{2.8b}$$

$$\nu = \theta \gamma + (1 - \theta) \hat{\mathfrak{Q}}_k \delta_x, \tag{2.8c}$$

$$\theta = \begin{cases} 1 & \text{if } \langle \delta_x, \gamma \rangle \geqslant 0.2 \langle \delta_x, \hat{\mathfrak{Q}}_k \delta_x \rangle, \\ \dfrac{0.8 \langle \delta_x, \hat{\mathfrak{Q}}_k \delta_x \rangle}{\langle \delta_x, (\hat{\mathfrak{Q}}_k \delta_x - \gamma) \rangle} & \text{if } \langle \delta_x, \gamma \rangle < 0.2 \langle \delta_x, \hat{\mathfrak{Q}}_k \delta_x \rangle. \end{cases} \tag{2.8, d}$$

For linearly constrained $\mathscr{R}$ and positive-definite $\hat{\mathfrak{Q}}_k$, (2.3) is a positive-definite quadratic programming problem with a unique solution x_k^{p}. In the initial stages of the algorithm, $\alpha_k > 1$ provides the option for choosing large steps. As discussed in Sections 3 and 4 below, α_k does not affect the convergence properties of the algorithm, provided it is chosen in the range $[1, \bar{\alpha}]$, with the sequence $\{\alpha_k\}$ converging to unity. An example is $\alpha_{k+1} = 2 - \tau_k$. It is shown below that $\tau_k > 0, \forall k$. Also, if $\tau_k = 1$ for $\alpha_k > 1$ then reducing α_{k+1} to unity makes it easier to maintain unit τ_k.

Numerous earlier versions of the GLP algorithm mostly assign the values $\tau_k = 1, \hat{\mathfrak{Q}}_k = I$ in (2.1)–(2.4) and suggest various strategies for choosing α_k. Goldstein (1964) and Levitin and Polyak (1966) have proposed the case in which α_k is chosen from a range bounded by a Lipschitz constant. The difficulty with estimating the Lipschitz constant was later resolved by McCormick and Tapia (1972) and Bertsekas (1976, 1982), who (with $\tau_k = 1$), proposed that α_k be determined by univariate minimisation and an Armijo-type rule respectively. Dunn (1981) discusses, under relaxed conditions, the convergence properties of Armijo strategies with $\hat{\mathfrak{Q}}_k = I, \tau_k = 1$. For $\hat{\mathfrak{Q}}_k = I, \alpha_k = 1$, Dunn (1979) discusses the convergence properties of Demyanov and Rubinov's (1971) approach, or its modification, for determining τ_k. Psenichny and Danilin (1978) and Dunn (1980) discuss the case with $\alpha_k = 1, \hat{\mathfrak{Q}}_k$ replaced by the exact Hessian of $\mathscr{F}(x)$, and τ_k determined with an Armijo-type strategy. Finally, the case with $\alpha_k = 1, \hat{\mathfrak{Q}}_k$ a positive-definite matrix and τ_k determined by an Armijo strategy is discussed by Allwright (1980). The traditional positive-definiteness condition on the Hessian of $\mathscr{F}(x)$ in the GLP algorithm has been replaced in Dunn (1980, 1981) by pseudoconvexity conditions and growth conditions of $\mathscr{F}(x)$.

In Sections 3 and 4 below, the stepsize strategies for α_k and τ_k are justified, and GLP algorithm (2.1)–(2.6) is shown to converge. The convergence of the stepsizes to unity and the Q-superlinear convergence rate of the algorithm are established, with a positive-definite Hessian approximation, even when the exact Hessian at the solution is not strictly positive-definite.

2.2 Parallel computation using proximal optimization

The parallelization of the algorithm can also be considered using the proximal optimization approach (Rockafeller, 1973; Bertsekas, 1982; Bertsekas and Tsitsiklis, 1989; Bertsekas and Tseng, 1994). We introduce an additional vector $y \in \mathbb{R}^n$ and formulate the following equivalent problem to (1.1):

$$\min \{\mathscr{F}(x) + c\|x - y\|_2^2 \mid x \in \mathscr{R}; y \in \mathbb{R}^n\}, \tag{2.9}$$

where c is a positive scalar. The problem can be solved by alternately minimizing the objective in (2.9) over $x \in \mathscr{R}$, while keeping y fixed, then minimizing the objective over $y \in \mathbb{R}^n$, while keeping x fixed, and repeating this procedure until convergence. The method is thus given by

$$x_{k+1} = \arg\min_{x\in\mathscr{R}} \{\mathscr{F}(x) + c\|x - y_k\|_2^2\},$$
$$y_{k+1} = x_{k+1},$$

which reduces to

$$x_{k+1} = \arg\min_{x\in\mathscr{R}} \{\mathscr{F}(x) + c\|x - x_k\|_2^2\}.$$

By appropriately assigning the minimizations of distinct partitions of x to different processors, Bertsekas and Tseng (1994) show that it is possible to parallelize the overall algorithm and consider its convergence properties.

3 CONVERGENCE OF THE ALGORITHM

In this section the stepsize strategies (2.5) and (2.6) are justified for symmetric positive-definite and $\hat{\mathfrak{Q}}_k$ and $\{\alpha_k\}$ satisfying the restrictions discussed in Section 2. It is shown that the algorithm is globally convergent. The main theorem of this section is (3.1) below, which justifies (2.5), (2.6) and establishes the existence of $\tau_k \in (0, 1]$ satisfying the stepsize rules. Most subsequent results are dependent on Theorem 3.1.

Lemma 3.1 (Descent property of $x_k^{\mathrm{p}} - x_k$)
Let $\alpha_k \in (0, \infty)$, $\hat{\mathfrak{Q}}_k$ be a bounded symmetric positive-definite matrix and x_k be given by (2.2). Then, x_k^{p} solving (2.3) satisfies

$$\langle \nabla\mathscr{F}(x_k), x_k^{\mathrm{p}} - x_k\rangle \leqslant -\frac{1}{\alpha_k}\|x_k^{\mathrm{p}} - x_k\|^2_{\hat{\mathfrak{Q}}_k}. \tag{3.1}$$

Proof
Since (2.3) is the projection of $\bar{x}$ on the convex region $\mathscr{R}$, we invoke Lemma 3.1.2 to express the inequality

$$\langle x_k^{\mathrm{p}} - \mathscr{X}, \hat{\mathfrak{Q}}_k(x_k^{\mathrm{p}} - \bar{x})\rangle \leqslant 0 \tag{3.2}$$

for any $\mathscr{X} \in \mathscr{R}$. Thus, we have

$$\|x_k^{\mathrm{p}} - x_k\|^2_{\hat{\mathfrak{Q}}_k} = \langle x_k^{\mathrm{p}} - x_k, \hat{\mathfrak{Q}}_k[x_k^{\mathrm{p}} + \alpha_k\hat{\mathfrak{Q}}_k^{-1}\nabla\mathscr{F}(x_k) - x_k]\rangle - \alpha_k\langle\nabla\mathscr{F}(x_k), x_k^{\mathrm{p}} - x_k\rangle.$$

The result follows, since the first term on the right may be formulated as (3.2), for $\mathscr{X} = x_k$. □

Lemma 3.2
Let $\tau_k \in [0,1], \hat{\mathfrak{Q}}_k$ be symmetric positive-definite and x_{k+1} be given by (2.4). Then

$$q_k(x_{k+1}) \leqslant \tau_k q_k(x_k^{\mathrm{p}}). \tag{3.3}$$

Proof
The result follows from the quadratic function

$$\begin{aligned} q_k(x_{k+1}) &= \tau_k \langle \nabla \mathscr{F}(x_k), x_k^{\mathrm{p}} - x_k \rangle + \frac{1}{2}\tau_k^2 \langle x_k^{\mathrm{p}} - x_k, \hat{\mathfrak{Q}}_k(x_k^{\mathrm{p}} - x_k) \rangle \\ &\leqslant \tau_k q_k(x_k^{\mathrm{p}}). \end{aligned}$$

□

Lemma 3.3
Let $\xi_{\bar{\alpha}} > 0, \alpha_k \in (0, \bar{\alpha}], \bar{\alpha} \in (0,2), \hat{\mathfrak{Q}}_k$ be bounded symmetric positive-definite and x_k be given by (2.2). Then, x_k^{p} solving (2.3) satisfies

$$q_k(x_k^{\mathrm{p}}) \leqslant \left(\frac{1}{2} - \frac{1}{\alpha_k}\right) \|x_k^{\mathrm{p}} - x_k\|_{\hat{\mathfrak{Q}}_k}^2 \leqslant 0, \tag{3.4}$$

$$\xi_{\bar{\alpha}} \|x_k^{\mathrm{p}} - x_k\|_{\hat{\mathfrak{Q}}_k}^2 \leqslant -q_k(x_k^{\mathrm{p}}). \tag{3.5}$$

Proof
Lemma 3.1 yields (3.4) immediately. Also, (3.5) holds for $\xi_{\bar{\alpha}} = 1/\bar{\alpha} - \frac{1}{2}$. □

Theorem 3.1
Let

(i) $\mathscr{R} \subseteq \mathbb{R}^n$ be a convex set and $\mathscr{F}(x) \in \mathbb{C}^2$;

(ii) $\hat{\mathfrak{Q}}_k$ in (2.1)–(2.3) satisfy

$$m\|v\|^2 \leqslant \langle v, \hat{\mathfrak{Q}}_k v \rangle \leqslant \mathcal{M}\|v\|^2, \quad \forall k, \qquad 0 \neq v \in \mathbb{R}^n, \qquad \infty > \mathcal{M} \geqslant m > 0;$$

(iii) $\{\alpha_k\}$ be any sequence converging to unity, with $\alpha_k \in (0, \bar{\alpha}], \bar{\alpha} \in [1,2)$.

Then there exists a stepsize $\tau_k \in (0,1]$ that satisfies (2.5) or (2.6), and hence the sequence $\{x_k\}$ computed by (2.4) generates a monotonically decreasing sequence $\{\mathscr{F}(x_k)\}$.

Proof
Using the second-order Taylor series expansion (see CN1.5),

$$\begin{aligned} \mathscr{F}(x_{k+1}) - \mathscr{F}(x_k) &= \langle \nabla \mathscr{F}(x_k), x_{k+1} - x_k \rangle + \frac{1}{2}\langle x_{k+1} - x_k, \hat{\mathfrak{Q}}_k(x_{k+1} - x_k) \rangle \\ &\quad + \int_0^1 (1-t)\langle x_{k+1} - x_k, [\mathfrak{Q}((x_k(t)) - \hat{\mathfrak{Q}}_k](x_{k+1} - x_k) \rangle \, \mathrm{d}t \qquad (3.7) \\ &\leqslant \tau_k q_k(x_k^{\mathrm{p}}) + \tau_k^2 \eta_k \|x_k^{\mathrm{p}} - x_k\|^2, \qquad (3.8) \end{aligned}$$

where

$$x_k(t) = x_k - t(x_k - x_{k+1}),$$

$\mathfrak{Q}(\cdot)$ is the Hessian of $\mathscr{F}$ at $(\cdot)$,

$$\eta_k = \int_0^1 (1-t)\|\mathfrak{Q}(x_k(t)) - \hat{\mathfrak{Q}}_k\|\,\mathrm{d}t,$$

and (3.3) is applied to obtain (3.8). From (3.5) and (3.8), we have

$$\mathscr{F}(x_{k+1}) - \mathscr{F}(x_k) \leqslant \tau_k q_k(x_k^{\mathrm{p}})\left(1 - \frac{\tau_k \eta_k}{m\xi_{\bar{\alpha}}}\right). \tag{3.9}$$

Since $0 < \rho_1 < 1$, there is a $\tau_k \in (0,1]$ such that

$$\rho_1 < 1 - \frac{\tau_k \eta_k}{m\xi_{\bar{\alpha}}} \leqslant 1. \tag{3.10}$$

By Lemma 3.3, $q_k(x_k^{\mathrm{p}}) \leqslant 0$. Thus, (2.5) holds for this τ_k. Suppose that τ^0 is the largest $\tau \in (0,1]$ satisfying inequality (2.5). All $\tau \leqslant \tau^0$ also satisfy this condition, and the selected $\tau_k \in [\bar{\tau}\tau^0, \tau^0]$. It follows that $\{\mathscr{F}(x_k)\}$ is monotonically decreasing.

To show the monotonic decrease of $\{\mathscr{F}(x_k)\}$ for (2.6), we use (3.1) with (3.7) to yield

$$\mathscr{F}(x_{k+1}) - \mathscr{F}(x_k) \leqslant \tau_k \langle \nabla\mathscr{F}(x_k), x_k^{\mathrm{p}} - x_k\rangle\left(1 - \frac{\tau_k \alpha_k}{2} - \frac{\tau_k \eta_k \alpha_k}{m}\right). \tag{3.11}$$

Since $0 < \rho_2 \leq 1 - \frac{1}{2}\bar{\alpha}$, there is a $\tau_k \in (0,1]$ such that

$$0 < \rho_2 \leqslant 1 - \frac{\tau_k \alpha_k}{2} - \frac{\tau_k \eta_k \alpha_k}{m} \leqslant 1 - \frac{\bar{\alpha}}{2}. \tag{3.12}$$

By Lemma 3.1, $\langle \nabla\mathscr{F}(x_k), x_k^{\mathrm{p}} - x_k\rangle \leqslant 0$. Thus, (2.6) holds for this τ_k. Suppose that τ^0 is the largest $\tau \in (0,1]$ satisfying inequality (2.6). All $\tau \leqslant \tau^0$ also satisfy this condition, and $\tau_k \in [\bar{\tau}\tau^0, \tau^0]$. It follows that $\{\mathscr{F}(x_k)\}$ is monotonically decreasing. □

A survey of the basic convergence concepts used in all subsequent results is given in CN.2.

Theorem 3.2
Let

(i) the assumptions of Theorem 3.1 be satisfied,

(ii) the set

$$\mathbb{F} \equiv \{x \in \mathscr{R} \mid \mathscr{F}(x) \leqslant \mathscr{F}(x_0)\}$$

be bounded.

Then

$$\lim_{k\to\infty} q_k(x_k^{\mathrm{p}}) = 0 \tag{3.13}$$

and

$$\lim_{k\to\infty} \langle \nabla\mathscr{F}(x_k), x_k^{\mathrm{p}} - x_k\rangle = 0. \tag{3.14}$$

Proof
Given $0 < \rho_1 < 1$, by (3.9), the choice

$$\tau^0 = \min\left\{1, \frac{m\xi_{\bar\alpha}(1-\rho_1)}{\eta_k}\right\}$$

always satisfies the stepsize strategy (2.5). Clearly, τ_k, chosen as $\tau_k = \bar\tau^{j_k}$, as discussed above, is in the range $\tau_k \in [\bar\tau\tau^0, \tau^0]$ and thereby also satisfies (2.5). As $\mathscr{F}(x) \in \mathbb{C}^2$, and $\mathbb{F}$ is compact, there is a scalar $\bar M < \infty$ such that $\eta_k \leqslant \bar M$. Similarly, given $0 < \rho_2 < 1 - \frac{1}{2}\bar\alpha$, (2.6) is satisfied by

$$\tau^0 = \min\left\{1, (1-\rho_2)\left[\bar\alpha\left(\frac{1}{2} + \frac{\eta_k}{m}\right)^{-1}\right]\right\}.$$

As $m, \bar\alpha, \xi_{\bar\alpha} \geqslant 0$, we have established that there is a $\hat\tau > 0$ such that the stepsizes determined by (2.5) or (2.6) satisfy $\tau_k \geqslant \hat\tau > 0, \forall k$.

The boundedness of $\mathscr{F}$ on $\mathbb{F}$ and $q_k(x_k^{\mathrm{p}}) \leqslant 0$ imply, in the case of (2.5), that

$$0 \leqslant \rho_1 \sum_k \tau_k |q_k(x_k^{\mathrm{p}})| \leqslant \sum_k [\mathscr{F}(x_k) - \mathscr{F}(x_{k+1})] < \infty.$$

Since $\tau_k \geqslant \epsilon > 0$, this yields (3.13). The equality (3.14) is established similarly by using (2.6) and

$$\langle \nabla\mathscr{F}(x_k), x_k^{\mathrm{p}} - x_k\rangle \leqslant 0.$$

□

Lemma 3.4
If (3.13) or if (3.14) are satisfied then

$$\lim_{k\to\infty} \|x_k^{\mathrm{p}} - x_k\| = 0. \tag{3.15}$$

Proof
The result follows from Lemmas 3.3 and 3.4, for (3.13) and (3.14) respectively.

□

Theorem 3.3
Let

(i) the assumptions of Theorem 3.1 be satisfied;

(ii) the assumptions of Theorem 3.2 be satisfied;

(iii) $h(x)$ be once-differentiable;

(iv) the active constraint gradients at x_k^p be linearly independent (see CN.3).

Then,

(a) the algorithm in Section 2, with stepsize strategy (2.5) or (2.6), generates a sequence $\{x_k\}$ that converges to x_*;

(b) if, furthermore, strict complementarity holds at the solution of subproblem (2.3), for large k, μ_k predicts the active inequality constraints at x_*.

Proof

By Theorem 3.1, the algorithm ensures the decrease of $\mathscr{F}(x)$ at each iteration, thereby ensuring $x_k \in \mathbb{F}$, for $\mathbb{F}$ compact. Hence, $\{x_k\}$ has a limit point x_*. Without loss of generality, we can take $\{x_k\} \to x_*$. To show that x_* also satisfies the necessary conditions for optimality for problem (1.1), we consider the optimality of subproblem (2.3). Let μ_{k+1} and ∇h_k^p denote respectively the multipliers of the subproblem and the matrix whose columns are the constraint gradients at x_k^p. Using (2.2) and (2.3), the conditions for (2.3) are given by

$$\nabla \mathscr{F}(x_k) + \frac{1}{\alpha_k} \hat{\mathfrak{Q}}_k (x_k^p - x_k) + \nabla h_k^p \mu_{k+1} = 0, \tag{3.16a}$$

$$h(x_k^p) \leqslant 0, \quad \langle h(x_k^p), \mu_{k+1} \rangle = 0, \quad \mu_{k+1} \geqslant 0. \tag{3.16b}$$

Using (3.15), we have $\lim_{k\to\infty} x_k = \lim_{k\to\infty} x_k^p = x_*$ and $\nabla \mathscr{F}(x_*) + \nabla h_* \mu_* = 0$.

To show (b), we note that, in view of the last two conditions in (3.16b), for sufficiently large k and with strict complementarity holding, none of the inactive constraints, i.e.

$$h_k^j < 0, \quad \mu_{k+1}^j = 0, \tag{3.10}$$

are predicted to be active at $x_*, h_*^j(x_*) < 0, \mu_*^j = 0$.

4 UNIT STEPSIZES AND SUPERLINEAR CONVERGENCE RATES

In this section, we consider the convergence of τ_k to unity. Coupled with $\alpha_k \to 1$, this leads to the Q- superlinear convergence rate of the algorithm (see CN.4). It is shown that both the convergence $\{\tau_k\} \to 1$ and superlinear convergence are dependent on the Hessian approximation $\hat{\mathfrak{Q}}_k$. The algorithm uses strictly positive-definite $\hat{\mathfrak{Q}}_k$ even when the Hessian at the solution is not strictly positive-definite. First, we establish the subspace in which $\hat{\mathfrak{Q}}_k$ needs to approximate the Hessian of $\mathscr{F}(x)$. This is used in Theorem 4.1 to establish the attainment of unit stepsizes and in Theorem 4.2 to establish the convergence rate. Finally, a similar convergence rate result is also discussed as a corollary to a general nonlinear programming algorithm in Chapter 8. However, in that case, $\hat{\mathfrak{Q}}_k$ refers to an approximation to the Hessian of the Lagrangian (see Corollary 8.3.1)

A consequence of Theorem 3.3 is that, for sufficiently large k, with strict complementarity holding, μ_k predicts the constraints active at x_*. Thus, for large k, the constraints satisfied as strict inequalities at x_* do not affect the computation of x_k. At that stage, it would make no difference if the constraints satisfied as equalities at x_* were treated as equality constraints. Hence, for large k, $h^j(x_*) = 0$ implies $h^j(x_k) = 0$, and $h^j(x_*) < 0$ implies $h^j(x_k) < 0$. Let h^* denote the vector constraints satisfied as equalities at x_*. Thus, we have $h^*(x_k) = 0$, for large k.

In order to characterize the subspace for approximating the Hessian of $\mathscr{F}(x)$, we need to invoke a special case of the mean value theorem that holds for h^* (see CN.5). Provided each element of the vector h^*, with the jth element given by $h^{*j} : \mathbb{H} \subset \mathbb{R}^n \to \mathbb{R}^1$, is differentiable on an open convex set $\mathbb{H}_0 \subset \mathbb{H}$, for any two points $x_{k+1}, x_k \in \mathbb{H}_0$, there exist $t^1, t^2, \ldots, t^j, \ldots \in (0, 1)$ such that

$$\mathscr{H}(x_{k+1}, x_k) \equiv [\nabla h^{*1}(x_k + t^1(x_{k+1} - x_k)) \vdots \nabla h^{*^2}(x_k + t^2(x_{k+1} - x_k)) \vdots \ldots], \quad (4.1)$$

where ∇h_j^* denotes the gradient of the jth element of h^*, and

$$h^*(x_{k+1}) - h^*(x_k) = \mathscr{H}^{\mathrm{T}}(x_{k+1}, x_k)(x_{k+1} - x_k). \quad (4.2)$$

Let the matrix $\mathscr{H}_k$ have the same rank as $\mathscr{H}(x_{k+1}, x_k)$ and let a linearly independent subset of the columns of $\mathscr{H}(x_{k+1}, x_k)$ form the columns of $\mathscr{H}_k$. Thus $\mathscr{H}_k$ is of full rank, and we can form the operator $P_k = I - \mathscr{H}_k(\mathscr{H}_k^{\mathrm{T}} \mathscr{H}_k)^{-1} \mathscr{H}_k^{\mathrm{T}}$, which satisfies $P_k P_k = P_k$ and

$$P_k(x_{k+1} - x_k) = x_{k+1} - x_k. \quad (4.3)$$

As $\{x_k\} \to x_*$,

$$\{\mathscr{H}(x_{k+1}, x_k)\} \to [\nabla h^{*1}(x_*) \vdots \nabla h^{*2}(x_*) \vdots \ldots], \quad (4.4)$$

and P_k becomes the operator projecting vectors in $\mathbb{R}^n$ onto the active constraints at x_*.

Theorem 4.1

Let the assumptions of Theorem 3.3 be satisfied. Then, there is a $\sigma > 0$, such that

(a)

$$\frac{\|(\mathfrak{Q}_* - \hat{\mathfrak{Q}}_k)(x_{k+1} - x_k)\|}{\|x_{k+1} - x_k\|} \leqslant \sigma, \quad (4.5)$$

or,

(b) for large k, the two-sided projected Hessian satisfies

$$\frac{\|P_k(\mathfrak{Q}_* - \hat{\mathfrak{Q}}_k)P_k(x_{k+1} - x_k)\|}{\|x_{k+1} - x_k\|} \leqslant \sigma, \tag{4.6}$$

and we have $\{\tau_k\} \to 1$.

Remark

Inequality (4.5) requires $\hat{\mathfrak{Q}}_k$ to be close to $\mathfrak{Q}_*$. This may be satisfied for $\mathfrak{Q}_*$ and $\hat{\mathfrak{Q}}_k$ strictly positive-definite or for some $\hat{\mathfrak{Q}}_k$ positive-definite and $\mathfrak{Q}_*$ not positive-definite. If $\mathfrak{Q}_*$ is not positive-definite, and there is no $\hat{\mathfrak{Q}}_k$ satisfying (4.5), we need to consider (4.6). The latter demands only the projections of $\mathfrak{Q}_*, \hat{\mathfrak{Q}}_k$, on the constraints, to be close and the projection of $\mathfrak{Q}_*$ to be positive-definite.

Proof of Theorem 4.1

For stepsize (2.5), we can write (3.7) as

$$\begin{aligned}\mathscr{F}(x_{k+1}) - \mathscr{F}(x_k) &\leqslant \tau_k q_k(x_k^{\mathrm{p}}) \\ &\quad + \int_0^1 (1-t)\langle x_{k+1} - x_k, [\mathfrak{Q}((x_k(t)) - \mathfrak{Q}_* + \mathfrak{Q}_* - \hat{\mathfrak{Q}}_k](x_{k+1} - x_k)\rangle \mathrm{d}t \quad (4.7) \\ &\leqslant \tau_k q_k(x_k^{\mathrm{p}}) + \tau_k^2 \eta_k^* \|x_k^{\mathrm{p}} - x_k\|^2 + \frac{1}{2}\tau_k^2 \|x_k^{\mathrm{p}} - x_k\| \, \|(\mathfrak{Q}_* - \hat{\mathfrak{Q}}_k)(x_k^{\mathrm{p}} - x_k)\| \quad (4.8) \\ &\leqslant \tau_k q_k(x_k^{\mathrm{p}}) \left\{ 1 - \frac{\tau_k}{m\xi_{\bar{\alpha}}} \left[\eta_k^* + \frac{1}{2} \frac{\|(\mathfrak{Q}_* - \hat{\mathfrak{Q}}_k)(x_{k+1} - x_k)\|}{\|x_{k+1} - x_k\|} \right] \right\}, \quad (4.9)\end{aligned}$$

where

$$\eta_k^* = \int_0^1 (1-t)\|\mathfrak{Q}((x_k(t)) - \mathfrak{Q}_*\| \, \mathrm{d}t$$

and (4.9) is obtained by invoking Lemma 3.3. The scalar $\rho_1 \in (0, 1)$ in (2.5) requires τ_k to satisfy

$$\rho_1 \leqslant 1 - \frac{\tau_k}{m\xi_{\bar{\alpha}}} \left[\eta_k^* + \frac{1}{2} \frac{\|(\mathfrak{Q}_* - \hat{\mathfrak{Q}}_k)(x_{k+1} - x_k)\|}{\|x_{k+1} - x_k\|} \right] \leqslant 1. \tag{4.10}$$

As in (3.9) and (3.10), there exists a $\tau_k \in (0, 1]$, satisfying (4.9) and hence (2.5). If σ in (4.5) is such that

$$\frac{1}{m\xi_{\bar{\alpha}}} \left(\eta_k^* + \frac{1}{2}\sigma \right) \leqslant 1 - \rho_1 \tag{4.11}$$

(in view of $\{x_k\} \to x_*, \eta_k^* \to 0$, this defines the number σ) then (4.10) holds with $\tau_k = 1$, and therefore, because $q_k(x_k^{\mathrm{p}}) \leqslant 0$, (2.5) is satisfied with $\tau_k = 1$.

If (4.5) cannot be achieved because $\mathfrak{Q}_*$ is not positive-definite then the projection of $\mathfrak{Q}_* - \hat{\mathfrak{Q}}_k$ can be used, for large k, by invoking (4.4) in (4.8) to yield

$$\mathscr{F}(x_{k+1}) - \mathscr{F}(x_k) \leqslant \tau_k q_k(x_k^{\mathrm{p}}) \left\{ 1 - \frac{\tau_k}{m\xi_{\bar{\alpha}}} \left[\eta_k^* + \frac{1}{2} \frac{\|P_k(\mathfrak{Q}_* - \hat{\mathfrak{Q}}_k)P_k(x_{k+1} - x_k)\|}{\|x_{k+1} - x_k\|} \right] \right\}. \tag{4.12}$$

Using (4.6) and the same arguments as before, we establish that (2.5) is satisfied with $\tau_k = 1$.

For stepsize (2.6), (3.7) with Lemma 3.1 yields

$$\mathscr{F}(x_{k+1}) - \mathscr{F}(x_k) \leqslant \tau_k \langle \nabla \mathscr{F}(x_k), x_k^{\mathrm{p}} - x_k \rangle$$
$$x \left\{ 1 - \frac{\bar{\alpha}\tau_k}{2} - \frac{\bar{\alpha}\tau_k}{m} \left[\eta_k^* + \frac{1}{2} \frac{\|(\mathfrak{Q}_* - \hat{\mathfrak{Q}}_k)(x_{k+1} - x_k)\|}{\|x_{k+1} - x_k\|} \right] \right\} \tag{4.13}$$

The scalar $\rho_2 \in (0, 1 - \frac{1}{2}\bar{\alpha})$ in (2.6) requires τ_k to satisfy

$$\rho_2 \leqslant 1 - \frac{\bar{\alpha}\tau_k}{2} - \frac{\bar{\alpha}\tau_k}{m} \left[\eta_k^* + \frac{1}{2} \frac{\|(\mathfrak{Q}_* - \hat{\mathfrak{Q}}_k)(x_{k+1} - x_k)\|}{\|x_{k+1} - x_k\|} \right] \leqslant 1 - \frac{\bar{\alpha}}{2}. \tag{4.14}$$

As in (3.11) and (3.12), there exists a $\tau_k \in (0, 1]$, satisfying (2.6). If σ in (4.5) is such that

$$\frac{\bar{\alpha}}{m} \left(\eta_k^* + \frac{1}{2}\sigma \right) \leqslant 1 - \frac{1}{2}\bar{\alpha} - \rho_1 \tag{4.15}$$

(since $\{x_k\} \to x_*, \eta_k^* \to 0$, this defines σ) then (4.14) holds with $\tau_k = 1$, and therefore, because $\langle \nabla \mathscr{F}(x_k), x_k^{\mathrm{p}} - x_k \rangle \leqslant 0$, (2.6) is satisfied with $\tau_k = 1$.

If (4.5) cannot be achieved because $\mathfrak{Q}_*$ is not positive-definite then the projection of $\mathfrak{Q}_* - \hat{\mathfrak{Q}}_k$ can be used, for large k, by invoking (4.4) to yield

$$\mathscr{F}(x_{k+1}) - \mathscr{F}(x_k) \leqslant \tau_k \langle \nabla \mathscr{F}(x_k), x_k^{\mathrm{p}} - x_k \rangle$$
$$\times \left\{ 1 - \frac{\bar{\alpha}\tau_k}{2} - \frac{\bar{\alpha}\tau_k}{m} \left[\eta_k^* + \frac{1}{2} \frac{\|P_k(\mathfrak{Q}_* - \hat{\mathfrak{Q}}_k)P_k(x_{k+1} - x_k)\|}{\|x_{k+1} - x_k\|} \right] \right\}. \tag{4.16}$$

Using (4.6) and the same arguments as before, we establish that (2.6) is satisfied with $\tau_k = 1$. □

Remark

The proof illustrates that it is 'easier' to attain $\tau_k = 1$ for smaller values of $\bar{\alpha}.\{\tau_k\}$ will accelerate towards unity as $\{\alpha_k\} \to 1$. Furthermore, if $\tau_k = 1$ while $\alpha_k > 1$, then reducing α_k to unity in subsequent iterations will increase $\xi_{\bar{\alpha}}$ to its largest value, $\xi_{\bar{\alpha}} = \frac{1}{2}$. Consider, for example, the effect of doing this in (4.11) or

(4.15). As α_k is reduced to unity, so can the bound $\bar{\alpha}$ be considered as reduced to unity. Thus, the bound on these inequalities and the bound on σ get more relaxed. Thus, conditions (4.5) and (4.6) become easier to satisfy, and this makes it easier to maintain $\tau_k = 1$.

The left sides of the bounds (4.5) and (4.6) are expected to approach zero if superlinear convergence is to be achieved. Theorem 4.2 below establishes these superlinear convergence conditions. In Chapter 8, this discussion is revisited and further generalized for quasi-Newton algorithms, following the earlier results of Dennis and Moré (1977), Han (1976) and Powell (1978a), among others.

Lemma 4.1
Let $\{x_k\} \to x_*$ *and* $\|x_* - x_{k+1}\| \leqslant \hat{e}\|x_* - x_k\|$, *for some* $\hat{e} \in [0, \infty)$. *Then,*

$$\|x_{k+1} - x_k\| \leqslant e_0 \|x_* - x_k\|, \tag{4.17}$$

where $e_0 = 1 + \hat{e}$.

Proof
The proof is immediate from the triangle inequality

$$\|x_{k+1} - x_* + x_* - x_k\| \leqslant \|x_{k+1} - x_*\| + \|x_* - x_k\| \leqslant (1 + \hat{e})\|x_k - x_*\|. \quad \square$$

Remark
The inequality $\|x_* - x_{k+1}\| \leqslant \hat{e}\|x_* - x_k\|$ is ensured whenever (4.17) is invoked below.

Lemma 4.2
Let $\{x_k\} \to x_*$. *Then* $\{x_k\}$ *is Q-superlinearly convergent, i.e.*

$$\|x_* - x_{k+1}\| \leqslant \beta_k^{(0)} \|x_* - x_k\|, \qquad \lim_{k \to \infty} \beta_k^{(0)} = 0, \tag{4.18}$$

if and only if $\|x_{k+1} - x_k\| \leqslant \beta_k \|x_k - x_{k-1}\|$, *with* $\lim_{k \to \infty} \beta_k = 0$

Proof
We have

$$\begin{aligned} \|x_* - x_k\| &\leqslant \lim_{t \to \infty} \sum_{j=k}^{t-1} \|x_{j+1} - x_j\| \\ &\leqslant \beta_k \|x_k - x_{k-1}\|(1 + \omega + \omega^2 + \omega^3 + \ldots) \\ &\leqslant \frac{\beta_k}{1 - \omega}(\|x_k - x_*\| + \|x_* - x_{k-1}\|) \end{aligned}$$

for some $\omega \in [0, 1)$ and k sufficiently large. As $\{\beta_k\} \to 0, \omega$ is chosen such that $1 > \beta_k + \omega, \forall k \geqslant K_0$. K_0 is an integer and is such that $\beta_k < 1, \forall k \geqslant K_0$. Rearranging the above expression yields the required result.

Suppose that $\|x_* - x_k\| \leqslant \beta_k^{(0)} \|x_* - x_{k-1}\|$, $\lim_{k\to\infty} \beta_k^{(0)} = 0$. The desired result is obtained using Lemma 2.1 and $\beta_k^{(0)} < 1$:

$$\begin{aligned}\|x_* - x_k\| &\leqslant \beta_k^{(0)} (\|x_* - x_k\| + \|x_k - x_{k-1}\|)\\ &\leqslant \frac{\beta_k^{(0)}}{1 - \beta_k^{(0)}} \|x_k - x_{k-1}\|.\end{aligned}$$

□

Theorem 4.2 (The two-sided projected Hessian condition for Q-superlinear convergence)
Let

(i) the assumptions of Theorem 3.3 be satisfied;

(ii) k be large, such that, by Theorem 4.1, $\alpha_k = \tau_k = 1$.

Then, the sequence $\{x_k\}$ generated by the GLP algorithm in Section 2 converges at a Q-superlinear rate if and only if the two-sided projected Hessian condition

$$\lim_{k\to\infty} \frac{\|P_k(\mathfrak{Q}_* - \hat{\mathfrak{Q}}_k)P_k(x_{k+1} - x_k)\|}{\|x_{k+1} - x_k\|} = 0 \tag{4.19}$$

is satisfied.

Proof
The first-order expansion of $\nabla\mathscr{F}(x)$ can be written using (4.4) as

$$\begin{aligned}\nabla\mathscr{F}(x_k) =& \nabla\mathscr{F}(x_{k-1}) + \hat{\mathfrak{Q}}_k(x_k - x_{k-1})\\ &+ \int_0^1 [\mathfrak{Q}(x_{k-1}(t)) - \hat{\mathfrak{Q}}_{k-1}]P_{k-1}(x_k - x_{k-1})\,\mathrm{d}t.\end{aligned} \tag{4.20}$$

The gradient of the objective function in (2.3) is given by the first two terms on the right in (4.20). Thus, for $x_{k+1} \in \mathscr{R}$, the inequality

$$\langle \nabla\mathscr{F}(x_{k-1}) + \hat{\mathfrak{Q}}_k(x_k - x_{k-1}), x_{k+1} - x_k\rangle = \langle q_{k-1}(x_k), x_{k+1} - x_k\rangle \geqslant 0 \tag{4.21}$$

follows from the optimality of x_k in (2.3) for

$$\min\left\{\frac{1}{2}\left\|x - x_{k-1} + \hat{\mathfrak{Q}}_{k-1}^{-1}\nabla\mathscr{F}(x_{k-1})\right\|^2_{\hat{\mathfrak{Q}}_{k-1}} \middle| x \in \mathscr{R}\right\}.$$

Also, the inequality

$$q_k(x_{k+1}) \geqslant \langle \nabla\mathscr{F}(x_k), x_{k+1} - x_k\rangle \tag{4.22}$$

follows from the positive-definiteness of $\hat{\mathfrak{Q}}_k$. For large k, using (2.4) with $\alpha_k = \tau_k = 1$, (4.20)–(4.22) and Lemma 3.3, we have

$$
\begin{aligned}
\frac{1}{2}m\|x_{k+1}-x_k\|^2 &\leqslant -q_k(x_{k+1})\\
&\leqslant -\langle\nabla\mathscr{F}(x_k), x_{k+1}-x_k\rangle\\
&\leqslant -\langle q_{k-1}(x_k), x_{k+1}-x_k\rangle\\
&\quad -\langle x_{k+1}-x_k, P_k(\mathfrak{Q}_*-\hat{\mathfrak{Q}}_{k-1})P_{k-1}(x_k-x_{k-1})\rangle\\
&\quad -\int_0^1\langle x_{k+1}-x_k, [\mathfrak{Q}((x_{k-1}(t))-\mathfrak{Q}_*](x_k-x_{k-1})\rangle\,\mathrm{d}t\\
&\leqslant\left[\bar{\eta}_k+\frac{\|P_k(\mathfrak{Q}_*-\hat{\mathfrak{Q}}_{k-1})P_{k-1}(x_k-x_{k-1})\|}{\|x_k-x_{k-1}\|}\right]\\
&\quad\times\|x_k-x_{k-1}\|\,\|x_{k+1}-x_k\| \qquad (4.23)
\end{aligned}
$$

where $\bar{\eta}_k=\int_0^1\|\mathfrak{Q}((x_{k-1}(t))-\mathfrak{Q}_*\|\,\mathrm{d}t$, and, as $\{x_k\}\to x_*, \{\bar{\eta}_k\}\to 0$. Hence, (4.23) yields

$$\|x_{k+1}-x_k\|\leqslant\beta_k\|x_k-x_{k-1}\|,$$

where

$$
\begin{aligned}
\beta_k=\frac{2}{m}\Bigg[\bar{\eta}_k&+\frac{\|P_{k-1}(\mathfrak{Q}_*-\hat{\mathfrak{Q}}_{k-1})P_{k-1}(x_k-x_{k-1})\|}{\|x_k-x_{k-1}\|}\\
&+\frac{\|(P_k-P_{k-1})(\mathfrak{Q}_*-\hat{\mathfrak{Q}}_{k-1})P_{k-1}(x_k-x_{k-1})\|}{\|x_k-x_{k-1}\|}\Bigg].
\end{aligned}
$$

If (4.19) is satisfied then $\lim_{k\to\infty}\beta_k=0$, and Lemma 4.2 yields the desired result.

Suppose, conversely, that $\{x_k\}$ converges Q-superlinearly, and thence, by Lemma 4.2,

$$\|x_{k+1}-x_k\|\leqslant\beta_k\|x_k-x_{k-1}\|,\qquad \lim_{k\to\infty}\beta_k=0.$$

By (4.23), we have (4.19). □

Powell (1978a) discusses condition (4.19) in relation to sequential quadratic programming algorithms that do not require $x_k\in\mathscr{R}, \forall k$. We discuss (4.19) further in Chapter 8, and show that it implies two-step Q-superlinear convergence for a general nonlinearly constrained optimisation algorithm.

The feasibility requirement in (2.3) is responsible for the stronger Q-superlinear convergence rate result in Theorem 4.2. In addition, we are able to show that (4.19) is a *necessary and sufficient* condition for Q-superlinear convergence of the GLP algorithm.

5 COMMENTS AND NOTES

CN.1 Convergence rates

Convergence rates indicate the speed of convergence to the solution. A general discussion is given in CN.4 below. In the case of the algorithm of this chapter, this is discussed in Section 4.

CN.2 Notes on convergence

A sequence of vectors $x_0, x_1, \ldots, x_k, \ldots$, in $\mathcal{S} \subseteq \mathbb{R}^n$, denoted by $\{x_k\}$, is said to *converge* to the limit $x_* \in \mathcal{S}$ if

$$\lim_{k\to\infty} \|x_k - x_*\| = 0$$

(that is, given $\epsilon > 0$, there is an N such that $k \geqslant N$ implies $\|x_k - x_*\| < \epsilon$). If $\{x_k\}$ converges to x, we write $\{x_k\} \to x_*$.

An *open ball* around a point x is a set of the form

$$\{y \in \mathbb{R}^n \mid \|y - x\| < \epsilon\}$$

for some $\epsilon > 0$. A subset $\mathcal{S} \subseteq \mathbb{R}^n$ is *open* if around every point $x \in \mathcal{S}$, there is an open ball contained in $\mathcal{S}$ (i.e. for $x \in \mathcal{S}$, there is an $\epsilon > 0$ such that $\|y - x\| < \epsilon$, with $y \in \mathcal{S}$). For example, the ball $\{y \in R^n \mid \|y\| < \epsilon\}$ is open. The *interior* of any set $\mathcal{S} \subseteq \mathbb{R}^n$ is the set of points $x \in \mathcal{S}$ that are at the centre of some ball contained in $\mathcal{S}$. A set is said to be *open* if all its points are interior points. Thus, the interior of a set is always open. The interior of the ball

$$\{y \in \mathbb{R}^n \mid \|y\| \leqslant \epsilon\}$$

is the open ball

$$\{y \in \mathbb{R}^n \mid \|y\| < \epsilon\}.$$

A point x is a *limit* of $\{x_k\}$ if there is a subsequence of $\{x_k\}$ convergent to x. The subsequence is denoted in terms of the subset $\mathcal{K}$ of the positive integers such that $\{x_k\} \to x$ for $k \in \mathcal{K}$. A related concept is an accumulation point: if $\mathcal{S} \subseteq \mathbb{R}^n$ and $x \in \mathbb{R}^n$ then x is called an *accumulation point* of $\mathcal{S}$ if every ball around x contains at least one point of $\mathcal{S}$ distinct from x. For example, the set of numbers $1/n, n = 1, 2, 3, \ldots$, has zero as an accumulation point.

A set $\mathcal{S}$ is *closed* if its complement $\mathbb{R}^n - \mathcal{S}$ is open. Alternatively, a set $\mathcal{S}$ is *closed* if every point arbitrarily close to $\mathcal{S}$ is a member of $\mathcal{S}$ of (i.e. $\mathcal{S}$ is closed if $\{x_k\} \in \mathcal{S}$ and $\{x_k\} \to x$ imply that $x \in \mathcal{S}$). For example, the ball $\{y \in \mathbb{R}^n \mid \|y\| \leqslant \epsilon\}$ is closed. The *closure* of any set $\mathcal{S} \subseteq \mathbb{R}^n$ is the smallest closed set containing $\mathcal{S}$. The *boundary* of a set is that part of the set that is not its interior.

A set is *bounded* if it lies entirely within a ball of some radius $\epsilon > 0$,

$$\{y \in \mathbb{R}^n \mid \|y - x\| \leqslant \epsilon\}.$$

A set is *compact* if and only if it is closed and bounded.

For a compact set $\mathscr{S}$, the Bolzano–Weierstrass theorem (Apostol, 1981) can be used to show that every infinite subset of $\mathscr{S}$ has an accumulation point in $\mathscr{S}$. Thus, if $\{x_k\} \subset \mathscr{S}$ (i.e. each member of the sequence is in $\mathscr{S}$,) then $\{x_k\}$ has a *limit point in* $\mathscr{S}$. This establishes that there is a subsequence of $\{x_k\}$ converging to a point in $\mathscr{S}$. (For example, the sequence defined by $x_{k+1} = -x_k$, $x_0 = 1$, has a subsequence $\{x_{2k}\}$ with limit point 1 and a subsequence $\{x_{2k+1}\}$ with limit point -1. Both sequences and their limits are clearly in the compact set $\{x \mid \| x \| \leqslant 1\}$.)

The convergence of the algorithms in this chapter and in Chapters 7–9 and 12 depends on the Bolzano–Weierstrass theorem. We are concerned with algorithms that aim to solve for a fixed optimal point while generating a sequence $\{x_k\}$ within a compact set $\mathscr{S}$ such that a certain descent property is satisfied for a merit function (e.g. $f(x_{k+1}) < f(x_k)$ and for x_* solving the problem, $f(x_*) \leqslant f(x_k), \forall k$). There need not be a unique accumulation point for each subsequence. However, each accumulation point would need to satisfy the optimality condition of the problem.

CN.3 Bounded multipliers

Alternatively, instead of the linear independence condition, it can be assumed that the multipliers associated with (2.3) at x_k^{P} are bounded (Fiacco and McCormick, 1968).

CN.4 Q-rates of convergence of a sequence $\{x_k\} \rightarrow x_*$

If there exists a constant $\gamma \in [0, 1)$, and an integer $\hat{k} \geqslant 0$ such that, $\forall k \geqslant \hat{k}$,

$$\|x_* - x_{k+1}\| \leqslant \gamma \|x_* - x_k\|$$

then $\{x_k\}$ is said to be convergent to x_* at a *Q-linear rate*. If for some sequence $\{\gamma_k\}$,

$$\lim_{k \to \infty} \gamma_k = 0$$

and

$$\|x_* - x_{k+1}\| \leqslant \gamma_k \|x_* - x_k\|,$$

then $\{x_k\}$ is said to be convergent to x_* at a *Q-superlinear rate*. If, instead of this rate, for some integer j, we have

$$\|x_* - x_{k+j}\| \leqslant \gamma_k \|x_* - x_k\|$$

then $\{x_j\}$ is said to be convergent to x_* at a *j-step Q-superlinear rate*. If there exist constant values $i > 1, \gamma \geqslant 0$ and an integer $\hat{k} \geqslant 0$ such that, $\forall k \geqslant \hat{k}$,

$$\|x_* - x_{k+1}\| \leqslant \gamma \|x_* - x_k\|^i$$

then $\{x_k\}$ is said to be converge to x_* *with a Q-order at least i*. For $i = 2$, the rate is said to be *Q-quadratic*. An authoritative discussion of general rates of convergence is given by Ortega and Rheinboldt (1970, Chap. 9).

CN.5 A mean value theorem for $f : \mathbb{L} \subset \mathbb{R}^n \to \mathbb{R}^m, m > 1$

We note that mean value theorems usually apply to mappings $f : \mathbb{F} \subset \mathbb{R}^n \to \mathbb{R}^1$, and do not hold in general for mappings $f : \mathbb{F} \subset \mathbb{R}^n \to \mathbb{R}^m, m > 1$. We construct the result by treating each element of the vector individually (see Ortega and Rheinboldt, 1970, Theorem 3.2.2).

REFERENCES

Allwright, J.C. (1980). A feasible direction algorithm for convex optimization: global convergence rates. *J. Optim. Theory Applic.* **28**, 1–18.

Apostol, T.M. (1981). *Mathematical Analysis*, 2nd edn. Addison-Wesley, Reading, MA.

Bertsekas, D.P. (1976). On the Goldstein–Levitin–Polyak gradient projection method. *IEEE Trans. Autom. Control* **AC-21**, 174–183.

Bertsekas, D.P. (1979). Convexification procedures and decomposition methods for non-convex optimisation problems. *J. Optim. Theory Applic.* **29**, 169–197.

Bertsekas, D.P. and P. Tseng (1994). Partial proximal minimisation algorithms for convex programming. *SIAM J Optim.* **4**, 551–572.

Bertsekas, D. and J.N. Tsitsiklis (1989). *Parallel and Distributed Computation—Numerical Results*. Prentice-Hall, New York.

Broyden C.G. (1969). A new method for solving nonlinear simultaneous equations. *Comput. J.* **12**, 95–100.

Broyden C.G. (1970). The convergence of a class of double-rank minimisation algorithms 2. The new algorithm. *J. Inst. Math. Applic.* **6**, 222–231.

Darlington, J., C.C. Pantelides, B. Rustem and B.A. Tanyi (1996). An efficient approximate algorithm for optimal robust decisions under uncertainty. Department of Computing, Imperial College.

Demyanov, V.F. and A.M. Rubinov (1971). *Approximate Methods in Optimization Problems*. Elsevier, New York.

Dennis, J.E. and J.J. Moré (1977). Quasi-Newton methods, motivation and theory. *SIAM Rev.* **19**, 46–89.

Dunn, J.C. (1979). Rates of convergence for conditional gradient algorithms near singular and nonsingular extremals. *SIAM J Control Optim.* **17**, 187–211.

Dunn, J.C. (1980). Newton's method and the Goldstein step-length rule for constrained minimization problems. *SIAM J Control Optim.* **18**, 659–674.

Dunn, J.C. (1981). Global and asymptotic convergence rate estimates for a class of projected gradient processes. *SIAM J Control Optim.* **19**, 368–400.

Fiacco, A.V. and G.P. McCormick (1968). *Nonlinear Programming: Sequential Unconstrained Minimization Techniques*. Wiley, New York.

Fletcher, R. (1970). A new approach to variable metric algorithms. *Comput. J.* **13**, 317–322.

Gafni, E. and D. Bertsekas (1984). Two-metric projection methods for constrained optimization. *SIAM J. Control Optim.* **22**, 936–964.

Goldfarb D. (1970). A family of variable metric algorithms derived by variational means. *Math. Comput.* **24**, 23–26.

Goldstein, A.A. (1964). Convex programming in Hilbert space. *Bull. Am. Math. Soc.* **70**, 709–710.

Han, S.-P. (1976). Superlinearly convergent variable metric algorithms for general nonlinear programming problems. *Math. Programming* **11**, 263–282.

Levitin, E.S. and B.T. Polyak (1966). Constrained minimization methods. *USSR Comput. Math. Math. Phys.* **6**, 1–50.

McCormick, G.P. and R.A. Tapia (1972). The gradient projection method under mild differentiability conditions. *SIAM J. Control Optim.* **10**, 93–98.

Ortega, J.M. and W.C. Rheinboldt (1970). *Iterative Solution of Nonlinear Equations in Several Variables.* Academic Press, London.

Polak, E. (1971). *Computational Methods in Optimization.* Academic Press, London.

Powell, M.J.D. (1978a). The convergence of variable metric methods for nonlinearly constrained optimization calculations. In *Nonlinear Programming* (J.B. Rosen, O.L. Mangasarian, K. Ritter, eds), pp. 27–63. Academic Press, New York.

Powell, M.J.D. (1978b). A fast algorithm for nonlinearly constrained optimization calculations. In *Numerical Analysis Proceedings, Dundee 1977* (G.A. Watson, ed.). Springer-Verlag, Berlin.

Psenichny, B.N. and Y.M. Danilin (1978). *Numerical Methods in Extremal Problems.* Mir, Moscow.

Rockafeller, R.T. (1976). Monotone operators and the proximal point algorithm. *SIAM J. Control Optim.* **14**, 877–898.

Rustem, B. (1984). A class of superlinearly convergent projection algorithms with relaxed stepsizes. *Appl. Math. Optim.* **12**, 29–43.

Shanno, D.F. (1970). Conditioning of quasi-Newton methods for function minimization. *Math. Comput.* **24**, 647–654.

CHAPTER 7

Nonlinear Optimization with Equality and Inequality Constraints

The approach to nonlinear optimization in Chapter 6 is restricted to convex feasible regions. A powerful approach to optimization with general nonlinear objective functions and nonlinear constraints is provided by the sequential quadratic programming algorithm. Assuming that all functions are differentiable—twice in theory, only once computationally—the algorithm constructs quadratic programming approximations to the original problem and solves these iteratively until convergence is attained. Each quadratic programming solution yields a search direction along which the current iterate can be improved. A step is taken along the search direction, according to a stepsize strategy, that ensures sufficient improvement in that iteration.

In this chapter, we consider mainly two algorithms for nonlinear equality- and inequality-constrained optimization. Both algorithms utilize stepsize strategies based on *differentiable penalty functions* and quadratic programming subproblems. The essential difference between the algorithms is in the stepsize strategies used. The objective function in the quadratic subproblems includes a linear term that is dependent on the penalty functions. The quadratic term in this objective function utilizes an approximate Hessian of the Lagrangian augmented by the penalty functions. In this approximation, it is possible to ignore the second-derivative terms arising from the constraints in the penalty functions.

The penalty parameter is determined using a strategy—slightly different for each algorithm—that ensures boundedness as well as a descent property. In particular, the boundedness follows since the strategy is always satisfied for finite values of the parameter.

These properties ensure global convergence and the condition under which unit stepsizes are achieved. There is also a compatibility between the quadratic objective function and the stepsize strategy. This maintains the consistency of the properties for unit steps and subsequent convergence rates studied in Chapter 8.

A variant of the algorithm that ensures the satisfaction of the inequality constraints with a barrier function is also discussed. The barrier function ensures that the algorithm operates within the interior of the feasible region of the inequality constraints. This is a simple interior point variant that overcomes the necessity of solving inequality constrained subproblem at every iteration.

1 NONLINEAR PROGRAMMING: AUGMENTED LAGRANGIAN SQP ALGORITHM

In this chapter, we consider algorithms for solving the equality- and inequality-constrained optimization problem

$$\min \{f(x) \mid g(x) = 0; h(x) \leqslant 0\}, \tag{1.1}$$

where $f : \mathbb{R}^n \to \mathbb{R}^1, g : \mathbb{R}^n \to \mathbb{R}^e, h : \mathbb{R}^n \to \mathbb{R}^i (e, i \leqslant n)$. We consider an extension of the well-known penalty $\|g(x)\|_2^2$ to ensure feasibility and adopt an augmented Lagrangian *sequential quadratic programming* (SQP) framework (Rustem, 1986, 1993). The penalty is adapted to ensure progress towards the solution. This chapter provides a detailed study of the inequality-constrained case. The penalty parameter, which is often a source of numerical problems, is shown not to increase indefinitely.

Associated with (1.1), we consider the Lagrangian L: $\mathbb{R}^n \times \mathbb{R}^e \times \mathbb{R}^i_+ \to \mathbb{R}$:

$$L(x, \lambda, \mu) = f(x) + \langle g(x), \lambda \rangle + \langle h(x), \mu \rangle. \tag{1.2}$$

A local solution x_* of (1.1) and the corresponding multipliers λ_*, μ_* are such that $\nabla_x L(x_*, \lambda_*, \mu_*) = 0$ and $\mu_* \geqslant 0$. We also define the penalty function by l: $\mathbb{R}^n \times \mathbb{R}_+ \times \mathbb{R}^i_+ \to \mathbb{R}$:

$$l(x, c, \pi) \equiv f(x) + \frac{c}{2} \|g(x)\|_2^2 + \frac{1}{2c} \left\| [ch(x) + \pi]_+ \right\|_2^2, \tag{1.3}$$

where $0 \leqslant c \in \mathbb{R}_+$. $0 \leqslant \pi \in \mathbb{R}^1_+$ and $(\cdot)_+$ is such that its jth element,

$$(\cdot)^j_+ = \max\{(\cdot)^j, 0\}$$

The augmented Lagrangian for (1.1), $L^a : \mathbb{R}^n \times \mathbb{R}^e \times \mathbb{R}^i_+ \times \mathbb{R}_+ \times \mathbb{R}^i_+ \to \mathbb{R}$, is given by

$$L^{\mathrm{a}}(x, \lambda, \mu, c, \pi) = l(x, c, \pi) + \langle g(x), \lambda \rangle + \langle h(x), \mu \rangle. \tag{1.4}$$

1.1 Convexification

The augmented Lagrangian ensures that the local convexity assumed to establish the optimality conditions in Chapter 1 can, to some extent, be extended beyond this locality. This convexification is based on results established in Arrow, Gould and Howe (1973) and Rockafeller (1973). We establish the following result due to Finsler (1937) (see also Arrow, Gould and Howe, 1973; Bellman, 1995) while taking account of the difference in the penalty function used in (1.3).

Lemma 1.1 (Finsler's lemma)
Let Q be a real $n \times n$ matrix and let J^{T} be a real $n \times m$ matrix. Suppose $\langle x, Qx \rangle > 0$ for all $x \neq 0$ such that $J^{\mathrm{T}} x = 0$. Then, for all $x \neq 0$ and for c sufficiently large, we have

$$\langle x . (Q + cJJ^{\mathrm{T}}) \times \rangle > 0.$$

That is, if Q is positive-definite on the null space of J^{T} then, for c sufficiently large, $Q + cJJ^{\mathrm{T}}$ is positive-definite on the whole space.

Proof
Let $K = \{x | \langle x, x \rangle = 1\}$ and $\langle J^T x, J^T x \rangle \geqslant 0$, for all $x \in K$. Then, we shall show that the following two statements are equivalent in K:

(a) $\langle x, Qx \rangle > 0$ whenever $\langle J^T x, J^T x \rangle = 0$;

(b) for all c sufficiently large, $\langle x, Qx \rangle + c\langle J^T x, J^T x \rangle > 0$, for all x.

By assumption, (a) is satisfied. Hence, from (b), for all c sufficiently large,

$$\langle x, (Q + cJJ^T)x \rangle > 0$$

for $\langle x, x \rangle = 1$, and hence whenever $x \neq 0$.

To establish the equivalence of (a) and (b), we note that (b) trivially implies (a). We consider the converse. Let K' be a subset of K on which $\langle x, Qx \rangle \leqslant 0$. K' is compact. Since

$$\langle J^T x, J^T x \rangle \geqslant 0$$

on K, we have

$$\langle x, Qx \rangle + c\langle J^T x, J^T x \rangle \geqslant \langle x, Qx \rangle > 0, \quad \forall x \in K - K' \text{ and } \forall c \geqslant 0.$$

If $K' = \emptyset$, the proof would be finished. If $K' \neq \emptyset$, let

$$\zeta_Q = \min \{\langle x, Qx \rangle \mid x \in K'\},$$
$$\zeta_J = \min \{\langle J^T x, J^T x \rangle \mid x \in K'\}$$

and $\zeta_J = \langle J^T x', J^T x' \rangle$, for some $x' \in K'$. We have $\langle J^T x, J^T x \rangle \geqslant 0, \forall x \in K'$. But if

$$\langle J^T x', J^T x' \rangle = 0$$

then, by (a) $\langle x', Qx' \rangle > 0$, contrary to the definition of K'. Hence,

$$\zeta_J = \langle J^T x', J^T x' \rangle > 0,$$

and clearly, for all c sufficiently large, we have

$$\zeta_Q + c\zeta_J > 0.$$

Then, $\forall x \in K'$, by definition of a minimum, the inequality

$$\langle x, Qx \rangle + c\langle J^T x, J^T x \rangle \geqslant \zeta_Q + c\zeta_J > 0$$

follows for all c sufficiently large. We have thus established the result for $x \in K - K'$ and $x \in K'$. □

Let the jth element of the vector π be defined by

$$\pi^j(x) = \begin{cases} -ch^j(x) & \text{if} \quad h^j(x) \leqslant 0, \\ \kappa & \text{if} \quad h^j(x) > 0, \end{cases}$$

where $\kappa \geqslant 0$ is a constant and $c \geqslant 0$. We note that Algorithms 2.1 and 2.2 below iteratively adjust π, such that, at convergence to the solution x_*, we have $\pi_* = -ch(x_*)$.

Theorem 1.1

Suppose

(i) *π is determined as above, with $\pi_* = -ch(x_*)$;*

(ii) *x_*, λ_*, μ_* satisfy the second-order sufficiency conditions of Theorem 1.3.3 for x_* to be a strict local solution to (1.1);*

(iii) *strict complementarity holds;*

(iv) *c is sufficiently large; and*

(v) *for all i and j, the expressions*

$$\frac{\partial^2\left(\frac{1}{2c}\left\|(ch(x)+\pi_*)_+\right\|_2^2\right)}{\partial(h^i)^2} \quad \text{and} \quad \frac{\partial^2\left(\frac{c}{2}\|g(x)\|_2^2\right)}{\partial(g^j)^2}$$

are continuous at the points $(h^i(x_*), c)$ and *$(g^j(x_*), c)$ respectively, where* $h^i(x_*) \leqslant 0$ and $g^j(x_*) = 0$.

Then, the function $L^{\mathrm{a}}(x, \lambda_*, \mu_*, c, \pi_*)$ has an unconstrained strict local minimum at x_*; and, for all (λ, μ) sufficiently close to (λ_*, μ_*), $L^{\mathrm{a}}(x, \lambda, \mu, c, \pi)$ is strictly convex in x close to x_*.

Proof

For any $c > 0$, we have, by the first-order necessary conditions of optimality of (1.1) (see Section 1.3),

$$\begin{aligned}\nabla L^{\mathrm{a}}(x_*, \lambda_*, \mu_*, c, \pi_*) &= \nabla f(x_*) + \nabla g(x_*)[cg(x_*)+\lambda_*] + \nabla h(x_*)[(ch(x_*)+\pi_*)_+ + \mu_*] \\ &= \nabla f(x_*) + \nabla g(x_*)\lambda_* + \nabla h(x_*)\mu_* \\ &= \nabla L(x_*, \lambda_*, \mu_*) \\ &= 0.\end{aligned}$$

The Hessian of L^{a} is given by

$$\begin{aligned}\nabla^2 L^{\mathrm{a}}(x_*, \lambda_*, \mu_*, c, \pi_*) &= \nabla^2 f(x_*) + c\left\{\sum_{j=1}^{e} \nabla^2 g^j(x_*) g^j(x_*) + \nabla g(x_*)\nabla g^{\mathrm{T}}(x_*)\right\} \\ &\quad + \sum_{j=1}^{e} \nabla^2 g^j(x_*)\lambda_*^j + \sum_{i=1}^{i} \nabla^2 h^i(x_*)\left(ch^i(x_*) + \pi_*^i\right)_+ \\ &\quad + c\nabla h(x_*)\nabla h^{\mathrm{T}}(x_*) + \sum_{i=1}^{i} \nabla^2 h^i(x_*)\mu_*^i \\ &= \nabla^2 L(x_*, \lambda_*, \mu_*) + c[\nabla g(x_*)\nabla g^{\mathrm{T}}(x_*) + \nabla h(x_*)\nabla h^{\mathrm{T}}(x_*)].\end{aligned}$$

Defining the index set

$$\mathscr{I}_0 = \{i \mid h^i(x_*) = 0, \mu^i > 0\}$$

and the Jacobian matrix of active constraints

$$J = [\nabla g^1(x_*) \vdots \nabla g^2(x_*) \vdots \ldots \vdots \nabla g^e(x_*) \vdots \nabla h^i(x_*), \ldots; i \in \mathcal{I}_0],$$

by Theorem 1.3.3 we have

$$\langle v, \nabla^2 L(x_*, \lambda_*, \mu_*) v \rangle > 0, \quad \forall v \neq 0, \quad J^{\mathrm{T}} v = 0.$$

By Lemma 1.1, $\nabla^2 L(x_*, \lambda_*, \mu_*) + \sigma J J^{\mathrm{T}}$ is positive-definite for some scalar σ sufficiently large. Now, if c is sufficiently large, we have $c > \sigma$, and

$$\begin{aligned}\nabla^2 L^{\mathrm{a}}(x_*, \lambda_*, \mu_*, c, \pi_*) =& \nabla^2 L(x_*, \lambda_*, \mu_*) + \sigma J J^{\mathrm{T}} \\ &+ \{c[\nabla g(x_*) \nabla g^{\mathrm{T}}(x_*) + \nabla h(x_*) \nabla h^{\mathrm{T}}(x_*)] - \sigma J J^{\mathrm{T}}\}\end{aligned}$$

is positive definite, since

$$\begin{aligned}\langle v, [\nabla^2 L(x_*, \lambda_*, \mu_*) + \sigma J J^{\mathrm{T}} + \{c[\nabla g(x_*) \nabla g^{\mathrm{T}}(x_*) + \nabla h(x_*) \nabla h^{\mathrm{T}}(x_*)] - \sigma J J^{\mathrm{T}}\}] v \rangle \\ \geqslant \langle v, [\nabla^2 L(x_*, \lambda_*, \mu_*) + \sigma J J^{\mathrm{T}}] v \rangle \\ > 0.\end{aligned}$$

The inequality constraints, not in $\mathcal{I}_0$, add a nonnegative term to the right-hand side of the above inequality. By continuity, $\nabla^2 L^{\mathrm{a}}(x, \lambda, \mu, c, \pi)$ is positive-definite for (x, λ, μ) sufficiently close to (x_*, λ_*, μ_*). □

The next theorem is a saddle-point result, which is largely a consequence of Theorem 1.1 and the discussion in Section 1.3. It is used in later chapters as well.

Theorem 1.2
Let the conditions of Theorem 1.1 be satisfied. Then, we have the saddle point

$$L^{\mathrm{a}}(x, \lambda_*, \mu_*, c, \pi_*) \geqslant L^{\mathrm{a}}(x_*, \lambda_*, \mu_*, c, \pi_*) \geqslant L^{\mathrm{a}}(x_*, \lambda, \mu, c, \pi_*)$$

for every x in some neighbourhood of x_* and every point $(\lambda, \mu) \in \mathbb{R}^e \times \mathbb{R}^i_+$. Furthermore, for all (λ, μ) sufficiently close to (λ_*, μ_*), $L^{\mathrm{a}}(x, \lambda, \mu, c, \pi)$ is strictly convex in x sufficiently close to x_*, and, for all $x \in \mathbb{R}^n$, $L^{\mathrm{a}}(x, \lambda, \mu, c, \pi)$ is strictly concave in $(\lambda, \mu) \in \mathbb{R}^e \times \mathbb{R}^i_+$.

Proof
The left inequality follows directly from Theorem 1.1. Also, the way π is determined ensures $\pi_* = -ch(x_*)$. The right inequality follows from the fact that $g(x_*) = 0$ and $h(x_*) \leqslant 0$ and any deviation of the multipliers at their optimal value will result in a value of $L^{\mathrm{a}}(x_*, \lambda, \mu, c, \pi_*)$ that is either the same as, or inferior to (i.e. less than), $L^{\mathrm{a}}(x_*, \lambda_*, \mu_*, c, \pi_*)$. □

1.2 The quadratic programming subproblem

It should be noted that neither the augmented Lagrangian nor the penalty function (1.3) possess continuous Hessians. In order to overcome the difficulties arising from this, we shall, at appropriate junctures, consider the effect of replacing the Hessian of (1.3) by that of

$$\bar{L}^{\mathrm{a}}(x, \lambda, \mu, c, \pi) = \bar{l}(x, c, \pi) + \langle g(x), \lambda \rangle + \langle h(x), \mu \rangle, \tag{1.5}$$

where

$$\bar{l}(x, c, \pi) = f(x) + \frac{c}{2}\|g(x)\|_2^2 + \frac{1}{2c}\left\|ch(x) + \pi\right\|_2^2. \tag{1.6}$$

$\bar{l}$ and l differ only in their last terms, corresponding to the inequality constraints. In contrast to (1.4), the Hessian of $\bar{L}^{\mathrm{a}}$ exists, provided f, g and h are suitably differentiable. The role of π in replacing the Hessian of (1.3) by that of (1.5) is crucial, and resembles the *smearing* procedure adopted by Polak and Tits (1980). The difference is in the choice of π and c. It is shown below that this choice allows the treatment of the active constraints *locally* as equality constraints.

The SQP quasi-Newton algorithm considered in this chapter involves the basic iteration

$$x_{k+1} = x_k + \tau_k d_k, \tag{1.7}$$

where τ_k is the stepsize factor and d_k solves the *quadratic programming subproblem* (QPS) (see CN.1)

$$\min \{q(d, c_k, \pi_k) \mid \nabla g_k^{\mathrm{T}} d + g_k = 0, \quad \nabla h_k^{\mathrm{T}} d + h_k \leqslant 0\}; \tag{1.8}$$

the multipliers of this QPS, corresponding to d_k, are denoted by λ_{k+1}, μ_{k+1}. The objective function is given by

$$q(d, c, \pi_k) = \langle d, \nabla l(x_k, c, \pi_k)\rangle + \frac{1}{2}\langle d, \hat{\mathfrak{Q}}_k d\rangle, \tag{1.9}$$

with $\nabla l = \nabla_x l. \hat{\mathfrak{Q}}_k$ is a quasi-Newton approximation to the Hessian $\mathfrak{Q}_k = \nabla_x^2 \bar{L}^{\mathrm{a}}(x_k, \lambda_k, \mu_k, c_k, \pi_k)$. $\hat{\mathfrak{Q}}_k$ is clearly different from the Hessian of the ordinary Lagrangian (1.2). Furthermore, it is possible to exclude from $\hat{\mathfrak{Q}}$ the second-derivative terms arising from the constraints in the penalty term (Tapia, 1986). Since this term vanishes at a feasible solution, all subsequent results are applicable if it is ignored in $\hat{\mathfrak{Q}}$.

The determination of the stepsize τ_k is an important factor in the performance of quasi-Newton algorithms (Dixon, 1972). In Section 2, we discuss two strategies for computing τ_k along with the rules for determining π_k and c_k. The following assumptions are invoked throughout the chapter.

Assumptions

(1) $f, g, h \in \mathbb{C}^2(\mathbb{R}^n)$;

(2) $\exists x_*, \lambda_*, \mu_*$ satisfying the first-order conditions of (1.1);

(3) the QPS (1.8) has a feasible solution at each iteration.

The quadratic objective function (1.9) includes ∇l instead of the more common ∇f only (see e.g. Han, 1977; Powell, 1978, 1983; Powell and Yuan, 1986). The reason for this is somewhat convoluted. As discussed in Theorem 4.1 below, the condition for $\{\tau_k\} \rightarrow 1$ is dependent on the accuracy of $\hat{\mathfrak{Q}}_k$.[†] In order to establish the convergence rate results with this $\hat{\mathfrak{Q}}_k, \nabla l$ is required in (1.9). Thus, ∇l is incorporated in order to ensure the *consistency* of the algorithm when τ_k is approaching unity, with the subsequent convergence rate studied in Chapter 8.

The convergence of τ_k to unity can also be shown if ∇l is replaced by ∇f in the QPS. Once $\{\tau_k\} \rightarrow 1$ is established, the superlinear convergence rates of both algorithms are shown in Chapter 8 to depend on the accuracy of the approximate Hessian used in the QPS. The latter is equally applicable if $\hat{\mathfrak{Q}}_k$ is chosen to approximate the Hessian of the ordinary (1.2) or the augmented (1.4) Lagrangian, with or without the above second-order terms. However, to achieve $\{\tau_k\} \rightarrow 1$, the algorithm uses $\hat{\mathfrak{Q}}_k$ as an approximation of the augmented Lagrangian Hessian and, for this $\hat{\mathfrak{Q}}_k, \nabla l$ is used in (1.9). The inclusion of the equality constraint penalty term in the quadratic objective function has also been considered, for different reasons, elsewhere in the literature (e.g. Tapia, 1977, 1986).

SQP algorithms have proved to be effective in solving problem (1.1) (see e.g. Fan, Sarkar and Lasdon, 1988). Following the original suggestion by Wilson (1963), several variants have been proposed. One common feature of all these algorithms is the formulation of a QPS whose objective function is based on a linear term depending on ∇f_k and a quadratic term depending on a quasi-Newton approximation of the Hessian of the Lagrangian (1.2). A second common feature among SQP algorithms is a strategy that determines the stepsize τ_k. However, it is essentially in the design of this strategy that most algorithms seem to differ from each other. Specifically, in Bartholomew-Biggs (1974, 1978, 1981, 1987, 1988) an equality-constrained QPS is formulated and the stepsize strategy is based on a penalty function that, in the case of inequality constraints, is differentiable only once. The penalty parameter updating strategy of Bartholomew-Biggs is somewhat different from the one adopted in this chapter. For the equality-constrained case, the relationship of both strategies is discussed in Rustem (1986). Han (1977), Mayne and Polak (1982) and Powell (1978) consider a stepsize based on a nondifferentiable penalty function. Mayne and Polak (1982) also have an adjustment strategy for the penalty parameter, which is shown not to grow indefinitely. Polak and Mayne (1981) consider two stepsize strategies: one to ensure the decrease in the objective function and another, nondifferentiable strategy, to ensure feasibility.

[†]With or without the second-order terms arising from the constraints in the penalty term.

Gabay (1982) considers, for the equality-constrained problem, a reduced-order algorithm that uses an equality-constrained QPS based on the reduced Hessian of the Lagrangian (1.2) and two stepsize strategies: one based on a nondifferentiable penalty function and another based on a norm condition to ensure improved feasibility and subsequent Q-superlinear convergence rate. The reduced order is obtained by considering a null-space approach, similar to the concept discussed in Chapter 2. Schittkowski (1981a, b, 1983) suggests the solution of an inequality-constrained QPS and a stepsize based on an augmented Lagrangian function. Gill *et al.* (1986) discuss a stepsize strategy based on an augmented Lagrangian, and demonstrate $\{\tau_k\} \to 1$. In this approach, the penalty parameters c and π are determined using different criteria than those discussed below. In addition, the stepsize strategy used alters x, λ, μ, π instead of just x and π, as discussed below. Finally, Burke and Han (1989) discuss an algorithm that allows the possibility for the QPS to be infeasible. A stepsize strategy and a penalty parameter updating scheme are utilized to ensure convergence. Both the stepsize and penalty parameter determination schemes are somewhat different from the approach adopted below. The latter extends the approach discussed in Rustem (1984, 1986) to ensure the attainment of unit stepsizes and superlinear convergence.

For some of the stepsize strategies above, it has not been possible to establish $\{\tau_k\} \to 1$. Local convergence rate results have been obtained *assuming* that there exists a stage (i.e. $K \geqslant 0$) after which, due to attraction towards the solution, $\tau_k = 1, \forall k \geqslant K$ (see e.g. Bartholomew-Biggs, 1978; Boggs, Tolle and Wang, 1982; Chiang and Tapia 1989; Coleman and Conn, 1984; Fontecilla, Steihaug and Tapia, 1987; Han, 1976; Palomares and Mangasarian, 1976; Powell, 1978; Stoer and Tapia, 1986). However, Chamberlain (1979) and Maratos (1978) have provided counterexamples illustrating that strategies similar to those proposed by Han (1977) and Powell (1978) may not converge to unity. In order to force convergence in such situations, Chamberlain *et al.* (1982) have proposed the 'watchdog technique', which allows, whenever it becomes necessary, the relaxation of a given sufficient decrease criterion. Similar strategies with an additional correction step have been proposed, for example, by Bartholomew-Biggs and Nguyen (1990), Coleman and Conn (1982), Dixon (1987) and Mayne and Polak (1982) (see CN.2). For the remaining strategies, unit stepsizes can be established by invoking additional strategies or assumptions, which ensure that the multipliers λ_k, μ_k generated by the QPS (1.8), as well as $\{x_k\}$ and the Hessian approximations generated by the algorithms, are sufficiently close to their respective values at the solution of (1.1) (see e.g. Schittkowski, 1981, Lemma 6.1, Theorem 6.4). In the context of an exact penalty function, Dixon (1980) establishes conditions for unit-stepsize achievement. One of these conditions is related to the accuracy of the one-sided projected Hessian approximation.

Another aspect is that, although all the above stepsize strategies are well defined around the solution of (1.1), no strategy for inequality-constrained problems is, in general, twice continuously differentiable. Indeed, some strategies are not even once-differentiable. Differentiability is not always important, but if it can be assured, convergence to unit stepsizes can be demonstrated with otherwise reasonably relaxed conditions. In Algorithms 2.1 and 2.2 below, twice continuous differentiability is maintained. For equality constraints, the chosen penalty function ensures this, whereas for inequality constraints the smearing technique proposed by Polak and Tits (1980) is adapted and incorporated in the algorithms as a simple rule that maintains twice continuous differentiability.

The basic algorithm in Section 2 is formulated in full, rather than reduced dimension. The equality-constrained case could be formulated in reduced dimension by adopting the approach of Gabay (1982) for reducing the dimensionality of the QPS and approximating the Hessian in this dimension. On the other hand, if slack variables are not to be used, the inequality-constrained case is easier to treat in full dimension. As in Bartholomew- Biggs (1974, 1981) and Powell (1978), Algorithms 2.1 and 2.2 use a positive-definite approximation to the Hessian. Departing from these, as in Tapia (1977), the Hessian approximates the augmented Lagrangian (1.5), with or without the second-order terms arising from the penalty function. The stepsize strategies are based on this penalty function.

The descent property of the direction d_k generated by the QPS is ensured by a simple procedure. It is shown that the penalty parameter c does not grow indefinitely. The convergence of the algorithms, as shown in Theorems (3.1) and (3.2) below, can be demonstrated for any uniformly bounded symmetric positive definite choice for the Hessian approximation. This choice is, however, important in the achievement of unit stepsizes and superlinear rates of convergence (Rustem, 1994). The latter is considered in Chapter 8. The fact that $\hat{\mathfrak{Q}}_k$ incorporates an approximation to the Hessian of the penalty function is important for establishing the condition for $\{\tau_k\} \to 1$. It is at this stage that maintaining the existence of $\mathfrak{Q}_k$ is important, particularly for inequality constraints. Theorem 4.1 relates $\{\tau_k\} \to 1$ to the accuracy of the Hessian of (1.5) used in the QPS.

2 THE SQP ALGORITHMS

The main difference between Algorithms 2.1 and 2.2 is the stepsize strategy and its implications for the penalty parameter. We first define the way in which the vector π is determined in both algorithms. Let the jth element of the vectors h and π be respectively denoted by h^j and π^j.
Let

$$\pi^j_{k+1/2} = \begin{cases} -c_{k+1} h^j_k & \text{if } h^j_k \leqslant 0, \\ \pi^j_k & \text{if } h^j_k > 0 \end{cases} \tag{2.1a}$$

and

$$\pi_{k+1}^{j} = \begin{cases} -c_{k+1}h_{k+1}^{j} & \text{if} \quad h_{k+1}^{j} \leqslant 0, \\ \pi_{k+1/2}^{j} & \text{if} \quad h_{k+1}^{j} > 0, \end{cases} \tag{2.1b}$$

where π_0 is set in **Step 0** with $c_0 \in (0, \infty]$ given as one of the inputs to the program choice and c_{k+1} is set in **Step 3** below. The reason for distinguishing between π_{k+1} and $\pi_{k+1/2}$ is to ensure, in practice, that the first inequality in Corollary 2.1 below is satisfied in general. It is shown below that the penalty parameter c_k is not altered at the later stages of the algorithms and that $c_k = c_*$. For such a constant c_*, it can be verified from (2.1) that $\pi_{k+1/2} = \pi_k$, since

$$\pi_k^j = \pi_{k+1/2}^j = \begin{cases} -c_* h_k^j & \text{if} \quad h_k^j \leqslant 0, \\ \pi_k^j & \text{if} \quad h_k^j > 0, \end{cases}$$

For both algorithms, a possible scheme to approximate the Hessian $\hat{\mathfrak{Q}}_k$ is Powell's (1978) modification to the BFGS (Broyden–Fletcher–Goldfarb–Shanno) formula in Section 6.2, with γ in (6.2.8b) replaced by

$$\gamma = \nabla_x \bar{L}^{\mathrm{a}}(x_{k+1}, \lambda_{k+1}, \mu_{k+1}, c_{k+1}, \pi_{k+1}) - \nabla_x \bar{L}^{\mathrm{a}}(x_k, \lambda_{k+1}, \mu_{k+1}, c_{k+1}, \pi_{k+1/2}).$$

Algorithm 2.1

Step 0: Given$x_0, c_0 \in \mathbb{R}_+$, a small positive $\delta \in (0, \infty), \rho_1 \in (0, 1 - 1/2\epsilon_1)$, $\epsilon_1 \in (\frac{1}{2}, 1], \tilde{\tau} \in (0, 1), \hat{\mathfrak{Q}}_0$, set $k = 0, \pi_0^j = \max\{-c_0 h^j(x_0), 0\}, j = 1, \ldots, i$.

Step 1: Compute $\nabla f_k, \nabla g_k, \nabla h_k$. Solve the QPS (1.8) to obtain d_k and the associated multiplier vectors λ_{k+1}, μ_{k+1}.

Step 2: Test for optimality: if optimality is achieved, **stop**. Else **go to Step 3.**

Step 3: If

$$\langle d_k, \nabla f_k + \epsilon_1 \hat{\mathfrak{Q}}_k d_k \rangle - \langle (h_k)_+, \pi_k \rangle - c_k[\|g_k\|_2^2 + \|(h_k)_+\|_2^2] \leqslant 0 \tag{2.2}$$

then $c_{k+1} = c_k$. Else set

$$c_{k+1} = \max\left\{ \frac{\langle d_k, \nabla f_k + \epsilon_1 \hat{\mathfrak{Q}}_k d_k \rangle - \langle (h_k)_+, \pi_k \rangle}{\|g_k\|_2^2 + \|(h_k)_+\|_2^2}, c_k + \delta \right\}. \tag{2.3}$$

Calculate $\pi_{k+1/2}$ using (2.1b).

Step 4: Find the smallest nonnegative integer j_k and $\tau_k = \bar{\tau}^{j_k}$ such that the inequality

$$l(x_{k+1}, c_{k+1}, \pi_{k+1}) - l(x_k, c_{k+1}, \pi_{k+1/2}) \leqslant \tau_k \rho_1 \langle \nabla l(x_k, c_{k+1}, \pi_{k+1/2}), d_k \rangle \tag{2.4}$$

is satisfied with $x_{k+1} = x_k + \tau_k d_k$. π_{k+1} is computed using (2.1c).

Step 5: Update $\hat{\mathfrak{Q}}_k$ to compute $\hat{\mathfrak{Q}}_{k+1}$.

Step 6: Set $k = k + 1$ and **go to Step 1.**

Algorithm 2.2

Steps 0–2: Given $x_0, c_0 \in \mathbb{R}_+$, a small positive $\delta \in (0, \infty), \rho_2 \in (0, 1)$, $\epsilon_2 \in (0, \frac{1}{2}), \bar{\tau} \in (0, 1), \hat{\mathfrak{Q}}_0$, proceed as in Algorithm 2.1.

Step 3: If

$$\langle d_k, \nabla f_k^{\mathrm{T}} + \left(\frac{1}{2} + \epsilon_2\right)\hat{\mathfrak{Q}}_k d_k\rangle - \langle (h_k)_+, \pi_k\rangle - c_k[\|g_k\|_2^2 + \|(h_k)_+\|_2^2] \leqslant 0 \tag{2.5}$$

then $c_{k+1} = c_k$. Else set

$$c_{k+1} = \max\left\{\frac{\langle d_k, \nabla f_k^{\mathrm{T}} + \left(\frac{1}{2} + \epsilon_2\right)\hat{\mathfrak{Q}}_k d_k\rangle - \langle (h_k)_+, \pi_k\rangle}{\|g_k\|_2^2 + \|(h_k)_+\|_2^2}, c_k + \delta\right\}. \tag{2.6}$$

Calculate $\pi_{k+1/2}$ using (2.1b).

Step 4: Line search: choose the stepsize $\tau_k = \bar{\tau}^{j_k}$ such that the inequality

$$l(x_{k+1}, c_{k+1}, \pi_{k+1}) - l(x_k, c_{k+1}, \pi_{k+1/2}) \leqslant \tau_k \rho_2 q(d_k, c_{k+1}, \pi_{k+1/2}) \tag{2.7}$$

is satisfied with $x_{k+1} = x_k + \tau_k d_k$ π_{k+1} is computed using (2.1).

Steps 5, 6: As in Algorithm 2.1.

Remark

An alternative to δ in **Step 3** of both algorithms is $\delta = \beta c_k$ for some $\beta \in (0, \infty)$. The next few results are concerned with the properties of $\pi_{k+1}, \pi_{k+1/2}$.

Lemma 2.1

Let $\kappa \geqslant 0$ be a constant, $c_{k+1} \geqslant 0$, and

$$\pi_{k+1}^j(x) = \begin{cases} -c_{k+1} h^j(x) & \text{if } h^j(x) \leqslant 0, \\ \kappa & \text{if } h^j(x) > 0. \end{cases} \tag{2.1a}$$

We have

$$(c_{k+1} h^j(x) + \pi_{k+1}^j(x))_+ = c_{k+1} h^j(x) + \pi_{k+1}^j(x) \geqslant 0. \tag{2.8a}$$

Proof

When $h^j(x) \leqslant 0, \pi_{k+1}^j(x) = -c_{k+1} h^j(x) \geqslant 0$ and when $h^j(x) > 0, \pi_{k+1}^j(x) = \kappa \geqslant 0$. Thus, we have

$$h^j(x) \leqslant 0 \Rightarrow (c_{k+1} h^j(x) + \pi_{k+1}^j(x))_+ = c_{k+1} h^j(x) + \pi_{k+1}^j(x) = 0, \tag{2.8b}$$

$$h^j(x) > 0 \Rightarrow (c_{k+1}h^j(x) + \pi^j_{k+1}(x))_+ = c_{k+1}h^j(x) + \pi^j_{k+1}(x) \geqslant 0, \tag{2.8c}$$

and (2.8a) follows directly from (2.8b,c). □

Lemma 2.2
We have

$$(h_k)^{\mathrm{T}}_+ \pi_{k+1/2} = (h_k)^{\mathrm{T}}_+ \pi_k. \tag{2.9}$$

Proof
The result follows from (2.1) since $\pi^j_{k+1/2} \neq \pi^j_k$ if $h^j_k \leqslant 0$, in which case $(h^j_k)_+ = 0$.
□

The corollary below ensures that the descent discussion in Lemma 2.3 and Theorem 4.1 can always be related to the stepsize strategy in **Step 4**, when the penalty parameter is not necessarily a constant. The twice differentiability of $\bar{l}$ leads to the convergence results of the next section.

Corollary 2.1
We have

$$\begin{aligned} l(x_{k+1}, c_{k+1}, \pi_{k+1}) - l(x_k, c_{k+1}, \pi_{k+1/2}) &\leqslant l(x_{k+1}, c_{k+1}, \pi_{k+1/2}) - l(x_k, c_{k+1}, \pi_{k+1/2}) \\ &\leqslant \bar{l}(x_{k+1}, c_{k+1}, \pi_{k+1/2}) - l(x_k, c_{k+1}, \pi_{k+1/2}) \\ &= \bar{l}(x_{k+1}, c_{k+1}, \pi_{k+1/2}) - \bar{l}(x_k, c_{k+1}, \pi_{k+1/2}). \end{aligned}$$

Proof
Given

$$l(x_{k+1}, c_{k+1}, \pi_{k+1}) = f_{k+1} + \frac{c_{k+1}}{2}\|g_{k+1}\|_2^2 + \frac{1}{2c_{k+1}}\left\|(c_{k+1}h_{k+1} + \pi_{k+1})_+\right\|_2^2,$$

the first inequality arises, since, by (2.1a, b),

$$0 = (c_{k+1}h^j_{k+1} + \pi^j_{k+1})_+ \leqslant (c_{k+1}h^j_{k+1} + \pi^j_{k+1/2})_+, \quad \text{for } h^j_{k+1} \leqslant 0,$$

and, by (2.1a, b), $\pi^j_{k+1} = \pi^j_{k+1/2}$

$$(c_{k+1}h^j_{k+1} + \pi^j_{k+1})_+ = (c_{k+1}h^j_{k+1} + \pi^j_{k+1/2})_+, \quad \text{for } h^j_{k+1} > 0.$$

Thus, we have, in general,

$$(c_{k+1}h^j_{k+1} + \pi^j_{k+1})_+ \leqslant (c_{k+1}h^j_{k+1} + \pi^j_{k+1/2})_+.$$

The second inequality arises from $(\cdot)^2_+ \leqslant (\cdot)^2$. As in Lemma (2.1), the equality arises from (2.1), and thus

$$(c_{k+1}h^j_k + \pi^j_{k+1/2})_+ = (c_{k+1}h^j_k + \pi^j_{k+1/2}).$$

□

Lemma 2.3 (The descent property of d_k)
*Let $g_k \neq 0$ and $(h_k)_+ \neq 0$ (the case when $g_k = 0$ and $(h_k)_+ = 0$ is discussed in Lemma 2.4). The direction d_k satisfying the constraints of the QPS and the penalty parameter c_{k+1} chosen as in **Step 3** of Algorithm 2.1 ensure the inequality*

$$\langle \nabla l(x_k, c_{k+1}, \pi_{k+1/2}), d_k \rangle \leqslant -\epsilon_1 \langle d_k, \hat{\mathfrak{Q}}_k d_k \rangle, \qquad \epsilon_1 \in (\tfrac{1}{2}, 1], \tag{2.10}$$

*and d_k with c_{k+1} chosen as in **Step 3** of Algorithm 2.2 ensure*

$$q(d_k, c_{k+1}, \pi_{k+1/2}) \leqslant -\epsilon_2 \langle d_k, \hat{\mathfrak{Q}}_k d_k \rangle, \qquad \epsilon_2 \in (0, \tfrac{1}{2}]. \tag{2.11}$$

Remark
Following the discussions in Chapters 1 and 2, the first-order optimality conditions of the QPS (1.8) are given by

$$\hat{\mathfrak{Q}}_k d_k + \nabla f_k + \nabla g_k (c_k g_k + \lambda_{k+1}) + \nabla h_k [(c_k h_k + \pi_k)_+ + \mu_{k+1}] = 0 \tag{2.12a}$$

$$\langle \mu_{k+1}, (\nabla h_k^{\mathrm{T}} d_k + h_k) \rangle = 0, \qquad \mu_{k+1} \geqslant 0, \tag{2.12b}$$

$$\nabla h_k^{\mathrm{T}} d_k + h_k \leqslant 0, \qquad \nabla g_k^{\mathrm{T}} d_k + g_k = 0. \tag{2.12c}$$

Proof of Lemma 2.3
In order to establish the required result, d_k is only required to satisfy the linearized constraints (2.12c). The other optimality conditions are required for subsequent results. Using (2.12c), (2.8a) in Lemma 2.1 and (2.9) in Lemma 2.2, we can write

$$\begin{aligned}
&\langle d_k, [\nabla f_k + \epsilon_1 \hat{\mathfrak{Q}}_k d_k + c_{k+1} \nabla g_k g_k + \nabla h_k (c_{k+1} h_k + \pi_{k+1/2})_+] \rangle \\
&\quad \leqslant \langle d_k, \nabla f_k + \epsilon_1 \hat{\mathfrak{Q}}_k d_k \rangle - c_{k+1} \|g_k\|_2^2 - \langle h_k, (c_{k+1} h_k + \pi_{k+1/2})_+ \rangle \\
&\quad = \langle d_k, \nabla_k^f + \epsilon_1 \hat{\mathfrak{Q}}_k d_k \rangle - c_{k+1} [\|g_k\|_2^2 \|(h_k)_+\|_2^2] - \langle (h_k)_+, \pi_k \rangle.
\end{aligned} \tag{2.13}$$

In order to ensure (2.10), (2.13) must be nonpositive, and c_{k+1} chosen in **Step 3** of Algorithm 2.1 ensures the latter. The corresponding result for (2.11) can be obtained by replacing ϵ_1 in (2.13) by $\epsilon_2 + \frac{1}{2}$. □

We consider the effect of the penalty parameter on the direction and Lagrange multipliers generated by the QPS. It is well known that, *for a given value of* $\hat{\mathfrak{Q}}_k$, the direction generated by the algorithms, for equality constraints only, is independent of c_k (see Powell, 1980; Tapia, 1977). For certain choices of the Hessian, the independence from c_k can be further relaxed. Consider the Newton version of the algorithms for solving the *equality-constrained problem only*. The necessary optimality condition of the subproblem becomes

$$\left[\nabla^2 f(x_k) + c_k \nabla g_k \nabla g_k^{\mathrm{T}} + \sum_i \nabla^2 g^i(x_k)(c_k g_k^i + \lambda_k^i) \right] d_k + c_k \nabla g_k g_k + \nabla g_k \lambda_{k+1} = 0 \tag{2.14}$$

and $\nabla g_k^{\mathrm{T}} d_k + g_k = 0$. If the second-derivative term, $\nabla^2 g^i(x_k)(c_k g_k^i)$, due to the constraint curvatures, is ignored then it can be verified that the penalty parameter does not affect the multiplier. However, this result is only clear when the Hessian, or its approximation, used in the QPS does not include the term $\nabla^2 g^i(x_k)(c_k g_k^i)$. For other choices of the Hessian used in the QPS and also for inequality-constrained problems, this conclusion may not apply, and hence *c may affect the direction and the Lagrange multipliers* of the QPS.

It is shown in the next Lemma that (2.13) is nonpositive and that the descent property of Lemma 2.3 also holds when $g_k = 0$ and $(h_k)_+ = 0$.

Lemma 2.4 (The finiteness of the penalty parameter †)
Let f, g and h be once-differentiable and let $d_k, \lambda_{k+1}, \mu_{k+1}$ generated by the QPS be bounded. Then we have the following.

(i) If for some or all k the constraints of (1.1) are satisfied, i.e. $g(x_k) = 0$ and $h(x_k) \leqslant 0$, then the descent properties of (2.10) and (2.11) of Lemma 2.3 are satisfied for any choice of $c_k \in [0, \infty)$.

(ii) Let $c_k \in [0, \infty)$ and $g_k = 0$ and $h_k \leqslant 0$, for some k. Then, ***Steps 3*** *of both algorithms choose $c_{k+1} = c_k$. For this choice, the direction d_k ensures the descent property of Lemma 2.3.*

(iii) $\forall k, \exists c_{k+1} \in [0, \infty)$ satisfying ***Steps 3*** *of both algorithms.*

(iv) Let the sequence $\{x_k\}$ generated by either algorithm be bounded. Then c_k is increased finitely often and $\exists$ an integer $k_* \geqslant 0$ and $\exists c_* \in [0, \infty)$ such that the algorithms choose $c_k = c_*, \forall k \geqslant k_*$.

Proof
To show (i), consider the optimality conditions (2.12), with (2.1) as the rule for setting π, for $g_k = 0, h_k \leqslant 0$:

$$\begin{aligned}\langle d_k, [\nabla l(x_k, c_k, \pi_k) + \epsilon_1 \hat{\mathfrak{Q}}_k d_k]\rangle &\leqslant \langle d_k, [\nabla l(x_k, c_k, \pi_k) + \hat{\mathfrak{Q}}_k d_k]\rangle \\ &= -\langle d_k, [\nabla g_k \lambda_{k+1} + \nabla h_k \mu_{k+1}]\rangle \\ &= \langle g_k, \lambda_{k+1}\rangle + \langle h_k, \mu_{k+1}\rangle \\ &\leqslant 0,\end{aligned}$$

which establishes the case for (2.10). The same can also be shown for (2.11). Therefore, the descent conditions (2.10) and (2.11) are satisfied. To show (ii), consider (2.2) and (2.5), which are satisfied $\forall c_k \in [0, \infty)$ since the descent conditions (2.10) and (2.11) are satisfied for $g_k = 0$ and $h_k \leqslant 0$. Thus, both algorithms choose $c_{k+1} = c_k$. For this choice, it can simply be verified that the descent property of Lemma 2.3 holds. (iii) follows simply from (i) and (ii), since **Steps 3** of both algorithms generate at each iteration a finite c_{k+1}. To

† c_k is the value used in the quadratic subproblem when computing d_k, λ_{k+1}, μ_{k+1}.

establish (iv), we note that at each iteration k, **Steps 3** of both algorithms generate a $c_{k+1} \in [0, \infty)$ such that the descent conditions of Lemma 2.3 hold. If c_k is increased infinitely often then, by **Step 3**, we have $\{c_{k+1}\} \to \infty$. As there exists a $c_{k+1} \in [0, \infty)$ that the algorithms can choose at every iteration, then there exists a $c_* \in [0, \infty)$ such that the test of **Step 3** is always satisfied for $c_k = c_*, \forall k \geqslant k_*$. k_* is the iteration at which the algorithms reach c_*. □

The algorithms increase the penalty parameter until c_* is attained. An increase in the penalty parameter is not required after this, since the condition in **Step 3** is satisfied with $c_k = c_*$. The numerical behaviour of the penalty parameter is discussed in Rustem (1993).

3 GLOBAL CONVERGENCE OF THE ALGORITHMS

We first establish that the stepsize strategies (2.4) and (2.7) can always be satisfied. If the satisfaction of these strategies were assumed then the inequalities (3.6a,b) below could be replaced by (2.4) and (2.7), and the monotonicity result of Theorem 4.1 below would follow from Lemmas 2.3 and 2.4(i).

It can be observed from Lemma 2.3 that

$$\langle d_k, \nabla l(x_k, c_{k+1}, \pi_{k+1/2})\rangle \leqslant q(x_k, c_{k+1}, \pi_{k+1/2}) < 0, \quad \forall d_k \neq 0.$$

The next few results confirm that $\tau_k > 0, \forall d_k \neq 0$.

Lemma 3.1
The inequality $q(\tau d, c, \pi) \leqslant \tau q(d, c, \pi)$ holds for all bounded d, c and π and bounded and positive-semidefinite $\hat{\mathfrak{Q}}_k$ and $\tau \in [0, 1]$.

Proof
The result follows from the inequality

$$q(\tau d, c, \pi) = \tau\langle d, [\nabla l(x_k, c, \pi) + \tau \frac{1}{2} \hat{\mathfrak{Q}}_k d]\rangle \leqslant \tau q(d, c, \pi).$$ □

Remark
As shown in Lemmas 2.3 and 2.4, **Steps 3** of both algorithms are always satisfied for finite c_k, and there exists a $c_* \in [0, \infty)$ for which the test in **Step 3** is satisfied $\forall k \geqslant k_*$. The integer k_* is the iteration at which c_* is attained by the algorithms.

Theorem 3.1
Let

(*i*) f, g, h be twice continuously differentiable functions;

(*ii*) *the approximate Hessian be such that, for each k,*

$$m\|y\|_2^2 \leqslant \langle y, \hat{\mathfrak{Q}}_k y\rangle \leqslant \mathcal{M}\|y\|_2^2, \qquad \mathcal{M} \geqslant m > 0, \quad 0 \neq y \in \mathbb{R}^n; \tag{3.1}$$

(*iii*) *for each k, there exists a bounded Karush–Kuhn–Tucker triple* $(d_k, \lambda_{k+1}, \mu_{k+1}) \in \mathbb{R}^n \times \mathbb{R}^e \times \mathbb{R}^i$ *of the QPS (see CN.3);*

(*iv*) *there exist an integer* k_* *and a scalar* $c_* \geqslant 0$ *such that the penalty parameter restrictions (2.2) and (2.5) are satisfied, for Algorithms 2.1 and 2.2 respectively,* $\forall k \geqslant k_*$ *with* $c_{k+1} = c_k = c_*$.

Then the stepsize computed in Step 4 of either algorithm is such that $\tau_k \in (0, 1]$, *and hence the sequence* $\{x_k\}$, *computed by the algorithms, generates a corresponding sequence* $\{l(x_k, c_*, \pi_k)\}$ *that is monotonically decreasing for* $k \geqslant k_*$.

Remark

The above result can be extended to show that there exists a $\tau_k \in (0, 1]$ that satisfies the stepsize strategy in **Step 4** *before* the algorithms achieve a constant c_* value for c_k. This can be done by considering

$$\bar{l}(x_{k+1}, c_{k+1}, \pi_{k+1/2}) - \bar{l}(x_k, c_{k+1}, \pi_{k+1/2})$$

and the expansion of $\bar{l}(x, c_{k+1}, \pi_{k+1/2})$ instead of $\bar{l}(x, c_*, \pi_k)$ in (3.3) below. The assumption of $c_k = c_*$ is required for the overall monotonicity of l.

Proof of Theorem 3.1

We note that for $c_{k+1} = c_k$, (2.1) yields $\pi_{k+1/2} = \pi_k$. Using this in Corollary 2.1 and the definition of $\bar{l}$, we have

$$l(x_{k+1}, c_*, \pi_{k+1}) - l(x_k, c_*, \pi_{k+1/2}) \quad \leqslant \bar{l}(x_{k+1}, c_*, \pi_k) - \bar{l}(x_k, c_*, \pi_k). \tag{3.2}$$

Using the second-order expansion of $\bar{l}(x, c_*, \pi_k)$, (3.2) can be written as

$$\begin{aligned} l(x_{k+1}, c_*, \pi_{k+1}) - l(x_k, c_*, \pi_{k+1/2}) \leqslant \tau_k \langle d_k, [\nabla \bar{l}(x_k, c_*, \pi_k) \\ + \frac{1}{2}\tau_k \hat{\mathfrak{Q}}_k d_k]\rangle + \tau_k^2 \bar{\eta}_k \|d_k\|_2^2, \end{aligned} \tag{3.3}$$

where

$$\bar{\eta}_k = \int_0^1 (1-t)\|\nabla_x^2 \bar{l}(x_k + t\tau_k d_k, c_*, \pi_k) - \hat{\mathfrak{Q}}_k\| \mathrm{d}t. \tag{3.4}$$

For π_k determined by (2.1), Lemma 2.1 ensures that

$$\nabla \bar{l}(x_k, c_{k+1}, \pi_{k+1/2}) = \nabla l(x_k, c_{k+1}, \pi_{k+1/2}), \tag{3.5a}$$

and thence

$$\nabla \bar{l}(x_k, c_*, \pi_k) = \nabla l(x_k, c_* \pi_k). \tag{3.5b}$$

The relations (3.3), (3.5) and Lemmas 2.3 and 3.1 yield the inequalities

$$l(x_{k+1}, c_*, \pi_{k+1}) - l(x_k, c_*, \pi_k) \leqslant \tau_k \langle d_k, \nabla \bar{l}(x_k, c_*, \pi_k) \rangle \left[1 - \frac{\tau_\kappa}{\epsilon_1} \left(\frac{1}{2} + \frac{\bar{\eta}_\kappa}{m} \right) \right], \quad (3.6a)$$

$$l(x_{k+1}, c_*, \pi_{k+1}) - l(x_k, c_*, \pi_k) \leqslant \tau_k q(d_k, c_*, \pi_k) \left(1 - 2\tau_k \frac{\bar{\eta}_\kappa}{m\epsilon_2} \right). \quad (3.6b)$$

The scalar $\rho_1 \in (0, 1 - 1/2\epsilon_1)$ in the stepsize strategy (2.4) clearly determines τ_k such that

$$\rho_1 \leqslant 1 - \frac{\tau_\kappa}{\epsilon_1} \left(\frac{1}{2} + \frac{\bar{\eta}_\kappa}{m} \right) \leqslant 1 - \frac{1}{2\epsilon_1}.$$

Since, by (2.10), $d_k^{\mathrm{T}} \nabla l(\cdot) \leqslant 0$, there exists a $\tau_k \in (0, 1]$ to ensure that (3.6a), and thence (2.4), holds. Suppose that τ^0 is the largest $\tau \in (0, 1]$ satisfying inequalities (3.6a) and (2.4). It follows that all $\tau \leqslant \tau^0$ also satisfy these conditions and that the strategy in **Step 4** of the algorithm selects a $\tau_k \in [\bar{\tau}\tau^0, \tau^0]$. Since, by (2.10), $\langle d_k, \nabla l(\cdot) \rangle \leqslant 0$, it follows that $\{l(x_k, c_*, \pi_k)\}$ is monotonically decreasing for Algorithm 2.1.

For Algorithm 2.2, $\rho_2 \in (0, 1)$ in (2.7) determines τ_k such that

$$\rho_2 \leqslant 1 - 2\tau_k \frac{\bar{\eta}_k}{m\epsilon_2} \leqslant 1.$$

Since, by (2.11), $q(d_k, c_*, \pi_k) \leqslant 0$, there exists a $\tau_k \in (0, 1]$ such that (3.6b), and thence (2.7), holds. Using the same arguments as above, it follows that $\{l(x_k, c_*, \pi_k)\}$ is monotonically decreasing. □

The Hessian of $\bar{l}(x, c_*, \pi_k)$ is given by

$$\nabla_x^2 \bar{l}(x_k, c_*, \pi_k) = \nabla^2 f_k + c_* \left(\sum_{j=1}^{e} \nabla^2 g_k^j g_k^j + \nabla g_k \nabla g_k^{\mathrm{T}} \right) + \sum_{j=1}^{i} \nabla^2 h_k^j (c_* h_k^j + \pi_k^j)_+ + c_* \nabla h_k \nabla h_k^{\mathrm{T}}. \quad (3.7)$$

Since the way in which π_k is determined ensures, by Lemma 2.1, that $(\cdot) = (\cdot)_+$, and since for a feasible point $x_k, (\cdot)_+ = 0$, we note that for such points the second and fourth terms on the right of (3.7) vanish. This fact may clearly be used in the approximate Hessian $\hat{\mathfrak{Q}}_k$ of $\bar{L}^a$ in (1.5). The results below are not affected by ignoring these terms (see Tapia, 1986).

Lemma 3.2

Let the assumptions of Theorem 3.1 be satisfied and let, for some $k_0 \geqslant 0$ and $k \geqslant k_0$, the set

$$\mathbb{F} = \{x \in \mathbb{R}^n \mid l(x, c_*, \pi_k(x)) \leqslant l(x_{k_0}, c_*, \pi_{k_0}(x_{k_0}))\}$$

be bounded. Then, we have

$$\lim_{k \to \infty} q(d_k, c_*, \pi_k) = 0, \tag{3.8a}$$

$$\lim_{k \to \infty} \langle \nabla l(x_k, c_*, \pi_k), d_k \rangle = 0. \tag{3.8b}$$

Proof

By (2.8a) of Lemma 2.1, we have

$$l(x, c_*, \pi_k(x)) = \bar{l}(x, c_*, \pi_k(x)), \quad \bar{l}(x_{k_0}, c_*, \pi_{k_0}(x_{k_0})) = l(x_{k_0}, c_*, \pi_{k_0}(x_{k_0})),$$

and hence the set

$$\bar{\mathbb{F}} = \{x \in \mathbb{R}^n \mid \bar{l}(x, c_*, \pi_k(x)) \leqslant \bar{l}(x_{k_0}, c_*, \pi_{k_0}(x_{k_0}))\}$$

is also bounded.

Given $\rho_1 \in (0, 1 - 1/2\epsilon_1)$, by (3.6a), the choice

$$\tau^0 = \min \left\{ 1, \frac{1 - \rho_1}{\dfrac{1}{\epsilon_1}\left(\dfrac{1}{2} + \dfrac{\bar{\eta}_\kappa}{m}\right)} \right\}$$

always satisfies the stepsize strategy (2.4). Clearly, τ_k, chosen as $\tau_k = \bar{\tau}^{j_k}$, as discussed above, is in the range $\tau_k \in [\bar{\tau}\tau^0, \tau^0]$, and thereby also satisfies (2.4). Since $\bar{l}(x, c_*, \pi_k)$ is twice continuously differentiable, and $\bar{\mathbb{F}}$ is compact, there is a scalar $\bar{M} < \infty$ such that $\bar{\eta}_k \leqslant \bar{M}$. Thus, since $m > 0$, we have $\tau_k \geqslant \hat{\tau} > 0, \forall k$, for the stepsize strategy (2.4) and for some positive $\hat{\tau}$.

Given $\rho_2 \in (0, 1)$, by (3.6b), the choice

$$\tau^0 = \min \left\{ 1, \frac{(1 - \rho_2) m \epsilon_2}{2\bar{\eta}_\kappa} \right\}$$

always satisfies the stepsize strategy (2.7). As above, it can be established that $\tau_k \geqslant \hat{\tau} > 0, \forall k$. Hence, the boundedness of $l(x, c_*, \pi_k(x))$ on $\mathbb{F}$ and $q(d_k, c_*, \pi_k) \leqslant 0$ imply, in the case of (2.7),

$$0 \leqslant \rho_2 \sum_h \tau_k \mid q(d_k, c_*, \pi_k) \mid \leqslant \sum [l(x_k, c_*, \pi_k) - l(x_{k+1}, c_*, \pi_{k+1})] < \infty,$$

which gives (3.8a). The equality (3.8b) is shown similarly, using the above argument with (2.4) and $d_k^{\mathrm{T}} \nabla l(x_k, c_*, \pi_k) \leqslant 0$. □

Lemma 3.3

Inequalities (2.10) or (2.11) and Lemma 2.3 imply that $\lim_{\kappa \to \infty} \|d_k\| = 0$.

Proof

For (3.8a), inequality (2.11), and, for (3.8b), inequality (2.10) provide the result. □

Theorem 3.2

Let the assumptions of Theorem 3.1 be satisfied. Then, we have the following.

(*i*) *Algorithms 2.1 and 2.2 either terminate at a Kuhn–Tucker point (satisfying the first-order necessary conditions) of (1.1) or they generate an infinite sequence* $\{x_k\}$ *in which there exists a subsequence* $\{x_k\}$ *with* $k_* \leqslant k \in K \subset \{0, 1, \ldots\}$ *such that* $\{\|d_k\|\} \rightarrow 0$, *and thus every accumulation point* x_* *of the infinite sequence* $\{x_k\}$ *is a Kuhn–Tucker point of (1.1).*

(*ii*) *If, furthermore, strict complementarity holds at the solution of the QPS* (1.8), *for large* k, $k_* \leqslant k \in K$, μ_k *predicts the active inequality constraints at* x_*.

Proof

By Lemmas 2.4, 3.2 and 3.3, there exists a subsequence $\{x_k\}$, $k_* \leqslant k \in K$ such that $\|d_k\| \rightarrow 0$. Let there exist $x_* \in \mathbb{R}^n, \lambda_* \in \mathbb{R}^e, \mu_* \in \mathbb{R}^i$ such that $\{x_k\} \rightarrow x_*$, $\{\lambda_k\} \rightarrow \lambda_*, \{\mu_k\} \rightarrow \mu_*, k_* \leqslant k \in K$. The existence of such points is ensured since, by Theorem 3.1, the algorithms decrease $l(x_k, c_*, \pi_k)$ sufficiently at each iteration, thereby ensuring $x_k \in \mathbb{F}$, with $\mathbb{F}$ compact.

In order to show that the triple (x_*, λ_*, μ_*) satisfies the necessary conditions of optimality for (1.1), we consider the optimality conditions (2.12) of the QPS (1.8) and apply Lemma 3.3. Letting $k \rightarrow \infty, k_* \leqslant k \in K$, we have

$$g(x_*) = 0, \qquad h(x_*) \leqslant 0, \qquad \langle \mu_*, h(x_*) \rangle = 0, \qquad \mu_* \geqslant 0, \tag{3.9a}$$

$$\begin{aligned} 0 &= \nabla f(x_*) + \nabla g(x_*)[c_* g(x_*) + \lambda_*] + \nabla h(x_*)[(c_* h(x_*) + \pi_*)_+ + \mu_*] \\ &= \nabla f(x_*) + \nabla g(x_*)\lambda_* + \nabla h(x_*)\mu_*. \end{aligned} \tag{3.9b}$$

The last equality follows from (2.1), since, for $h(x_*) \leqslant 0, \pi_* = -c_* h(x_*)$. Expressions (3.9a,b) are the first-order optimality conditions of (1.1).

To show (ii), we note that, in view of the last two conditions in (3.9a), for sufficiently large k and with strict complementarity holding, none of the inactive constraints, i.e.

$$h_k^j + d_k^{\mathrm{T}} \nabla h_k^j < 0, \qquad \mu_{k+1}^j = 0, \tag{3.10}$$

are predicted to be active at $x_*, h_*^j(x_*) < 0, \mu_*^j = 0$. □

As a consequence of (ii) in the above theorem, for sufficiently large k, say $\bar{K}$, the inactive constraints at x_* do not affect the computation of x_k. At that stage, it would make no difference if the active constraints were treated as equality constraints. Hence, for $k \geqslant \bar{K}, h^j(x_*) = 0$ implies $h_k^j + d_k^{\mathrm{T}} \nabla h_k^j = 0$. If $\bar{h}$ denotes the vector of active constraints at x_* then

$$\bar{h}_k + d_k^{\mathrm{T}} \nabla \bar{h}_k = 0, \quad \forall k \geqslant \bar{K}, \tag{3.11}$$

where $\nabla \bar{h}_k$ is the matrix whose columns are the normals of $\bar{h}$ at x_k. With strict complementarity, the multiplier μ is such that $\mu_*^j > 0$ implies $\mu_k^j > 0$, and $\mu_*^j = 0$ implies $\mu_k^j = 0$.

4 CONVERGENCE TO UNIT STEPSIZES

We consider the conditions that ensure $\{\tau_k\} \to 1$. Convergence to unit steps can be ensured using two different approaches leading to two different conditions. The first approach relates $\{\tau_k\} \to 1$ to the accuracy of the Hessian approximation $\hat{\mathfrak{Q}}$ of $\bar{L}^{\mathrm{a}}$ in (1.5). This is discussed in Theorem 4.1. The second approach relates $\{\tau_k\} \to 1$ to the accuracy of the approximate Hessian of $\bar{l}$. This is discussed, for the equality-constrained case, in Rustem (1986). Its extension to the inequality-constrained case is straightforward, and will not be stated explicitly in this study. The advantage of the latter approach is that it does not involve any discussion of multipliers. However, it has two disadvantages. The first is that it involves the Hessian of $\bar{l}$, and this is only indirectly approximated by the algorithm since it constitutes part of the Hessian of $\bar{L}^{\mathrm{a}}$. The second disadvantage is that it invokes an assumption concerning the positive-semidefiniteness of the approximate Hessian, with respect to x, of $\langle \lambda, g(x) \rangle + \langle \mu, h(x) \rangle$.

We shall first establish, in Lemma 4.1, two properties of a set of alternative multipliers, $\tilde{\lambda}, \tilde{\mu}$, used in Theorem 4.1. It should be noted that Algorithms 2.1 and 2.2 use the multipliers λ, μ and *not* $\tilde{\lambda}, \tilde{\mu}$.

The result in Theorem 4.1 requires a comparison of λ, μ and $\tilde{\lambda}, \mu$. To derive the conventional multipliers λ, μ for the constraints active at the solution of (1.1), let the matrix of active constraint gradients be given by

$$\mathcal{G}_k = [\nabla g_k \vdots \nabla \bar{h}_k], \tag{4.1}$$

and assume that the columns of $\mathcal{G}_k$ are linearly independent. For $k \geqslant \bar{K}$, the Kuhn–Tucker triple $(d_k, \lambda_{k+1}, \bar{\mu}_{k+1})$ is given by the solution of

$$\min \left\{ q(d, c_k, \bar{\pi}_k) \middle| \mathcal{G}_k^{\mathrm{T}} d + \begin{bmatrix} g_k \\ \bar{h}_k \end{bmatrix} = 0 \right\} \tag{4.2}$$

and

$$d_k = -P_k[\hat{\mathfrak{Q}}_k]\hat{\mathfrak{Q}}_k^{-1}\nabla f_k - \hat{\mathfrak{Q}}_k^{-1}\mathcal{G}_k\left(\mathcal{G}_k^{\mathrm{T}}\hat{\mathfrak{Q}}_k^{-1}\mathcal{G}_k\right)^{-1}\begin{bmatrix} g_k \\ \bar{h}_k \end{bmatrix}, \tag{4.3}$$

$$P_k[\hat{\mathfrak{Q}}_k] = I - \hat{\mathfrak{Q}}_k^{-1}\mathcal{G}_k\left(\mathcal{G}_k^{\mathrm{T}}\hat{\mathfrak{Q}}_k^{-1}\mathcal{G}_k\right)^{-1}\mathcal{G}_k^{\mathrm{T}}, \tag{4.4}$$

$$\begin{bmatrix} \lambda_{k+1} \\ \bar{\mu}_{k+1} \end{bmatrix} = (\mathcal{G}_k^{\mathrm{T}}\hat{\mathfrak{Q}}_k^{-1}\mathcal{G}_k)^{-1}\left(\mathcal{G}_k^{\mathrm{T}}\hat{\mathfrak{Q}}_k^{-1}\nabla f_k - \begin{bmatrix} g_k \\ \bar{h}_k \end{bmatrix}\right) - \begin{bmatrix} c_k g_k \\ (c_k\bar{h}_k + \bar{\pi}_k)_+ \end{bmatrix}, \tag{4.5}$$

where $\bar{\mu}$ and $\bar{\pi}$ denote the vectors derived from μ and π corresponding to the inequality constraints, active at the solution of (1.1), given by $\bar{h}$. From

(4.3)–(4.5), and also (2.14), it can be confirmed that, for a given value of $\hat{\mathfrak{Q}}_k$, the parameters c and π do not affect d_k. Hence, a change in c_k to c_{k+1} and π_k to $\pi_{k+1/2}$ would not affect d_k. In general, however, the multipliers may be affected (see (4.5) and the discussion related to (2.14)).

We now define a new set of multipliers, which are only used in the following lemma and theorem. For $\tilde{c}_{k+1} \in (-\infty, +\infty)$, consider $\tilde{\lambda}, \tilde{\mu}$ given by

$$\begin{bmatrix} \tilde{\lambda}_{k+1} \\ \tilde{\mu}_{k+1} \end{bmatrix} = (\mathscr{G}_k^{\mathrm{T}} \hat{\mathfrak{Q}}_k^{-1} \mathscr{G}_k)^{-1} \left(\mathscr{G}_k^{\mathrm{T}} \hat{\mathfrak{Q}}_k^{-1} \nabla f_k - \begin{bmatrix} g_k \\ \bar{h}_k \end{bmatrix} \right) - \tilde{c}_{k+1} \begin{bmatrix} g_k \\ \bar{h}_k \end{bmatrix}. \tag{4.6}$$

It should be noted that the algorithms use $c_k \geqslant 0$ and that c_k also influences $\hat{\mathfrak{Q}}_k \cdot \ \bar{c}_{k+1}$ is not necessarily the same value as c_k. The effect of $\bar{c}$ in (4.6) is well defined, since an alteration of the penalty parameter *after* the QPS is solved *does* affect the multipliers.

Lemma 4.1
Let the assumptions of Theorem 3.2 be satisfied, and, in particular, let the columns of $\mathscr{G}_k$ be linearly independent and strict complementarity hold at the solution of the QPS. Then
(i) for $\tilde{\lambda}_{k+1}, \tilde{\mu}_{k+1}$ given by (4.6), $\exists \tilde{c}_{k+1} \in (-\infty, +\infty)$ such that

$$\langle g_k, \tilde{\lambda}_{k+1} \rangle + \langle \bar{h}_k, \tilde{\mu}_{k+1} \rangle \geqslant 0; \tag{4.7}$$

(ii) as $k \to \infty$ and $\|d_k\| \to 0$, we have

$$\left\| \begin{matrix} \lambda_{k+1} - \tilde{\lambda}_{k+1} \\ \bar{\mu}_{k+1} - \tilde{\mu}_{k+1} \end{matrix} \right\| \to 0. \tag{4.8}$$

Proof
To show (i), consider (4.6) and let

$$a_k = \left\langle \begin{bmatrix} g_k \\ \bar{h}_k \end{bmatrix}, (\mathscr{G}_k^{\mathrm{T}} \hat{\mathfrak{Q}}_k^{-1} \mathscr{G}_k)^{-1} \left\{ \mathscr{G}_k)^{\mathrm{T}} \hat{\mathfrak{Q}}_k^{-1} \nabla f_k - \begin{bmatrix} g_k \\ \bar{h}_k \end{bmatrix} \right\} \right\rangle,$$

$$b_k = \left\| \begin{matrix} g_k \\ \bar{h}_k \end{matrix} \right\|_2^2,$$

$$\left\langle \begin{bmatrix} g_k \\ \bar{h}_k \end{bmatrix}, \begin{bmatrix} \tilde{\lambda}_{k+1} \\ \tilde{\mu}_{k+1} \end{bmatrix} \right\rangle = a_k - \tilde{c}_{k+1} b_k,$$

$$\tilde{c}_{k+1} = \begin{cases} c_* & \text{if } \ a_k - c_* \, b_k \geqslant 0, \\ a_k / b_k & \text{if } \ a_k - c_* b_k < 0. \end{cases} \tag{4.9}$$

By definition, $b_k \geqslant 0$. Also, if $b_k = 0$, we have $a_k = 0$. Thus, this choice of $\tilde{c}$ ensures (4.7). A natural initial choice is $\tilde{c}_0 = 0$. As $\{b_k\} \to 0$, it is clear from (4.9) that $\{\tilde{c}_k\} \to c_*$. To show (ii), consider (4.5) and (4.6). According to Lemma 3.3 and Theorem 3.2, for $k \geqslant k_*$, we have

$$\left\| \begin{matrix} \lambda_{k+1} & -\tilde{\lambda}_{k+1} \\ \bar{\mu}_{k+1} & -\tilde{\mu}_{k+1} \end{matrix} \right\| = \left\| \begin{bmatrix} c_* g_k \\ (c_* \bar{h}_k + \bar{\pi}_k)_+ \end{bmatrix} - \bar{c}_{k+1} \begin{bmatrix} g_k \\ \bar{h}_k \end{bmatrix} \right\| \to 0,$$

since $g_k \to 0$, $\bar{h}_k \to 0$ and $\bar{\pi}_k = -c_* \bar{h}_k \to 0$. □

Theorem 4.1

Let the assumptions of Lemma 4.1 be satisfied. For $k \geqslant k_*$, there is a number $\sigma > 0$ such that if $\Psi_k \leqslant \sigma$ then

$$\Psi_k = \frac{\|P_k^{\mathrm{T}}(\mathfrak{Q}_* - \hat{\mathfrak{Q}}_k)P_k d_k\|}{\|d_k\|} + \frac{\left\| \left\langle \begin{bmatrix} g_k \\ \bar{h}_k \end{bmatrix}, \bar{\mathscr{G}}_k^{\mathrm{T}}(\mathfrak{Q}_* - \hat{\mathfrak{Q}}_k)\left(\bar{\mathscr{G}}_k \begin{bmatrix} g_k \\ \bar{h}_k \end{bmatrix} - 2P_k d_k \right) \right\rangle \right\|}{\|d_k\|^2}, \tag{4.10}$$

with $P_k, \bar{\mathscr{G}}_k$ given either by $P_k = P_k[\hat{\mathfrak{Q}}_k]$ (see 4.4) and $\bar{\mathscr{G}}_k = \hat{\mathfrak{Q}}_k^{-1}\mathscr{G}_k(\mathscr{G}_k^{\mathrm{T}}\hat{\mathfrak{Q}}_k^{-1}\mathscr{G}_k)^{-1}$, or by $P_k = P_k[I]$ and $\bar{\mathscr{G}}_k = \mathscr{G}_k(\mathscr{G}_k^{\mathrm{T}}\mathscr{G}_k)^{-1}$. Then the step-size strategies (2.4) and (2.7) are satisfied for $\tau_k = 1$.

Proof

Consider the second-order expansion of $\bar{L}^{\mathrm{a}}$ given by (1.5):

$$\begin{aligned} &\bar{L}^{\mathrm{a}}(x_{k+1}, \lambda_{k+1}, \bar{\mu}_{k+1}, c_*, \bar{\pi}_k) - \bar{L}^{\mathrm{a}}(x_k, \lambda_{k+1}, \bar{\mu}_{k+1}, c_*, \bar{\pi}_k) \\ &\quad \leqslant \left\langle d_k, \left\{ \nabla \bar{l}(x_k, c_*, \bar{\pi}_k) + \frac{1}{2}(\hat{\mathfrak{Q}}_k + \mathfrak{Q}_* - \hat{\mathfrak{Q}}_k)d_k + \mathscr{G}_k \begin{bmatrix} \lambda_{k+1} \\ \bar{\mu}_{k+1} \end{bmatrix} \right\} \right\rangle + \eta_k \|d_k\|_2^2, \end{aligned} \tag{4.11}$$

where

$$\eta_k = \int_0^1 (1-t)\|\nabla_x^2 \bar{L}^{\mathrm{a}}(x_k + t d_k, \lambda_{k+1}, \bar{\mu}_{k+1}, c_*, \bar{\pi}_k) - \mathfrak{Q}_*\| \,\mathrm{d}t$$

and $\mathfrak{Q}_*$ denotes $\nabla_x^2 \bar{L}^{\mathrm{a}}$ evaluated at the solution of (1.1). Using (1.5), (3.2), (2.12b,c) and (4.7), we can write (4.11) as

$$\begin{aligned} l(x_{k+1}, c_*, \bar{\pi}_{k+1}) - l(x_k, c_*, \bar{\pi}_k) \leqslant &\left\langle d_k, \left\{ \nabla \bar{l}(x_k, c_*, \bar{\pi}_k) + \frac{1}{2}(\hat{\mathfrak{Q}}_k + \mathfrak{Q}_* - \hat{\mathfrak{Q}}_k)d_k \right\} \right\rangle \\ &+ \left\langle s_k, \begin{bmatrix} g_{k+1} \\ \bar{h}_{k+1} \end{bmatrix} \right\rangle + \eta_k \|d_k\|_2^2, \end{aligned} \tag{4.12}$$

where

$$s_k = \begin{bmatrix} \tilde{\lambda}_{k+2} & -\lambda_{k+2} & +\lambda_{k+2} & -\lambda_{k+1} \\ \tilde{\mu}_{k+2} & -\bar{\mu}_{k+2} & +\bar{\mu}_{k+2} & -\bar{\mu}_{k+1} \end{bmatrix}.$$

We use

$$\{\|\lambda_{k+2} - \lambda_{k+1}\|\} \to 0, \qquad \{\|\bar{\mu}_{k+2} - \bar{\mu}_{k+1}\|\} \to 0$$

and Lemma 4.1(ii) to deduce that $\{\|s_k\|\} \to 0$. Using the second-order expansions of g and $\bar{h}$ with (2.12c) and (3.11), we have

$$\left\| \begin{matrix} g_{k+1} \\ \bar{h}_{k+1} \end{matrix} \right\|_2 \leqslant \left\| \begin{matrix} g_k & +\nabla g_k^{\mathrm{T}} d_k \\ \bar{h}_k & +\nabla \bar{h}_k^{\mathrm{T}} d_k \end{matrix} \right\|_2 + e_1 \|d_k\|_2^2 = e_1 \|d_k\|_2^2, \tag{4.13}$$

for some $e_1 \geqslant 0$. Also, d_k satisfies

$$d_k = P_k d_k - \bar{\mathscr{G}}_k \begin{bmatrix} g_k \\ \bar{h}_k \end{bmatrix}. \tag{4.14}$$

Relations (4.12)–(4.14), (3.5) and Lemma 2.3 yield the inequalities

$$l(x_{k+1}, c_*, \bar{\pi}_{k+1}) - l(x_k, c_*, \bar{\pi}_k) \leqslant \langle d_k, \nabla l(x_k, c_*, \bar{\pi}_k) \rangle \Bigg[1 - \frac{1}{2\epsilon_1} - \frac{1}{m\epsilon_1} \left(\eta_k + \frac{1}{2} e_1 \|s_k\|_2 + \frac{1}{2} \psi_k \right) \Bigg], \tag{4.15}$$

$$l(x_{k+1}, c_*, \bar{\pi}_{k+1}) - l(x_k, c_*, \bar{\pi}_k) \leqslant q(d_k, c_*, \bar{\pi}_k) \left[1 - \frac{1}{m\epsilon_2} \left(2\eta_k + \frac{1}{2} e_1 \|s_k\|_2 + \frac{1}{2} \psi_k \right) \right]. \tag{4.16}$$

Since, by Theorem 3.2, $\{x_k\} \to x_*$, $\{\lambda_k\} \to \lambda_*$, $\{\mu_k\} \to \mu_*$, we have $\{\eta_k\} \to 0$. Also, by (4.8), we have $\{\|s_k\|_2\} \to 0$. The scalar $\rho_1 \in (0, 1 - 1/2\epsilon_1)$ in (2.4) is bounded by

$$\rho_1 \leqslant 1 - \frac{1}{2\epsilon_1} - \frac{1}{m\epsilon_1} \left(\eta_k + \frac{1}{2} e_1 \|s_k\|_2 + \frac{1}{2} \psi_k \right). \tag{4.17}$$

If $\sigma > 0$ is such that

$$\frac{1}{m\epsilon_1} \left(\eta_k + \frac{1}{2} e_1 \|s_k\|_2 + \frac{1}{2} \sigma \right) \leqslant 1 - \frac{1}{2\epsilon_1} - \rho_1$$

(in view of $\eta_k \to 0, s_k \to 0$, this defines the number σ for the stepsize strategy (2.4)) then (4.17) holds, and therefore, because $d_k^{\mathrm{T}} \nabla l(x_k, c_*, \bar{\pi}_k) \leqslant 0$, (2.4) is satisfied with $\tau_k = 1$.

For the stepsize (2.7), consider inequality (4.16). It can be argued, as above, that if σ is such that

$$\frac{1}{m\epsilon_2} \left(2\eta_k + \frac{1}{2} e_1 \|s_k\|_2 + \frac{1}{2} \sigma \right) \leqslant 1 - \rho_2$$

(again, in view of $\eta_k \to 0, s_k \to 0$, this defines σ for (2.7)) then, because $q(d_k, c_*, \bar{\pi}_k) \leqslant 0$, (2.7) is satisfied with $\tau_k = 1$. □

Corollary 4.1 (Applications to existing algorithms)
With the same assumptions as in Theorem 4.1, the stepsize strategies (2.4) and (2.7) are satisfied for $\tau_k = 1$ *if*

(i) for $k \geqslant k_*, \Psi_k \leqslant \sigma$, *where*

$$\Psi_k = \frac{\|(\mathfrak{Q}_* - \hat{\mathfrak{Q}}_k)d_k\|}{\|d_k\|};$$

(ii) for $k \geqslant k_*$,

$$\left\| \begin{matrix} g_k \\ \bar{h}_k \end{matrix} \right\| = 0 \quad \text{and} \quad \Psi_k \leqslant \sigma,$$

where

$$\Psi_k = \frac{\|P_k^{\mathrm{T}}(\mathfrak{Q}_* - \hat{\mathfrak{Q}}_k)P_k d_k\|}{\|d_k\|}; \tag{4.18}$$

(iii) for $k \geqslant k_*$

$$\lim_{k\to\infty} \left\| \begin{matrix} g_k \\ \bar{h}_k \end{matrix} \right\| \Big/ \|d_k\| = 0 \quad \text{and} \quad \Psi_k \leqslant \sigma,$$

where Ψ_k *is given by (4.18).*

Proof

To establish (i) we simply need to repeat the proof of Theorem 4.1 without invoking expression (4.14). (ii) and (iii) can be shown by noting that the above conditions on g_k and $\bar{h}_k$ reduce Ψ_k in (4.15) and (4.16) to expression (4.18). □

Case (i) in the above corollary is related to Algorithms 2.1 and 2.2. Case (ii) relates to the Goldstein–Levitin–Polyak algorithm, in which constraint satisfaction is ensured at every iteration (see Rustem, 1984). Case (iii) relates to the equality-constrained algorithm, in reduced dimensions, proposed by Gabay (1982).

5 AN INTERIOR POINT ALGORITHM

As an alternative to solving an inequality-constrained QPS like (1.8) at every iteration, it is possible to formulate a barrier function that ensures the eventual satisfaction of the inequality constraints (Fiacco and McCormick, 1968). For simplicity, we consider the problem

$$\min \{f(x) \mid g(x) = 0; x \geqslant 0\}, \tag{5.1}$$

and, with the terminology introduced in this chapter, and Section 2.4, we define the penalty barrier function

$$l_\eta(x, c) \equiv f(x) + \frac{1}{2}c\|g(x)\|_2^2 - \eta \sum_{j=1}^{n} \ln x^j, \quad c \geqslant 0, \quad \eta > 0. \tag{5.2}$$

Problem (5.1) is replaced by

$$\min \{l_\eta(x, c) \mid g(x) = 0\}. \tag{5.3}$$

We note that (5.3) imposes $x > 0$, and we need to solve a sequence of problems, for decreasing values of η, in order to approach the boundary of the inequalities and for increasing values of the penalty parameter c, in order to enforce the satisfaction of the equality constraints. Let $x(c, \eta)$ and $\lambda(c, \eta)$ solve (5.3) for a given pair of penalty and barrier parameters, $X = \operatorname{diag}(x(c, \eta))$ (i.e. the diagonal matrix whose ith element is the ith element of $x(c, \eta)$) and let $\mathit{1}$ denote the n-dimensional vector whose elements are all unity. It follows from Chapter 1 that the first-order necessary conditions for optimality for (5.1) are given by

$$\nabla f(x(c, \eta)) + \nabla g(x(c, \eta))[cg(x(c, \eta)) + \lambda(c, \eta)] - \eta X^{-1} \mathit{1} = 0, \tag{5.4a}$$

$$g(x(c, \eta)) = 0, \tag{5.4b}$$

or, by introducing the nonlinear transformation $z = \eta X^{-1} \mathit{1}$, with $Z = \operatorname{diag}(z)$,

$$\nabla f(x(c, \eta)) + \nabla g(x(c, \eta))[cg(x(c, \eta)) + \lambda(c, \eta)] + z = 0, \tag{5.5a}$$

$$X\, Z\, \mathit{1} = \eta \mathit{1}, \qquad g(x(c, \eta)) = 0. \tag{5.5b}$$

The application of the Newton algorithm to solve (5.4) or (5.5), for η decreasing gradually to zero, is the main aim. It should be noted that the Newton algorithm applied to either of these systems would not follow the same path due to the nonlinear transformation used in the latter. However, in either case, the paths end at a solution of the common system (El-Bakry *et al.*, 1996). Algorithms for solving these systems are considered by a number of authors (e.g. El-Bakry *et al.*, 1996; Yamashita and Yabe, 1996). The penalty parameter c can be adjusted to ensure that the algorithm generates a descent direction for $l_\eta(x, c)$, as well as $f(x) + \frac{1}{2} c \|g(x)\|_2^2$ (Akrotirianakis, 1998). Thus, if

$$c_{k+1} = \max \left\{ \frac{\langle d_k, \nabla f_k - \eta X^{-1} \mathit{1} + \epsilon_1 \hat{\mathfrak{Q}}_k d_k \rangle}{\|g_k\|_2^2}, c_k + \delta \right\}, \tag{5.6}$$

or if

$$c_{k+1} = \max \left\{ \frac{\langle d_k, \nabla f_k + \epsilon_1 \hat{\mathfrak{Q}}_k d_k \rangle}{\|g_k\|_2^2}, c_k + \delta \right\}, \tag{5.7}$$

then d_k generated by the quasi-Newton step for solving (5.4) or (5.5) is a descent direction for $l_\eta(x, c)$, and $f(x) + \frac{1}{2} c \|g(x)\|_2^2$ such that

$$\langle \nabla l_\eta(x_k, c_{k+1}), d_k \rangle \leqslant -\epsilon_1 \langle d_k \mathfrak{Q}_k d_k \rangle, \quad \epsilon_1 \in (\tfrac{1}{2}, 1], \tag{5.8}$$

$$\langle \nabla f(x_k) + c_{k+1} \nabla g_k g_k, d_k \rangle \leqslant -\epsilon_1 \langle d_k, \hat{\mathfrak{Q}}_k d_k \rangle, \quad \epsilon_2 \in (\tfrac{1}{2}, 1]. \tag{5.9}$$

The result follows directly from Lemma 2.3 with the use of the linearized equality constraints in (2.12c).

6 CONCLUDING REMARKS

Algorithms 2.1 and 2.2 use a Hessian approximation that is consistent with the quadratic subproblem and the stepsize strategies. The achievement of unit stepsizes can thereby be shown to depend on the accuracy of this approximation. An alternative Hessian approximation that could be utilized is one that ignores the second-derivative terms of the constraints arising from the penalty function. The desirable properties of this choice are discussed in Tapia (1986). As the algorithm converges to a feasible point, it can easily be shown that this approximation would also satisfy Theorem 4.1.

The existence of the Hessian of the augmented Lagrangian is essential for the above results. In the equality-constrained case, this is ensured by the differentiability of the objective and constraint functions. In the inequality-constrained case, however, this existence is ensured by a simple smearing process incorporated in the algorithm.

The penalty parameter c is adjusted to ensure a descent property. It is shown that any growth in c is not indefinite and that a bound is maintained.

7 COMMENTS AND NOTES

CN.1 Notation for sequences

An operator evaluated at x_* or at x_k is sometimes denoted without the argument, by using only the subscript $*$ or k. Thus, $g_* = g(x_*), h_k = h(x_k)$, $f_k = f(x_k)$. Similarly, $\nabla f_k = \nabla f(x_k)$, and the matrices $\nabla g_k = \nabla g(x_k)$ and $\nabla h_k = \nabla h(x_k)$ are such that their columns are the gradients, at x_k, of the elements of $g(x_k)$ and $h(x_k)$ respectively.

CN.2 Exact penalty approaches

Another exact penalty approach to overcome the Maratos effect is due to Dew (1985a, b). Discussions of exact penalty functions can be found in Bertsekas (1982) and di Pillo and Grippo (1989).

CN.3 Boundedness and the QPS

In Theorem 3.1, the boundedness conditions can be equivalently associated with the properties of the QPS. Along with assumptions (i) and (ii) in the theorem, boundedness is ensured when the active constraint normals at the solution of the QPS are linearly independent. Another condition implied by (iii) is that the feasible set of the QPS is nonempty.

REFERENCES

Akrotirianakis, I. (1998). An interior point algorithm for nonlinear programming. PhD Thesis, Department of Computing, Imperial College, London.

Bartholomew-Biggs, M.C.B. (1974). The development of a class of constrained minimization algorithms and their application to the problem of power scheduling. PhD Thesis, University of London.

Bartholomew-Biggs, M.C.B. (1978). On the convergence of some constrained minimization algorithms based on recursive quadratic programming. *J. Inst. Maths. Applic.* **21**, 67–81.

Bartholomew-Biggs, M.C.B. (1981). Line search procedures for nonlinear programming algorithms with quadratic programming subproblems. Numerical Optimization Centre, Report 116, Hatfield Polytechnic.

Bartholomew-Biggs, M.C.B. (1987). Recursive quadratic programming methods Based on the augmented Lagrangian. *Math. Programming Study* **31**, 21–41.

Bartholomew-Biggs M.C.B. (1988). A globally convergent version of REQP for constrained optimization. *IMA J. Numer. Anal.* **8**, 253–271.

Bartholomew-Biggs, M.C. and T.T. Nguyen (1990). Orthogonal and conjugate basis methods for solving equality constrained minimization problems. Numerical Optimization Centre, Report 222, Hatfield Polytechnic.

Bellman, R. (1995). *Introduction to Matrix Analysis*. SIAM, Philadelphia.

Bertsekas, D.P. (1982). *Constrained Optimization and Lagrange Multiplier Methods*. Academic Press, New York.

Boggs, P.T., J.W. Tolle and P. Wang (1982). On the local convergence of quasi-newton methods for constrained optimization. *SIAM J. Control Optim.* **20**, 161–171.

Burke, J.V. and S.-P. Han (1989). A robust sequential quadratic programming method. *Math. Programming* **43**, 277–303.

Chamberlain, R.M. (1978). Some examples of cycling in variable metric methods for constrained minimization. *Math. Programming* **16**, 378–383.

Chamberlain, R.M., M.J.D. Powell, C. Lemarechal and H.C. Pedersen (1982). The watchdog technique for forcing convergence in algorithms for constrained optimization. *Math. Programming Study* **16**, 1–17.

Chiang, J. and R. Tapia (1989). Convergence rates for the variable, the multiplier, and the pair in SQP methods. Technical Report 89–2, Rice University.

Coleman, T.F. and A.R. Conn (1982). Nonlinear programming via an exact penalty function: asymptotic analysis. *Math. Programming* **24**, 123–136.

Coleman, T.F. and A.R. Conn (1984). On the local convergence of a quasi-Newton method for the nonlinear programming problem. *SIAM J. Numer. Anal.* **21**, 755–769.

Dew, M.C. (1985a). An exact penalty function algorithm for accurate optimisation of industrial problems. PhD Thesis, Hatfield Polytechnic.

Dew, M.C. (1985b). An algorithm for nonlinear equality constrained optimisation using the di Pillo–Grippo exact penalty function. Numerical Optimization Centre, Report 159, Hatfield Polytechnic.

di Pillo, G. and L. Grippo (1989). Exact penalty functions in constrained optimization. *SIAM J Control Optim.* **27**, 1333–1360.

Dixon, L.C.W. (1972). The choice of step length, a crucial factor in the performance of variable metric algorithms. In *Numerical Methods for Non-Linear Optimization* (F.A. Lootsma, ed.). Academic Press, London.

Dixon, L.C.W. (1980). On the convergence properties of variable metric recursive quadratic programming algorithms. Proceedings of Xth International Symposium on Mathematical Programming, Montreal; Numerical Optimization Centre, Report 110, Hatfield Polytechnic.

Dixon, L.C.W. (1987). Conjugate bases for constrained optimisation. Proceedings of ICIAM 87, Paris; Numerical Optimization Centre, Report 181, Hatfield Polytechnic.

El-Bakry, A.S., R.A. Tapia, T. Tsuchiya and Y. Zhang (1996). On the formulation and theory of the Newton interior-point method for nonlinear programming. *J. Optim. Theory Applic.* **89**, 507–541.

Fan, Y, S. Sarkar and L. Lasdon (1988). Experiments with successive quadratic programming algorithms, *J. Optim. Theory Applic.* **56**, 359–383.

Fiacco, A.V. and G.P. McCormick (1968). *Nonlinear Programming: Sequential Unconstrained Minimization Techniques.* Wiley, New York.

Finsler, P. (1937). Über das Vorkommen definiter und semidefiniter Formen in schären quadratischen Formen. *Comment. Math. Helv.* **9**, 188–192.

Fontecilla, R., T. Steihaug and R.A. Tapia (1987). A convergence theory for a class of quasi-Newton methods for constrained optimization. *SIAM J. Numer Anal.* **24**, 1133–1151.

Gabay, D. (1982). Reduced quasi-Newton methods with feasibility improvement for nonlinearly constrained optimization. *Math. Programming Study* **16**, 18–44.

Gill, P.E., W. Murray, M.A. Saunders and M.H. Wright (1986). Some theoretical properties of an augmented Lagrangian merit function. Technical Report SOL 86-6R, Stanford University.

Han, S.-P. (1976). Superlinearly convergent variable metric algorithms for general nonlinear programming problems. *Math. Programming* **11**, 263–282.

Han, S.-P. (1977). A globally convergent method for nonlinear programming. *J. Optim. Theory Applic.* **22**, 297–309.

Maratos, N, (1978). Exact penalty function algorithms for finite dimensional and control optimization problems. PhD Thesis, University of London.

Mayne, D.Q. and E. Polak (1982). A superlinearly convergent algorithm for constrained minimization problems. *Math. Programming Study* **16**, 45–61.

Palomares, U.M.G. and O.L. Mangasarian (1976). Superlinearly convergent quasi-Newton algorithms for nonlinearly constrained optimization problems. *Math. Programming* **11**, 1–13.

Polak, E. and D.Q. Mayne (1981). A robust secant method for optimization problems with inequality constraints. *J. Optim. Theory Applic.* **33**, 463–477.

Polak, E. and A.L. Tits (1980). A globally convergent, implementable multiplier method with automatic penalty limitation. *Appl. Math. Optim.* **6**, 335–360.

Powell, M.J.D. (1978). A fast algorithm for nonlinearly constrained optimization calculations. In *Numerical Analysis Proceedings, Dundee 1977* (G.A. Watson, ed.). Springer-Verlag, Berlin.

Powell, M.J.D. (1980). Gradient conditions and Lagrange multipliers in nonlinear programming. In *Nonlinear Optimization* (L.C.W. Dixon, E. Spedicato and G.P. Szego, eds). Birkhäuser, Boston.

Powell, M.J.D. (1983). Variable metric methods for constrained optimization. In *Mathematical Programming. The State of the Art* (A. Bachem, M. Grötschel and B. Korte, eds). Springer-Verlag, Berlin.

Powell, M.J.D. and Y. Yuan (1986). A recursive quadratic programming algorithm that uses differentiable exact penalty functions. *Math. Programming* **35**, 265–278.

Rustem, B. (1984). A class of superlinearly convergent projection algorithms with relaxed stepsizes. *Appl. Math. Optim.* **12**, 29–43.

Rustem, B. (1986). Convergent stepsizes for constrained optimization algorithms. *J. Optim. Theory Applic.* **49**, 135–160.

Rustem, B. (1993). Equality and inequality constrained optimization algorithms with convergent stepsizes. *J. Optim. Theory Applic.* **76**, 429–453.

Rustem, B. (1994). The two-step and three-step Q-superlinear convergence of successive quadratic programming algorithms. *J. Optim. Theory Applic.* **83**, 613–619.

Schittkowski, K. (1981a). The nonlinear programming method of Wilson, Han and Powell with an augmented Lagrangian type line search function, Part 1: Convergence analysis. *Numer. Math.* **38**, 84–114.

Schittkowski, K. (1981b). The nonlinear programming method of Wilson, Han and Powell with an augmented Lagrangian type line search function, Part 2: An efficient implementation with linear least squares subproblems. *Numer. Math.* **38**, 115–127.

Schittkowski, K. (1983). On the convergence of a sequential quadratic programming method with an augmented Lagrangian line search function. *Math. Operationsforsch Statistik, Ser. Optim.* **14**, 197–216.

Stoer, J. and R.A. Tapia (1986). On the characterization of Q-superlinear convergence of quasi-Newton methods for constrained optimization. Technical Report 84–2, Rice University.

Tapia, R.A. (1977). Diagonalized multiplier methods and quasi- Newton methods for constrained optimization. *J. Optim. Theory Applic.* **22**, 135–194.

Tapia, R.A. (1986). On secant updates for use in general constrained optimization. Technical Report 84–3, Rice University.

Wilson, R.B. (1963). A simplical algorithm for concave programming. PhD Dissertation, Graduate School of Business Administration, Harvard University.

Yamashita, H. and H. Yabe (1996). Superlinear and quadratic convergence of some primal–dual interior point methods for constrained optimization. *Math. Programming* **75**, 377–406.

CHAPTER 8

Convergence Rates of SQP Algorithms

In this chapter, we investigate the rates of convergence of sequential quadratic programming algorithms discussed in Chapter 7. Convergence rates are important to assess the effectiveness and computational efficiency of algorithms. These rates can be evaluated in a neighbourhood of the optimum where the behaviour of the algorithm can be accurately predicted and measured. For example, as in the discussion leading to the achievement of unit steps, all inequality constraints active at the solution are treated as equality constraints in a neighbourhood of the solution. This is, nevertheless, a simplification for algorithms that have special augmented Lagrangian features that do not simply disappear in this neighbourhood. An appropriate treatment is discussed in Section 4.

We consider the convergence rates for the variable x and for the variable–multiplier pair (x, λ) in SQP methods for equality-constrained optimization and establish the condition for the Q-superlinear convergence of the sequence $\{x_k\}$. In Section 2, we show that this rate implies the two-step Q-superlinear convergence of the pair $\{x_k, \lambda_k\}$. In Section 3, we establish the condition for the two-step Q-superlinear convergence of $\{x_k\}$, and show that this implies the three-step Q-superlinear convergence of $\{x_k, \lambda_k\}$. The former is associated with the one-sided projected Hessian condition, in Section 2, and the latter is associated with the two-sided projected Hessian condition that we have already encountered in Chapter 6. In Section 4, we consider the effect of strategies for the inequality-constrained problem. We also show that these results can be used to analyse related algorithms discussed in Chapter 6 and sequential quadratic programming variants using a restoration strategy to maintain or improve feasibility.

1 INTRODUCTION TO THE CONVERGENCE RATES OF SQP ALGORITHMS

In this chapter, we consider first the local convergence properties of a generic class of SQP algorithms (see e.g. Rustem, 1986, 1993, 1994) that solve the equality-constrained nonlinear programming problem

$$\min \{f(x) \mid g(x) = 0\}, \tag{1.1}$$

where $f : \mathbb{R}^n \to \mathbb{R}, g : \mathbb{R}^n \to \mathbb{R}^e (e \leqslant n)$. We discussed such an algorithm extensively in Chapter 7. The equality-constrained case is of interest since, by Theorem 7.3.2(ii), in a neighbourhood of the optimum, the inequality constraints satisfied as equalities at the solution can be treated as equalities. The remaining inequality constraints do not affect the solution. The discussion applies to those SQP algorithms that employ unit steps and ensure progress towards feasibility without altering the QPS, as in (1.3) below. The algorithm in Chapter 7 does employ a complex strategy to ensure feasibility, which does affect the QPS even after unit steps have been achieved. We analyse the complications arising from this later in Section 4.

For Sections 1–3, the basic *sequential quadratic programming* (SQP) algorithm consists of the iterative procedure

$$x_{k+1} = x_k + d_k, \tag{1.2}$$

where d_k solves the QPS

$$\min \{q(d) \mid \mathscr{G}_k^{\mathrm{T}} d + g_k = 0\} \tag{1.3}$$

(see CN. 1 in Chapter 7). The objective function of this QPS is the quadratic approximation

$$q(d) = \langle d, \nabla \ell_*(x_k) \rangle + \frac{1}{2} \langle d, \hat{\mathfrak{Q}}_k d \rangle, \tag{1.4}$$

$$\ell_*(x) \equiv f(x) + \frac{1}{2} c_* \|g(x)\|_2^2, \tag{1.5}$$

with $\nabla \ell_* = \nabla_x \ell_*$ and a finite fixed value $c_* \geqslant 0$. The Lagrangian associated with (1.3) is given by

$$q(d) + \langle (\mathscr{G}_k^{\mathrm{T}} d + g_k), \lambda \rangle.$$

Let λ_{k+1} denote the multiplier corresponding to d_k. The ith column of the matrix $\mathscr{G}_k$ is the gradient of the ith element of $g(x)$, at x_k.$\hat{\mathfrak{Q}}_k$ is a bounded approximation ($\|\hat{\mathfrak{Q}}_k\| < \infty$) to the Hessian $\mathfrak{Q}_k$, with respect to x, at (x_k, λ_k), of the augmented Lagrangian $L^{\mathrm{a}} : \mathbb{R}^n \times \mathbb{R}^e \to \mathbb{R}$

$$L^{\mathrm{a}}(x, \lambda, c_*) = \ell_*(x) + \langle g(x), \lambda \rangle. \tag{1.6}$$

The second-derivative terms arising from the constraints in the penalty function in (1.6) may be ignored in $\hat{\mathfrak{Q}}_k$ (Tapia, 1986). The results below remain applicable in this case, since these terms vanish at the solution. The results in this chapter also do not invoke the inherent positive-definiteness of the augmented Lagrangian Hessian (Arrow *et al.*, 1973; Rockafeller, 1973).

1.1 Studying the convergence rates of the variable and variable–multiplier pair

Convergence rates were first encountered in Chapter 6, where it was established that the necessary and sufficient condition for Q-superlinear convergence is given by the two-sided projected Hessian condition. A study of the rate of convergence of $\{x_k\}$ generated by SQP algorithms is similarly of interest, since computational effectiveness depends on the speed of convergence. Furthermore, the variable and multiplier pair is generated simultaneously with, and related to, an associated optimal decision. Thus, the convergence rate of $\{x_k, \lambda_k\}$ is of interest in its own right despite the fact that the Q-superlinear convergence of $\{x_k, \lambda_k\}$ can imply only a R-superlinear rate for $\{x_k\}$ (see CN.1; Boggs, Tolle and Wang, 1982). One reason is that, although for a given value of $\hat{\mathfrak{Q}}_k$, the penalty parameter c_* does not affect the direction generated by the QPS (1.3) (see (7.4.3)), it does affect the value of λ_k. A further reason is that the multipliers may contain relevant information as shadow prices in decision making problems (see e.g. UNIDO, 1972). Therefore, for practical reasons related to decisions, it might be useful to have some insight concerning the convergence rate of $\{x_k, \lambda_k\}$, as well as $\{x_k\}$, when the algorithm is being terminated.

Another motivation for studying the convergence rates of $\{x_k, \lambda_k\}$ and $\{x_k\}$ is to highlight the local results required to establish them. The one-sided projected Hessian condition (see (2.11b) below) of Boggs, Tolle and Wang (1982) for the convergence rate of $\{x_k\}$ is discussed, among others, by Fontecilla, Steihaug and Tapia (1987), Nocedal and Overton (1985), Powell (1983), Stoer (1985) and Stoer and Tapia (1986) without a linear convergence rate assumption (see also Biggs, 1978; Dixon 1980). In these studies, the common assumption, in its weakest form, appears to be second-order sufficiency. As shown in Theorem 2.1 below, this only ensures two-step Q-superlinear convergence of $\{x_k, \lambda_k\}$ *unless* the following local result is invoked:

$$r\|x_* - x_k\| \leqslant \|\lambda_* - \lambda_k\| \leqslant R\|x_* - x_k\|, \qquad r, R > 0, \quad k \geqslant k_0 \tag{1.7}$$

(Tapia, 1977, Proposition 9.1). We do not use this inequality in this chapter, but note that, if (1.7) is utilized, it is a simple matter to show the same (one-or two-step) superlinear rate of convergence for $\{x_k\}$ and $\{x_k, \lambda_k\}$. If (1.7) is not invoked, it can be shown that the Q-superlinear convergence of $\{x_k, \lambda_k\}$ requires the accuracy of the *full* Hessian approximation, i.e.

$$\lim_{k\to\infty} \frac{\|(\mathfrak{Q}_* - \hat{\mathfrak{Q}}_k)d_k\|}{\|d_k\|} = 0,$$

(see CN.2), where $\mathfrak{Q}_*$ denotes the augmented Lagrangian Hessian at the solution (see e.g. Glad, 1979; Han, 1976; Tapia, 1977). Therefore, we need to consider the implication of this for the projected Hessian conditions discussed by Boggs, Tolle and Wang (1982) and Powell (1978, 1983). Some variants of the SQP algorithm, do not exhibit this difficulty and, therefore, do not require Tapia's result (Coleman and Conn, 1984; Gabay, 1982). In the case of Coleman and Conn (1984), the reason is the additional step for computing λ_{k+1} as a function of $\nabla f_{k+1}, \nabla g_{k+1}$, as opposed to $\nabla f_k, \nabla g_k$ in (2.3) below. Coleman and Conn (1984, Lemma 3.1) use this to establish $\|\lambda_* - \lambda_k\| \leqslant R\|x_* - x_k\|$ and consider the two-sided projected Hessian condition (see (3.5)) of Powell (1978, 1983) to demonstrate the two-step Q-superlinear convergence of $\{x_k\}$. Inequality $\|\lambda_* - \lambda_k\| \leqslant R\|x_* - x_k\|$ can be used to establish the two-step Q-superlinear convergence of $\{x_k, \lambda_k\}$ for Coleman and Conn's algorithm. In Gabay's (1982) SQP algorithm, on the other hand, the additional step used towards feasibility is shown to ensure the Q-superlinear convergence of $\{x_k\}$ and the two-step Q-superlinear convergence of $\{x_k, \lambda_k\}$ (see Corollary 3.1 below). Gabay (1982) establishes, for $\{x_k\}$, two-step Q-superlinear convergence *without*, and Q-superlinear convergence *with*, the additional step.

Powell's (1978, 1983) two-sided projected Hessian condition for the two-step Q-superlinear convergence of $\{x_k\}$ is also discussed by Coleman and Conn (1982), Gabay (1982), Nocedal and Overton (1985) and Stoer (1985). Some of these studies assume that the approximate Hessian utilized by the QPS is positive-definite (e.g. Powell, 1978). Others require that the approximation is positive definite in the null space spanned by the constraint gradients (e.g. Coleman and Conn, 1984; Gabay, 1982) or just that it is bounded (e.g. Powell, 1983, Theorem 3). For the two-sided projected Hessian condition, it is shown in Theorem 3.1 below that only an inferior rate of convergence can be obtained for $\{x_k, \lambda_k\}$. However, using (1.7) would ensure the two-step Q-superlinear convergence of $\{x_k, \lambda_k\}$.

1.2 An outline of the convergence rate results

We begin with a derivation of Boggs, Tolle and Wang's (1982) one-sided projected Hessian condition. This is the *necessary and sufficient condition for the Q-superlinear convergence* of $\{x_k\}$, which also establishes the two-step Q-superlinear convergence of $\{x_k, \lambda_k\}$ (Theorem 2.1). The symmetric theorem establishing that the Q-superlinear convergence of $\{x_k, \lambda_k\}$ implies the two-step Q-superlinear convergence of $\{x_k\}$ is given in Chiang and Tapia (1989). Similarly, a derivation of Powell's (1978, 1983) two-step Q-superlinear convergence condition is established in Theorem 3.1 as an extension of Boggs, Tolle and Wang's (1982) result. It is shown that this rate implies the three-step Q-superlinear convergence of $\{x_k, \lambda_k\}$ (Rustem, 1994). The effects of inequality constraints are explored in Section 4.

2 Q-SUPERLINEAR RATE OF THE VARIABLE AND TWO-STEP Q-SUPERLINEAR RATE OF THE VARIABLE–MULTIPLIER PAIR

We consider the Q-superlinear convergence rates of $\{x_k\}$ and $\{x_k, \lambda_k\}$ for the basic SQP algorithm above. Apart from its mathematical relevance, the convergence rate of $\{x_k, \lambda_k\}$ is of interest because of the potential use of λ_* as the shadow price vector (see Chapter 4 and UNIDO, 1972). Also, the penalty parameter c_* does affect the values of the multipliers (see (2.3) below), and any possible implication of this on the convergence rates needs to be clarified.

Assumptions 2.1

(1) $f, g \in \mathbb{C}^2(\mathbb{R}^n)$.

(2) $\exists x_*, \lambda_*$ satisfying the first-order conditions of (1.1).

(3) For the sequence generated by (1.2) and (1.3), $\{x_k, \lambda_k\} \to (x_*, \lambda_*)$.

(4) $\langle v, \mathfrak{Q}_* v\rangle > 0,\ \forall v(\neq 0) \in \mathbb{R}^n, \mathcal{G}_*^T v = 0$, where $\mathcal{G}_*$ is the matrix value at x_*.

(5) The coefficient matrix $\begin{bmatrix} \hat{\mathfrak{Q}}_k & \mathcal{G}_k \\ \mathcal{G}_k^{\mathrm{T}} & 0 \end{bmatrix}$ is nonsingular (with $\|\hat{\mathfrak{Q}}_k\| < \infty$); or, equivalently, the QPS (1.3) has a feasible solution at each iteration (see CN.3).

Lemma 2.1

Let Assumptions 2.1 hold. Then, for some $e_0 \in [0, \infty)$,

$$\|d_k\| \leqslant e_0 \|x_* - x_k\|. \tag{2.1}$$

Proof

The first-order optimality condition of the QPS (1.3) can be written as

$$\begin{bmatrix} \hat{\mathfrak{Q}}_k & \mathcal{G}_k \\ \mathcal{G}_k^{\mathrm{T}} & 0 \end{bmatrix} \begin{bmatrix} d_k \\ \lambda_{k+1} - \lambda_* + c_* g_k \end{bmatrix} = -\begin{bmatrix} \nabla_x L(x_k, \lambda_*) \\ g_k \end{bmatrix},$$

where $L(x, \lambda)$ is the ordinary Lagrangian, given by $L(x, \lambda) = f(x) + \langle g(x), \lambda\rangle$. Using the first-order expansion of the right side of the above expression about (x_*, λ_*), by the nonsingularity of the coefficient matrix and by the first-order optimality condition $\nabla_x L(x_*, \lambda_*) = 0,\ g(x_*) = 0$, we have

$$\|d_k\| \leqslant \left\| \begin{matrix} d_k \\ \lambda_{k+1} - \lambda_* + c_* g_k \end{matrix} \right\| \leqslant \hat{e} \left\| \begin{bmatrix} \hat{\mathfrak{Q}}_k & \mathcal{G}_k \\ \mathcal{G}_k^T & 0 \end{bmatrix}^{-1} \right\| \|x_* - x_k\|,$$

for some $\hat{e} \in [0, \infty)$. Hence, (2.1) follows from the above expression. □

Lemma 2.2

Let $\{x_k\} \to x_*$. Then $\{x_k\}$ is Q-superlinearly convergent, i.e.

$$\|x_* - x_{k+1}\| \leqslant \beta_k^{(0)} \|x_* - x_k\|, \qquad \lim_{k\to\infty} \beta_k^{(0)} = 0,$$

if and only if

$$\|d_k\| \leqslant \beta_k \|d_{k-1}\|, \qquad \lim_{k\to\infty} \beta_k = 0. \tag{2.2}$$

Proof

We have

$$\begin{aligned} \|x_* - x_k\| &\leqslant \lim_{t\to\infty} \sum_{j=k}^{t-1} \|x_{j+1} - x_j\| \\ &\leqslant \beta_k \|d_{k-1}\| (1 + \omega + \omega^2 + \omega^3 + \cdots) \\ &\leqslant \frac{\beta_k}{1-\omega} (\|x_k - x_*\| + \|x_* - x_{k-1}\|) \end{aligned}$$

for some $\omega \in [0,1)$ and k sufficiently large. As $\{\beta_k\} \to 0, \omega$ is chosen such that $1 > \beta_k + \omega, \forall k \geqslant K_0$. K_0 is an integer and is such that $\beta_k < 1, \forall k \geqslant K_0$. Rearranging the above expression yields the required result.

Suppose that $\|x_* - x_k\| \leqslant \beta_k^{(0)} \|x_* - x_{k-1}\|$, $\lim_{k\to\infty} \beta_k^{(0)} = 0$, with $\beta_k^{(0)} < 1$. This yields the inequality

$$\begin{aligned}\|x_* - x_k\| &\leqslant \beta_k^{(0)}(\|x_* - x_k\| + \|d_{k-1}\|)\\ &\leqslant \frac{\beta_k^{(0)}}{1-\beta_k^{(0)}}\|d_{k-1}\|.\end{aligned}$$

Lemma 2.1 yields the desired result. □

The multiplier generated by the QPS (1.3), referred to in Lemma 2.3, is given by

$$\lambda_{k+1} = -[(\mathscr{G}_k^{\mathrm{T}}\mathscr{G}_k)^{-1}\mathscr{G}_k^{\mathrm{T}}(\nabla f_k + \hat{\mathfrak{Q}}_k d_k) + c_* g_k]. \tag{2.3}$$

Lemma 2.3

Let $\{x_k, \lambda_k\}$ converge to (x_*, λ_*), (2.2) be satisfied and $\nabla_x L(x, \lambda_*)$ be Lipschitz-continuous (see CN.1.8) in x. Then, we have

$$\|\lambda_* - \lambda_{k+1}\| \leqslant \beta_k^{(1)}\|x_* - x_{k-1}\|, \qquad \lim_{k\to\infty} \beta_k^{(1)} = 0. \tag{2.4}$$

Proof

Using (2.3), we can write

$$\lambda_{k+1} - \lambda_* = -[(\mathscr{G}_k^{\mathrm{T}}\mathscr{G}_k)^{-1}\mathscr{G}_k^{\mathrm{T}}(\nabla f_k + \mathscr{G}_k\lambda_* + \hat{\mathfrak{Q}}_k d_k) + c_* g_k]. \tag{2.5}$$

At the solution, the necessary condition for optimality requires the equality

$$\nabla_x L(x_*, \lambda_*) = \nabla f_* + \mathscr{G}_*\lambda_* = 0. \tag{2.6}$$

As $\nabla_x L(x, \lambda_*)$ is Lipschitz-continuous in x, we have, for $e_1 \in [0, \infty)$,

$$\|(\mathscr{G}_k^{\mathrm{T}}\mathscr{G}_k)^{-1}\mathscr{G}_k^{\mathrm{T}}\nabla_x L(x_k, \lambda_*)\| = \|(\mathscr{G}_k^{\mathrm{T}}\mathscr{G}_k)^{-1}\mathscr{G}_k^{\mathrm{T}}(\nabla f_k + \mathscr{G}_k\lambda_*)\| \leqslant e_1\|x_k - x_*\|. \tag{2.7}$$

Also, at the solution, $g(x_*) = 0$, and, for some $e_2 \in [0, \infty)$,

$$\|g_k\| \leqslant \|g_*\| + e_2\|x_k - x_*\| \leqslant e_2\|x_k - x_*\|. \tag{2.8}$$

Let $e_3 = \|(\mathscr{G}_k^T\mathscr{G}_k)^{-1}\mathscr{G}_k^{\mathrm{T}}\hat{\mathfrak{Q}}_k\| \in [0, \infty)$. Using (2.5)–(2.8), we have

$$\|\lambda_{k+1} - \lambda_*\| \leqslant (e_1 + e_2)\|x_k - x_*\| + e_3\|d_k\|, \tag{2.9}$$

which, using (2.1) of Lemma 2.1, yields

$$\|\lambda_{k+1} - \lambda_*\| \leqslant (e_1 + e_2 + e_3 e_0)\|x_k - x_*\|. \tag{2.10}$$

Given (2.2), Lemma 2.2 ensures the Q-superlinear rate for $\{x_k\}$ and thence the required result. □

Generically, inequality (2.10) is a known result (see e.g. Tapia, 1986, inequality (7.2)). We now establish the necessary and sufficient condition for the Q-superlinear rate of convergence of $\{x_k\}$. We also show that the Q-superlinear convergence of $\{x_k\}$ implies the two-step Q-superlinear convergence of $\{x_k, \lambda_k\}$. The symmetric result—that Q-superlinear convergence of $\{x_k, \lambda_k\}$ implies the two-step Q-superlinear convergence of $\{x_k\}$—is established in Chiang and Tapia (1989).

Theorem 2.1

(i) Let Assumptions 2.1 be satisfied and let

$$P_k = I - \mathcal{G}_k\left(\mathcal{G}_k^{\mathrm{T}}\mathcal{G}_k\right)^{-1}\mathcal{G}_k^{\mathrm{T}}. \tag{2.11a}$$

Then the sequence $\{x_k\}$ converges Q-superlinearly if and only if the one-sided projected Hessian condition

$$\lim_{k\to\infty} \frac{\|P_k(\mathfrak{Q}_* - \hat{\mathfrak{Q}}_k)d_k\|}{\|d_k\|} = 0 \tag{2.11b}$$

is satisfied.

(ii) The Q-superlinear rate of $\{x_k\}$ implies the two-step Q-superlinear rate of $\{x_k, \lambda_k\}$.

Proof

Consider the first-order expansion of $\nabla_x L^a$, where L^a is given by (1.6):

$$\nabla_x L^{\mathrm{a}}(x_k, \lambda_k, c_*) = \nabla_x L^{\mathrm{a}}(x_{k-1}, \lambda_k, c_*) + \hat{\mathfrak{Q}}_{k-1}d_{k-1} + \Delta\mathfrak{Q}(t)d_{k-1}, \tag{2.12}$$

with

$$\Delta\mathfrak{Q}(t) = \int_0^1 [\mathfrak{Q}(x_{k-1} + td_{k-1}) - \mathfrak{Q}_* + \mathfrak{Q}_* - \hat{\mathfrak{Q}}_{k-1}]\,\mathrm{d}t. \tag{2.13}$$

Clearly, as $\{x_k, \lambda_k\} \to (x_*, \lambda_*)$, we have $\{\int_0^1 \{\mathfrak{Q}(x_{k-1} + td_{k-1}) - \mathfrak{Q}_*\}\,\mathrm{d}t\} \to 0$.

The first-order optimality conditions for the QPS (1.3) at x_{k-1} ensure that in (2.12) $\nabla_x L^{\mathrm{a}}(x_{k-1}, \lambda_k, c_*) + \hat{\mathfrak{Q}}_{k-1}d_{k-1} = 0$. From the definition of L^a, we have

$$\nabla_x L^{\mathrm{a}}(x_k, \lambda_k, c_*) = \nabla f_k + \mathcal{G}_k(c_* g_k + \lambda_k). \tag{2.14}$$

The solution of the QPS at x_k satisfies the first-order condition

$$\nabla f_k + \hat{\mathfrak{Q}}_k d_k + \mathcal{G}_k(c_* g_k + \lambda_{k+1}) = 0. \tag{2.15}$$

Using (2.14) and (2.15) in (2.12), along with $\mathcal{G}_k^T d_k + g_k = 0$, yields

$$\begin{bmatrix} \hat{\mathfrak{Q}}_k & \mathcal{G}_k \\ \mathcal{G}_k^T & 0 \end{bmatrix} \begin{bmatrix} d_k \\ \lambda_{k+1} - \lambda_k \end{bmatrix} = - \begin{bmatrix} \Delta\mathfrak{Q}(t) d_{k-1} \\ g_k \end{bmatrix}. \tag{2.16}$$

Also, the second-order expansion of the constraints, with $g_{k-1} + \mathcal{G}_{k-1}^T d_{k-1} = 0$, yields, for some $e_4 \geqslant 0$,

$$\begin{aligned} \|g_k\| &= \left\| g_{k-1} + \mathcal{G}_{k-1}^{\mathrm{T}} d_{k-1} + \sum_i \int_0^1 (1-t)\langle d_{k-1}, \nabla^2 g^i(x_{k-1} + t d_{k-1}) d_{k-1}\rangle \, \mathrm{d}t \right\| \\ &\leqslant e_4 \|d_{k-1}\|^2. \end{aligned} \tag{2.17}$$

From (2.16), we have

$$\begin{bmatrix} \hat{\mathfrak{Q}}_k & \mathcal{G}_k \\ \mathcal{G}_k^T & 0 \end{bmatrix} \begin{bmatrix} d_k \\ \lambda_{k+1} - \lambda_k + \left(\mathcal{G}_{k-1}^{\mathrm{T}} \mathcal{G}_{k-1}\right)^{-1} \mathcal{G}_{k-1}^{\mathrm{T}} \Delta\mathfrak{Q}(t) d_{k-1} \end{bmatrix} = - \begin{bmatrix} P_{k-1} \Delta\mathfrak{Q}(t) d_{k-1} \\ g_k \end{bmatrix}. \tag{2.18}$$

Since the matrix on the left is assumed nonsingular, (2.11b), (2.18), (2.17) and $\{\|d_k\|\} \to 0$ yield $\|d_k\| \leqslant \beta_k \|d_{k-1}\|$, with $\lim_{k\to\infty} \beta_k = 0$. Thence, Lemma 2.2 yields the Q-superlinear rate of $\{x_k\}$.

Suppose, conversely, that $\{x_k\}$ converges Q-superlinearly and thence $\|d_k\| \leqslant \beta_k \|d_{k-1}\|$, with $\lim_{k\to\infty} \beta_k = 0$. Premultiplying the first block of (2.16) by P_k and using the identity

$$P_k d_k \equiv d_k - \mathcal{G}_k \left(\mathcal{G}_k^{\mathrm{T}} \mathcal{G}_k\right)^{-1} \mathcal{G}_k^{\mathrm{T}} d_k,$$

$$P_k \hat{\mathfrak{Q}}_k P_k d_k + P_k \hat{\mathfrak{Q}}_k \mathcal{G}_k \left(\mathcal{G}_k^{\mathrm{T}} \mathcal{G}_k\right)^{-1} \mathcal{G}_k^{\mathrm{T}} d_k = -(P_k - P_{k-1}) \Delta\mathfrak{Q}(t) d_{k-1} - P_{k-1} \Delta\mathfrak{Q}(t) d_{k-1}.$$

Assumption 2.1(5) ensures that $\|P_k \hat{\mathfrak{Q}}_k P_k\| \in (0, \infty)$. Also, as $\{\|d_k\|\} \to 0$, we have

$$\{\|P_k - P_{k-1}\|\} \to 0.$$

In view of $\mathcal{G}_k^{\mathrm{T}} d_k = -g_k$ and (2.17), dividing the above expression by $\|d_{k-1}\|$ yields (2.11b) as required.

Consider the convergence rate of $\{x_k, \lambda_k\}$. From (2.4), Lemma 2.2 and the Q-superlinear convergence of $\{x_k\}$, established above, we have

$$\begin{aligned} \left\| \begin{matrix} x_* - x_k \\ \lambda_* - \lambda_k \end{matrix} \right\| &\leqslant \left[\beta_k^{(1)} + \frac{\beta_k \beta_{k-1}}{(1-\omega-\beta_k)(1-\omega-\beta_{k-1})} \right] \|x_* - x_{k-2}\| \\ &\leqslant \left[\beta_k^{(1)} + \frac{\beta_k \beta_{k-1}}{(1-\omega-\beta_k)(1-\omega-\beta_{k-1})} \right] \left\| \begin{matrix} x_* - x_{k-2} \\ \lambda_* - \lambda_{k-2} \end{matrix} \right\|. \end{aligned}$$

As $\{\beta_k^{(1)}\}, \{\beta_k\} \to 0$, the two-step Q-superlinear convergence of $\{x_k, \lambda_k\}$ follows. □

3 TWO-STEP Q-SUPERLINEAR CONVERGENCE OF THE VARIABLE AND THREE-STEP Q-SUPERLINEAR CONVERGENCE OF THE VARIABLE-MULTIPLIER PAIR

In this section, we discuss the convergence rates corresponding to the two-sided projected Hessian condition first encountered in Chapter 6. Due to the feasibility of the sequence generated by the GLP algorithm, this condition ensured a much stronger result in Chapter 6. The basic techniques used below are similar to those in Section 2. Lemma 3.1 is essentially a corollary of Lemma 2.3, and is employed to establish the convergence rate of $\{x_k, \lambda_k\}$ when $\{x_k\}$ is two-step Q-superlinearly convergent.

Lemma 3.1

Let Assumptions 2.1 and the following condition be satisfied:

$$\|d_k\| \leqslant \beta_k^{(2)} \|d_{k-2}\|, \quad \text{with} \quad \lim_{k\to\infty} \beta_k^{(2)} = 0. \tag{3.1}$$

We then have,

$$\|\lambda_* - \lambda_k\| \leqslant \beta_k^{(3)} \|x_* - x_{k-3}\|, \quad \text{with} \quad \lim_{k\to\infty} \beta_k^{(3)} = 0. \tag{3.2}$$

Proof

We can conclude from (2.16) and (2.17) that, for bounded $\hat{\mathfrak{Q}}_k$, we have $e_5 \in [0, \infty)$,

$$\|d_k\| \leqslant e_5 \|d_{k-1}\|. \tag{3.3}$$

Also, using (2.16) and (2.17) with (3.3) yields

$$\|\lambda_{k+1} - \lambda_k\| \leqslant e_6 \|d_{k-1}\| + e_7 \|d_{k-1}\|^2 \leqslant (e_6 + e_7\|d_{k-1}\|)\|d_{k-1}\|,$$

with $e_6, e_7 \geqslant 0$. Using (3.1), with

$$\beta_k^{(4)} = \beta_{k-1}^{(2)}(e_6 + e_7\|d_{k-1}\|),$$

this can be rewritten as

$$\|\lambda_{k+1} - \lambda_k\| \leqslant \beta_k^{(4)} \|d_{k-3}\|. \tag{3.4}$$

Combining (3.4) with (2.1) and (3.3) leads, for some $\omega \in [0, 1)$, to

$$\begin{aligned}
\|\lambda_* - \lambda_k\| &\leqslant \lim_{t\to\infty} \sum_{j=k}^{t-1} \|\lambda_{j+1} - \lambda_j\| \\
&\leqslant \lim_{t\to\infty} \sum_{j=k}^{t-1} \beta_j^{(4)} \|d_{j-3}\| \\
&\leqslant \beta_k^{(4)} (\|d_{k-3}\| + \|d_{k-2}\|)(1 + \omega + \omega^2 + \omega^3 \ldots) \\
&\leqslant \frac{e_0(1 + e_5)\beta_k^{(4)}}{1 - \omega} \|x_* - x_{k-3}\|
\end{aligned}$$

□

Theorem 3.1

(i) Let Assumptions 2.1 be satisfied and let P_k be given by (2.11a). Then the sequence $\{x_k\}$ converges at a two-step Q-superlinear rate if one of the following two-sided projected Hessian conditions is satisfied:

$$\lim_{k\to\infty} \frac{\|P_k(\mathfrak{Q}_* - \hat{\mathfrak{Q}}_k)P_k d_k\|}{\|d_k\|} = 0 \tag{3.5a}$$

or

$$\lim_{k\to\infty} \frac{\|P_k(\mathfrak{Q}_* - \hat{\mathfrak{Q}}_k)P_k d_k\|}{\|d_{k-1}\|} = 0. \tag{3.5b}$$

(ii) The two-step Q-superlinear rate of $\{x_k\}$ implies the three-step Q-superlinear rate of $\{x_k, \lambda_k\}$.

Proof

Consider the equality derived using the linearized constraints:

$$\begin{aligned} d_{k-1} &= P_{k-1}d_{k-1} + \mathcal{G}_{k-1}(\mathcal{G}_{k-1}^{\mathrm{T}}\mathcal{G}_{k-1})^{-1}\mathcal{G}_{k-1}^{\mathrm{T}}d_{k-1} \\ &= P_{k-1}d_{k-1} - \mathcal{G}_{k-1}(\mathcal{G}_{k-1}^{\mathrm{T}}\mathcal{G}_{k-1})^{-1}g_{k-1}. \end{aligned}$$

Using this to replace d_{k-1} in (2.18) yields, for some $e_8 \in [0, \infty)$,

$$\|d_k\| \leqslant e_8[\|P_{k-1}\Delta\mathfrak{Q}(t)P_{k-1}d_{k-1}\| + \left\|P_{k-1}\Delta\mathfrak{Q}(t)\mathcal{G}_{k-1}(\mathcal{G}_{k-1}^{\mathrm{T}}\mathcal{G}_{k-1})^{-1}\right\|\|g_{k-1}\|]. \tag{3.6}$$

We can rewrite (3.6) by using (2.17) to express g_{k-1} in terms of $\|d_{k-2}\|^2$,

$$\begin{aligned} \|d_k\| \leqslant e_8 \Bigg[& \frac{\|P_{k-1}\Delta\mathfrak{Q}(t)P_{k-1}d_{k-1}\|}{\|d_{k-2}\|} \\ & + e_4\left\|P_{k-1}\Delta\mathfrak{Q}(t)\mathcal{G}_{k-1}(\mathcal{G}_{k-1}^{\mathrm{T}}\mathcal{G}_{k-1})^{-1}\right\|\|d_{k-2}\|\Bigg]\|d_{k-2}\|, \end{aligned} \tag{3.7}$$

or, alternatively, using (3.3),

$$\begin{aligned} \|d_k\| \leqslant e_8 \Bigg[& \frac{e_5\|P_{k-1}\Delta\mathfrak{Q}(t)P_{k-1}d_{k-1}\|}{\|d_{k-1}\|} \\ & + e_4\left\|P_{k-1}\Delta\mathfrak{Q}(t)\mathcal{G}_{k-1}(\mathcal{G}_{k-1}^{\mathrm{T}}\mathcal{G}_{k-1})^{-1}\right\|\|d_{k-2}\|\Bigg]\|d_{k-2}\|. \end{aligned} \tag{3.8}$$

Since $\{\|d_k\|\} \to 0$, expression (3.7) can be written as (3.1) provided (3.5b) holds, and expression (3.8) can be written as (3.1) provided (3.5a) holds.

We have established that (3.5) ensures (3.1). We now show that (3.1) and $\{x_k\} \to x_*$ entail the two-step Q-superlinear convergence of $\{x_k\}$. Using (3.1), (3.3) and (2.1), we can write

$$\begin{aligned}\|x_* - x_k\| &\leqslant \lim_{t\to\infty} \sum_{j=k}^{t-1} \|x_{j+1} - x_j\| \\ &\leqslant (\|d_k\| + \|d_{k+1}\|)(1 + \omega + \omega^2 + \omega^3 + \ldots) \\ &\leqslant \frac{1+e_5}{1-\omega}\beta_k^{(2)}\|d_{k-2}\| \qquad (3.9) \\ &\leqslant \frac{(1+e_5)e_0}{1-\omega}\beta_k^{(2)}\|x_* - x_{k-2}\| \qquad (3.10)\end{aligned}$$

for some $\omega \in [0, 1)$. Thus,

$$\lim_{k\to\infty} \frac{\|x_* - x_k\|}{\|x_* - x_{k-2}\|} = 0, \qquad (3.11)$$

which is the two-step Q-superlinear condition of $\{x_k\}$.

Combining (3.9) with (3.3), (3.2) and (2.1) yields

$$\begin{aligned}\left\|\begin{matrix} x_* - x_k \\ \lambda_* - \lambda_k \end{matrix}\right\| &\leqslant \beta_k^{(3)}\|x_* - x_{k-3}\| + \frac{1+e_5}{1-\omega}\beta_k^{(2)}\|d_{k-2}\| \\ &\leqslant \beta_k^{(3)}\|x_* - x_{k-3}\| + \frac{e_5(1+e_5)}{1-\omega}\beta_k^{(2)}\|d_{k-3}\|, \\ &\leqslant \left[\beta_k^{(3)} + \frac{e_0 e_5(1+e_5)}{1-\omega}\beta_k^{(2)}\right]\left\|\begin{matrix} x_* - x_{k-3} \\ \lambda_* - \lambda_{k-3} \end{matrix}\right\|\end{aligned}$$

which establishes the convergence rate of $\{x_k, \lambda_k\}$. □

Under certain circumstances, the convergence rate estimate (3.11) can be improved. The algorithms to which these improved estimates correspond are variants of the standard SQP algorithm.

Corollary 3.1

Let the assumptions of Theorem 3.1 hold and let

$$\lim_{k\to\infty} \frac{\|g_k\|}{\|d_k\|} = 0.$$

Then $\{x_k\}$ converges Q-superlinearly and $\{x_k, \lambda_k\}$ converges two-step Q-superlinearly.

Proof

Consider (3.6) written as

$$\|d_k\| \leqslant e_8 \left[\frac{\|P_{k-1}\Delta\mathfrak{Q}(t)P_{k-1}d_{k-1}\|}{\|d_{k-1}\|} + \left\|P_{k-1}\Delta\mathfrak{Q}(t)\mathcal{G}_{k-1}(\mathcal{G}_{k-1}^{\mathrm{T}}\mathcal{G}_{k-1})^{-1}\right\| \frac{\|g_{k-1}\|}{\|d_{k-1}\|}\right]\|d_{k-1}\|,$$

which yields $\|d_k\| \leqslant \beta_k \|d_{k-1}\|$, $\lim_{k\to\infty} \beta_k = 0$ and thence, using Lemma 2.2, the Q-superlinear convergence of $\{x_k\}$. Theorem 2.1 leads to the two-step Q-superlinear convergence of $\{x_k, \lambda_k\}$. □

Corollary 3.1 refers to the Goldstein–Levitin–Polyak algorithm in Chapter 6, which ensures feasibility (i.e. $g(x_k) = 0$) for all iterations and to the SQP algorithm of Gabay (1984) for equality constraints. Gabay's algorithm ensures the property

$$\lim_{k\to\infty} \frac{\|g_k\|}{\|x_k - x_*\|} = 0,$$

which, using results in Gabay (1984, expression (4.23)), can be shown to be equivalent to the above condition.

4 THE EFFECT OF INEQUALITY CONSTRAINTS

In this section, we consider the algorithm for solving (7.1.1) and the effect of inequality constraints. The framework is the same as that of Chapter 7, but we are particularly interested in the inequality constraints satisfied as equalities at the solution. When unit stepsizes have been attained, the algorithm consists of (1.2) with d_k solving

$$\min \{q(d, c_*, \pi_k) \mid \nabla g_k^{\mathrm{T}} d + g_k = 0; \nabla h_k^{\mathrm{T}} d + h_k \leqslant 0\} \tag{4.1}$$

and the multipliers corresponding to d_k denoted by λ_{k+1}, μ_{k+1}. The objective function of the QPS is given by

$$q(d, c_*, \pi_k) = \langle d, \nabla \ell(x_k, c_*, \pi_k)\rangle + \frac{1}{2}\langle d, \hat{\mathfrak{Q}}_k d\rangle, \tag{4.2}$$

with

$$\ell(x, c_*, \pi) \equiv f(x) + \frac{c_*}{2}\|g(x)\|_2^2 + \frac{1}{2c_*}\left\|(c_* h(x) + \pi)_+\right\|_2^2.$$

where $\nabla \ell = \nabla_x^{\ell}(x, c_*, \pi)$, π_k for the jth inequality constraint is determined using

$$\pi_k^j = \begin{cases} -c_* h_k^j & \text{if } h_k^j \leqslant 0, \\ \pi_{k-1}^j & \text{if } h_k^j > 0, \end{cases} \tag{4.3}$$

and $\pi_0^j = \min\{-c_* h_k^j, 0\}$.

As in Chapter 7, $c_* \in \mathbb{R}_+, \pi \in \mathbb{R}_+^i$, and $(\cdot)_+$ is such that its jth element $(\cdot)_+^j = \max\{(\cdot), 0\}$. The matrix $\hat{\mathfrak{Q}}_k$ is an approximation to the Hessian $\mathfrak{Q}_k$, with respect to x, at (x_k, λ_k, μ_k), of $\bar{L}^a : \mathbb{R}^n \times \mathbb{R}^e \times \mathbb{R}_+^i \times \mathbb{R}_+ \times \mathbb{R}_+^i \to \mathbb{R}$:

$$\bar{L}^{\mathrm{a}}(x, \lambda, \mu, c_*, \pi) = f(x) + \langle g(x), \lambda\rangle + \langle h(x), \mu\rangle + \frac{c_*}{2}\|g(x)\|_2^2 + \frac{1}{2c_*}\left\|c_* h(x) + \pi\right\|_2^2$$

which is expression (7.1.5) evaluated at c_*. (4.4)

Following the discussion in Section 1, there exists a stage beyond which the inequality constraints active at the solution of (7.1.1) can be treated as equality constraints. The rest of the inequality constraints, strictly satisfied at the solution, can be ignored (see e.g. Chapter 7 and Powell, 1978, 1983). Thus, the equality-constrained QPS to be considered hereinafter is given by

$$\min\{q(d, c_*, \pi_k) \mid \nabla g_k^{\mathrm{T}} d + g_k = 0; \quad \nabla \bar{h}_k^{\mathrm{T}} d + \bar{h}_k = 0\}, \tag{4.5}$$

where $\bar{h}$ denotes the subset of inequality constraints active at the solution. Despite the fact that $\bar{h}(x)$ is treated as a vector of equalities, its inequality origin cannot be ignored since the QPS utilizes $\nabla\ell$ and the approximate Hessian of $\bar{L}^a$.

The reason for using $\nabla\ell$ and $\bar{L}^a$ is to ensure consistency within the algorithm. The condition for achieving unit stepsizes, as in (1.2), is shown in Chapter 7 to be dependent on the accuracy of $\hat{\mathfrak{Q}}_k$. In order to establish the convergence rate results for the choice of $\hat{\mathfrak{Q}}_k$, $\nabla\ell$ is required in (4.2). Thus, $\hat{\mathfrak{Q}}_k$ and $\nabla\ell$ are chosen and used in the QPS to ensure the *consistency* of unit stepsize achievement with the convergence rate results. Therefore, the inequality origin of $\bar{h}$ has to be accounted for in $\nabla\ell$.

The SQP algorithm described in (1.2), (4.1) and (4.3) above differs from the *standard* SQP algorithm (i.e. without the penalty terms) in its explicit treatment of inequality constraints and use of $\nabla\ell$, instead of ∇f, and the Hessian of $\bar{L}^a$ in the QPS (4.5). The convergence results concerning unit stepsizes and the local convergence rate discussion below are made consistent for an inequality constrained algorithm.

Assumption 4.1
Let Assumptions 2.1 hold for $\mathscr{G}_k$ given by

$$\mathscr{G}_k = [\nabla g(x_k) \vdots \nabla \bar{h}(x_k)]$$

and the QPS given by (4.1) and (4.2).

A framework for inequality constraints is required by $\nabla\ell$ included in the QPS (4.1) and (4.2) because of the stepsize strategy. Thus, the *simple* treatment of the active constraints at the solution as equalities is not sufficient.

The distinction between $\{x_k, \lambda_k\}$ and $\{x_k, \lambda_k, \mu_k\}$ is made because the demonstration of the convergence rate of the former sequence is simpler. The main difficulty with the inequality constraints is the influence of π_k on μ_k. This is discussed below.

Lemma 4.1
Let $\{x_k\} \to x_*$ *and*

$$b_k = (c_* \bar{h}(x_k) + \pi_k)_+ - (c_* \bar{h}(x_k) + \pi_{k-1}).$$

Then, we have $\{b_k\} \to 0$.

Proof

Using the rule for setting π_k, given by (4.3), we have, for $\bar{h}^j(x_k) > 0$,

$$\pi_k^j = \pi_{k-1}^j \Rightarrow (c_*\bar{h}_k^j + \pi_k^j)_+ = (c_*\bar{h}_k^j + \pi_{k-1}^j)_+ = (c_*\bar{h}_k^j + \pi_{k-1}^j), \qquad b_k = 0, \quad (4.6)$$

and, for $\bar{h}_k^j \leqslant 0$,

$$\pi_k^j = -c_*\bar{h}_k^j \Rightarrow (c_*\bar{h}_k^j + \pi_k^j)_+ = 0; b_k = -(c_*\bar{h}_k^j + \pi_{k-1}^j). \qquad (4.7)$$

As $\{x_k\} \to x_*, \{\bar{h}_k^j\} \to 0$, and hence, by (4.3), $\{\pi_k^j\} \to -c_*\bar{h}_*^j = 0$ and $\{b_k\} \to 0$. □

The multiplier vector $(\lambda_{k+1}, \mu_{k+1})$ generated by the QPS (4.5) is given by

$$\begin{bmatrix} \lambda_{k+1} \\ \mu_{k+1} \end{bmatrix} = -\left\{ (\mathscr{G}_k^{\mathrm{T}}\mathscr{G}_k)^{-1}\mathscr{G}_k^{\mathrm{T}}(\nabla f_k + \hat{\mathfrak{Q}}_k d_k) + \begin{bmatrix} c_* g_k \\ (c_*\bar{h}_k + \pi_k)_+ \end{bmatrix} \right\} \qquad (4.8)$$

We consider the convergence rate of $\{x_k\}$ for both the inequality- and equality-constrained cases. The main complication in extending the results for $\{x_k, \lambda_k\}$ to $\{x_k, \lambda_k, \mu_k\}$ is due to the presence of π_k in $(c_*\bar{h}_k + \pi_k)_+$ in (4.8) above. Consider the alternative multiplier vector

$$\begin{bmatrix} \lambda_{k+1} \\ \mu_{k+1}^0 \end{bmatrix} = -\left[(\mathscr{G}_k^{\mathrm{T}}\mathscr{G}_k)^{-1}\mathscr{G}_k^{\mathrm{T}}(\nabla f_k + \hat{\mathfrak{Q}}_k d_k) + \begin{bmatrix} c_* g_k \\ 0 \end{bmatrix} \right]. \qquad (4.9)$$

As $\{x_k\} \to x_*, \{\bar{h}_k\} \to 0, \{g_k\} \to 0$, and the multipliers (4.8) and (4.9) converge to the same value (λ_*, μ_*) at the solution. Both (λ_k, μ_k) and (λ_k, μ_k^0) are equivalent when $(c_*\bar{h}_k + \pi_k)_+ = 0$ or when the penalty term for the inequality constraints is not included in the quadratic objective function in (4.5). The equivalence is exploited in the discussion below. We discuss the sequence $\{x_k, \lambda_k, \mu_k^0\}$ as a means of analysing the behaviour of the algorithm.

Lemma 4.2

Let Assumption 4.1 hold. Then, (2.1) holds for the inequality-constrained case.

Proof

The first order optimality condition of the QPS (4.5) can be written as

$$\begin{bmatrix} \hat{\mathfrak{Q}}_k & \mathscr{G}_k \\ \mathscr{G}_k^{\mathrm{T}} & 0 \end{bmatrix} \begin{bmatrix} d_k \\ \lambda_{k+1} - \lambda_* + c_* g_k \\ \mu_{k+1} - \mu_* + \left(c_*\bar{h}_k + \pi_k\right)_+ \end{bmatrix} = -\begin{bmatrix} \nabla_x L(x_k, \lambda_*, \mu_*) \\ g_k \\ \bar{h}_k \end{bmatrix},$$

where $L(x, \lambda, \mu)$ is the ordinary Lagrangian, given by $L(x, \lambda, \mu) = f(x) + \langle g(x), \lambda \rangle + \langle h(x), \mu \rangle$. Using the first-order expansion of the right side of the above expression about (x_*, λ_*, μ_*), by the nonsingularity of the coefficient matrix on the left and by the first-order optimality condition of (1.1),

$$\nabla_x L(x_*, \lambda_*, \mu_*) = 0, \qquad g(x_*) = 0, \qquad \bar{h}(x_*) = 0,$$

we have

$$\|d_k\| \leqslant \left\| \begin{matrix} d_k \\ \lambda_{k+1} - \lambda_* + c_* g_k \\ \mu_{k+1} - \mu_* + \left(c_* \bar{h}_k + \pi_k\right)_+ \end{matrix} \right\| \leqslant \hat{e}_0 \left\| \begin{bmatrix} \hat{\mathfrak{Q}}_k & \mathscr{G}_k \\ \mathscr{G}_k^{\mathrm{T}} & 0 \end{bmatrix}^{-1} \right\| \|x_* - x_k\|.$$

for some $\hat{e}_0 \in [0, \infty)$. Hence, (2.1) follows from the above expression. □

Lemma 4.3
Let $\{x_k\} \to x_*$. Then, Lemma 2.2 holds for the inequality-constrained case.

Proof
The proof is identical to that of Lemma 2.2, except that Lemma 4.2 is used, instead of Lemma 2.1, to establish the desired result. □

Lemma 4.4
Let Assumption 4.1 hold, $\{x_k, \lambda_k, \mu_k^0\}$ converge to (x_*, λ_*, μ_*) and (2.2) be satisfied. Then, we have

$$\left\| \begin{matrix} \lambda_* - \lambda_{k+1} \\ \mu_* - \mu_{k+1}^0 \end{matrix} \right\| \leqslant \beta_k^{(5)} \|x_* - x_{k-1}\|, \qquad \text{with} \quad \lim_{k \to \infty} \beta_k^{(5)} = 0. \tag{4.10}$$

Proof
Using (4.9), we can write

$$\begin{bmatrix} \lambda_{k+1} - \lambda_* \\ \mu_{k+1}^0 - \mu_* \end{bmatrix} = -\left\{ (\mathscr{G}_k^T \mathscr{G}_k)^{-1} \mathscr{G}_k^{\mathrm{T}} \left(\nabla f_k + \mathscr{G}_k \begin{bmatrix} \lambda_* \\ \mu_* \end{bmatrix} + \hat{\mathfrak{Q}}_k d_k \right) + \begin{bmatrix} c_* g_k \\ 0 \end{bmatrix} \right\} \tag{4.11}$$

At the solution, the necessary condition for optimality requires the equality

$$\nabla_x L(x_*, \lambda_*, \mu_*) = \nabla f_* + \mathscr{G}_* \begin{bmatrix} \lambda_* \\ \mu_*^0 \end{bmatrix} = 0. \tag{4.12}$$

Assuming that $\nabla_x L(x, \lambda_*, \mu_*)$ is Lipschitz-continuous in x, we have, for $e_9 \in [0, \infty)$,

$$\|\nabla_x L(x_k, \lambda_*, \mu_*)\| = \left\| \nabla f_k + \mathscr{G}_k \begin{bmatrix} \lambda_* \\ \mu_*^0 \end{bmatrix} \right\| \leqslant e_9 \|x_k - x_*\|. \tag{4.13}$$

Thus, using (2.8) and (4.11)–(4.13), we have, for $e_{10} \in [0, \infty)$,

$$\left\| \begin{matrix} \lambda_{k+1} - \lambda_* \\ \mu_{k+1}^0 - \mu_* \end{matrix} \right\| \leqslant (e_9 + e_2)\|x_k - x_*\| + e_{10}\|d_k\|, \tag{4.14}$$

which, using Lemma 4.2, yields

$$\left\| \begin{matrix} \lambda_{k+1} - \lambda_* \\ \mu_{k+1}^0 - \mu_* \end{matrix} \right\| \leqslant (e_9 + e_2 + e_{10}\hat{e}_0)\|x_k - x_*\|. \tag{4.15}$$

Given (2.2), Lemma 2.3 can be invoked to establish the Q-superlinear convergence of $\{x_k\}$. Hence, we have

$$\left\| \begin{matrix} \lambda_{k+1} - \lambda_* \\ \mu_{k+1}^0 - \mu_* \end{matrix} \right\| \leqslant (e_9 + e_2 + e_{10}\hat{e}_0)\|x_k - x_*\| \leqslant \beta_k^{(5)} \|x_* - x_{k-1}\|. \tag{4.16}$$

□

Lemma 4.5

Let Assumption 4.1 hold and (2.2) be satisfied. Then, (2.4) holds, and when $b_k = 0$, we have

$$\left\| \begin{matrix} \lambda_* - \lambda_k \\ \mu_* - \mu_k \end{matrix} \right\| \leqslant \beta_k^{(6)} \|x_* - x_{k-2}\|, \qquad \lim_{k\to\infty} \beta_k^{(6)} = 0. \tag{4.17}$$

Proof

Consider the first-order expansion of $\nabla_x \bar{L}^a$, where $\bar{L}^a$ is given by (4.4):

$$\nabla_x \bar{L}^{\mathrm{a}}(x_k, \lambda_k, \mu_k, c_*, \pi_{k-1}) = \nabla_x \bar{L}^{\mathrm{a}}(x_{k-1}, \lambda_k, \mu_k, c_*, \pi_{k-1}) + \hat{\mathfrak{Q}}_{k-1} d_{k-1} + \Delta\mathfrak{Q}(t) d_{k-1}, \tag{4.18}$$

with $\Delta\mathfrak{Q}(t)$ as in (2.13). The first-order optimality conditions for the quadratic subproblem (4.5) at x_{k-1} ensure that in (4.18),

$$\nabla_x \bar{L}^{\mathrm{a}}(x_{k-1}, \lambda_k, \mu_k, c_*, \pi_{k-1}) + \hat{\mathfrak{Q}}_{k-1} d_{k-1} = 0. \tag{4.19}$$

From the definition of $\bar{L}^a$, we have

$$\nabla_x \bar{L}^{\mathrm{a}}(x_k, \lambda_k, \mu_k, c_*, \pi_{k-1}) = \nabla f_k + \mathscr{G}_k \begin{bmatrix} c_* g_k + \lambda_k \\ c_* \bar{h}_k + \pi_{k-1} + \mu_k \end{bmatrix}. \tag{4.20}$$

The solution of the quadratic subproblem at x_k satisfies the first-order condition

$$\nabla f_k + \hat{\mathfrak{Q}}_k d_k + \mathscr{G}_k \begin{bmatrix} c_* g_k + \lambda_{k+1} \\ (c_* \bar{h}_k + \pi_k)_+ + \mu_{k+1} \end{bmatrix} = 0. \tag{4.21}$$

Using (4.20) and (4.21) in (4.18), along with $\mathscr{G}_k^T d_k + \begin{bmatrix} g_k \\ \bar{h}_k \end{bmatrix} = 0$, yields

$$\begin{bmatrix} \hat{\mathfrak{Q}}_k & \mathscr{G}_k \\ \mathscr{G}_k^{\mathrm{T}} & 0 \end{bmatrix} \begin{bmatrix} d_k \\ \lambda_{k+1} - \lambda_k \\ \mu_{k+1} - \mu_k + b_k \end{bmatrix} = - \begin{bmatrix} \Delta\mathfrak{Q}(t) d_{k-1} \\ g_k \\ \bar{h}_k \end{bmatrix}. \tag{4.22}$$

The second-order expansion of the constraints, with

$$\mathscr{G}_{k-1}^{\mathrm{T}} d_{k-1} + \begin{bmatrix} g_{k-1} \\ \bar{h}_{k-1} \end{bmatrix} = 0,$$

yields, for some $e_{11} \geqslant 0$,

$$\left\| \begin{matrix} g_k \\ \bar{h}_k \end{matrix} \right\| \leqslant e_{11} \|d_{k-1}\|^2. \tag{4.23}$$

From (4.22) and (4.23), we have

$$\left\| \begin{matrix} \lambda_{k+1} - \lambda_k \\ \mu_{k+1} - \mu_k + b_k \end{matrix} \right\| \leqslant e_{12} \|d_{k-1}\| + e_{13} \|d_{k-1}\|^2, \tag{4.24}$$

with $e_{12}, e_{13} \geqslant 0$. We can use (4.24) and (2.2) to obtain

$$\|\lambda_{k+1} - \lambda_k\| \leqslant \left\| \begin{matrix} \lambda_{k+1} - \lambda_k \\ \mu_{k+1} - \mu_k + b_k \end{matrix} \right\| \leqslant \beta_k^{(7)} \|d_{k-2}\|, \quad \text{where} \quad \lim_{k \to \infty} \beta_k^{(7)} = 0 \tag{4.25}$$

and $\beta_k^{(7)} = \beta_{k-1}(e_{12} + e_{13})$. Combining (4.25) with (2.1) leads to

$$\begin{aligned} \|\lambda_* - \lambda_k\| &\leqslant \lim_{t \to \infty} \sum_{j=k}^{t-1} \|\lambda_{j+1} - \lambda_j\| \\ &\leqslant \lim_{t \to \infty} \sum_{j=k}^{t-1} \beta_j^{(7)} \|d_{j-2}\| \\ &\leqslant \beta_k^{(7)} \|d_{k-2}\| (1 + \omega + \omega^2 + \omega^3 \ldots) \\ &\leqslant \frac{\hat{e}_0 \beta_k^{(7)}}{1 - \omega} \|x_* - x_{k-2}\| \end{aligned}$$

for some $\omega \in [0, 1)$. This yields (2.4) as required.

In order to establish (4.17), we note that for $b_k = 0$, (4.25) reduces to

$$\left\| \begin{matrix} \lambda_{k+1} - \lambda_k \\ \mu_{k+1} - \mu_k \end{matrix} \right\| \leqslant \beta_k^{(7)} \|d_{k-2}\|.$$

Using this inequality with the above argument yields (4.17). □

Theorem 4.1

Let Assumption 4.1 be satisfied. Then, the results of Theorem 2.1 hold for d_k *generated by the QPS (4.5) and* π_k *given by (4.3). Furthermore, the Q-superlinear rate of* $\{x_k\}$ *implies the two-step Q-superlinear rate of* $\{x_k, \lambda_k, \mu_k\}$.

Proof

Expression (4.22) can be rewritten as

$$\begin{bmatrix} \hat{\mathfrak{Q}}_k & \mathscr{G}_k \\ \mathscr{G}_k^{\mathrm{T}} & 0 \end{bmatrix} \begin{bmatrix} & d_k & \\ \begin{bmatrix} \lambda_{k+1} - \lambda_k \\ \mu_{k+1} - \mu_k + b_k \end{bmatrix} & + (\mathscr{G}_{k-1}^{\mathrm{T}} \mathscr{G}_{k-1})^{-1} \mathscr{G}_{k-1}^{\mathrm{T}} \Delta \mathfrak{Q}(t) d_{k-1} \end{bmatrix} = - \begin{bmatrix} P_{k-1} \Delta \mathfrak{Q}(t) d_{k-1} \\ g_k \\ \bar{h}_k \end{bmatrix}. \tag{4.26}$$

Since the matrix on the left is nonsingular, (2.11b), (4.26), (4.23) and $\{\|d_k\|\} \to 0$ yield $\|d_k\| \leqslant \gamma_k \|d_{k-1}\|$, with $\lim_{k\to\infty} \gamma_k = 0$. Thence, Lemma 4.3 yields the Q-superlinear convergence of $\{x_k\}$. The converse is demonstrated as in the proof of Theorem 2.1.

Consider the convergence rate of $\{x_k, \lambda_k, \mu_k\}$. When $\bar{h}_k^j > 0$, by (4.6), $b_k = 0$, and, from (4.24), we have

$$\left\| \begin{matrix} \lambda_{k+1} - \lambda_k \\ \mu_{k+1} - \mu_k \end{matrix} \right\| \leqslant \beta_k^{(6)} \|d_{k-2}\|,$$

which yields (4.17). Using (4.17) with the Q-superlinear convergence of $\{x_k\}$, the two-step Q-superlinear convergence of $\{x_k, \lambda_k, \mu_k\}$ follows. When $\bar{h}_k^j \leqslant 0$, by (4.3),

$$(c_* \bar{h}_k^j + \pi_k^j)_+ = c_* \bar{h}_k^j + \pi_k^j = 0.$$

Hence, the multiplier vector (4.8) reduces to (4.9). By (4.10) of Lemma 4.4, and using the Q-superlinear convergence result for $\{x_k\}$ established above, we have, for $\lim_{k\to\infty} \beta_k^{(8)} = 0$,

$$\left\| \begin{matrix} x_* - x_{k+1} \\ \lambda_* - \lambda_{k+1} \\ \mu_* - \mu_{k+1} \end{matrix} \right\| \leqslant \beta_k^{(8)} \left\| \begin{matrix} x_* - x_{k-1} \\ \lambda_* - \lambda_{k-1} \\ \mu_* - \mu_{k-1} \end{matrix} \right\|.$$

This establishes the two-step Q-superlinear convergence for $\{x_k, \lambda_k, \mu_k\}$. □

Lemma 4.6

Let the conditions of Lemma 3.1 be satisfied. Then, (3.2) follows for the inequality-constrained case, and, when $b_k = 0$, we have

$$\left\| \begin{matrix} \lambda_* - \lambda_k \\ \mu_* - \mu_k \end{matrix} \right\| \leqslant \beta_k^{(9)} \|x_* - x_{k-3}\|. \tag{4.27}$$

Proof

Consider inequality (4.24):

$$\left\| \begin{matrix} \lambda_{k+1} - \lambda_k \\ \mu_{k+1} - \mu_k + b_k \end{matrix} \right\| \leqslant (e_{12} + e_{13}\|d_{k-1}\|)\|d_{k-1}\|. \tag{4.28}$$

Using (3.1), this can be rewritten as

$$\|\lambda_{k+1} - \lambda_k\| \leqslant \beta_k^{(10)} \|d_{k-3}\|, \tag{4.29}$$

where $\beta_k^{(10)} = \beta_{k-1}^{(2)}(e_{12} + e_{13}\|d_{k-1}\|)$. Combining (4.29) with Lemma 4.2 leads to (3.2) as in Lemma 3.1.

In order to establish (4.27), we note that for $b_k = 0$, (4.28) becomes

$$\left\| \begin{matrix} \lambda_{k+1} - \lambda_k \\ \mu_{k+1} - \mu_k \end{matrix} \right\| \leqslant (e_{12} + e_{13}\|d_{k-1}\|)\|d_{k-1}\| \leqslant \beta_k^{(10)} \|d_{k-3}\|.$$

Using this inequality with the above argument yields (4.27). □

Lemma 4.7
Let Assumption 4.1 be satisfied, $\{x_k, \lambda_k, \mu_k^0\}$ converge to (x_*, λ_*, μ_*) and

$$\lim_{k\to\infty} \frac{\|x_* - x_k\|}{\|x_* - x_{k-2}\|} = 0.$$

Then

$$\left\| \begin{matrix} \lambda_* - \lambda_k \\ \mu_* - \mu_k^0 \end{matrix} \right\| \leqslant \beta_k^{(11)} \|x_* - x_{k-3}\| \quad \text{with} \quad \lim_{k\to\infty} \beta_k^{(11)} = 0.$$

Proof
Consider inequality (4.15):

$$\left\| \begin{matrix} \lambda_* - \lambda_k \\ \mu_* - \mu_k^0 \end{matrix} \right\| \leqslant (e_9 + e_2 + e_{10}\hat{e}_0)\|x_{k-1} - x_*\|,$$

which, given the two-step Q-superlinear convergence of $\{x_k\}$, yields the required result.

Theorem 4.2
Let Assumption 4.1 be satisfied. Then, the results of Theorem 3.1 hold for d_k generated by the QPS (4.5) and π_k given by (4.3). Furthermore, the two-step Q-superlinear rate $\{x_k\}$ implies the three-step Q-superlinear rate of $\{x_k, \lambda_k, \mu_k\}$.

Proof
We can conclude from (4.22) that, for some $e_{14} \in [0, \infty)$,

$$\|d_k\| \leqslant e_{14}\|d_{k-1}\|.$$

Consider the equality derived using the linearized constraints:

$$\begin{aligned} d_{k-1} &= P_{k-1}d_{k-1} + \mathscr{G}_{k-1}(\mathscr{G}_{k-1}^{\mathrm{T}}\mathscr{G}_{k-1})^{-1}\mathscr{G}_{k-1}^{\mathrm{T}}d_{k-1} \\ &= P_{k-1}d_{k-1} - \mathscr{G}_{k-1}(\mathscr{G}_{k-1}^{\mathrm{T}}\mathscr{G}_{k-1})^{-1}\begin{bmatrix} g_{k-1} \\ \bar{h}_{k-1} \end{bmatrix}. \end{aligned}$$

Using this to replace d_{k-1} in (4.26) yields, for some $e_{15} \in [0, \infty)$,

$$\|d_k\| \leqslant e_{15}\left[\|P_{k-1}\nabla\mathfrak{L}(t)P_{k-1}d_{k-1}\| + \|\mathscr{G}_{k-1}(\mathscr{G}_{k-1}^{\mathrm{T}}\mathscr{G}_{k-1})^{-1}\|\left\| \begin{matrix} g_{k-1} \\ \bar{h}_{k-1} \end{matrix} \right\|\right].$$

The remaining part of the proof for $\{x_k\}$ is the same as in Theorem 3.1.

Consider the convergence rate of $\{x_k, \lambda_k, \mu_k\}$. When $\bar{h}_k^j > 0$, by (4.6), $b_k = 0$, and we can apply (4.27) of Lemma 4.6 and use the above argument to establish the convergence rate of the sequence for the case $b_k = 0$:

$$\left\| \begin{matrix} x_* - x_k \\ \lambda_* - \lambda_k \\ \mu_* - \mu_k \end{matrix} \right\| \leqslant \left[\beta_k^{(9)} + \frac{\hat{e}_0 e_{14}(1 + e_{14})}{1 - \omega} \beta_k^{(2)} \right] \left\| \begin{matrix} x_* - x_{k-3} \\ \lambda_* - \lambda_{k-3} \\ \mu_* - \mu_{k-3} \end{matrix} \right\|.$$

When $\bar{h}_k^j \leqslant 0$, by (4.3),

$$(c_* \bar{h}_k^j + \pi_k^j)_+ = c_* \bar{h}_k^j + \pi_k^j = 0.$$

Hence, the multiplier vector (4.8) reduces to (4.9), and by invoking Lemma 4.7 with the two-step Q-superlinear convergence of $\{x_k\}$, established above, we have the convergence rate of the sequence for $\{x_k, \lambda_k, \mu_k\}$:

$$\left\| \begin{matrix} x_* - x_k \\ \lambda_* - \lambda_k \\ \mu_* - \mu_k \end{matrix} \right\| \leqslant \left[\beta_k^{(11)} + \frac{\hat{e}_0 e_{14}(1 + e_{14})}{1 - \omega} \beta_k^{(2)} \right] \left\| \begin{matrix} x_* - x_{k-3} \\ \lambda_* - \lambda_{k-3} \\ \mu_* - \mu_{k-3} \end{matrix} \right\|. \qquad \square$$

5 CONCLUDING REMARKS

In this chapter, the superlinear convergence of $\{x_k\}, \{x_k, \lambda_k\}$ and $\{x_k, \lambda_k, \mu_k\}$ has been demonstrated. The two important factors are the inequality constraints and the penalty function incorporated in the linear term of the objective of the QPS. The principal reason for including the penalty term is the consistency requirement between the conditions that ensure the attainment of unit stepsizes in Chapter 7 and the convergence rate discussion in this chapter.

As Boggs, Tolle and Wang (1982) point out, the Q-superlinear convergence of either or both of $\{x_k, \lambda_k\}$ and $\{x_k, \lambda_k, \mu_k\}$ does not, in general, imply the same rate for $\{x_k\}$. On the other hand, Theorems 2.1 and 4.1 show that the Q-superlinear convergence of $\{x_k\}$ implies the two-step Q-superlinear convergence of $\{x_k, \lambda_k\}$ and $\{x_k, \lambda_k, \mu_k\}$ respectively. Theorems 3.1 and 4.2 show that the two-step Q-superlinear convergence of $\{x_k\}$ implies the three-step Q-superlinear convergence of $\{x_k, \lambda_k\}$ and $\{x_k, \lambda_k, \mu_k\}$. The slower rate of convergence of $\{x_k, \lambda_k\}$ and $\{x_k, \lambda_k, \mu_k\}$ may be important if the multipliers are used as shadow prices in a decision process.

6 COMMENTS AND NOTES

CN.1 Q- and R-superlinear convergence rates

Q-convergence rates are discussed in CN.4 in Chapter 6. R- convergence rate is a weaker rate than Q-convergence of $\{x_k\} \to x_*$. The sequence $\{x_k\}$ is said to be *R-superlinearly convergent* if

$$\limsup_{k\to\infty} \|x_k - x_*\|^{1/k} = 0$$

(Ortega and Rheinboldt, 1970, Chap. 9).

CN.2 Characterization of Q-superlinear convergence

The Q-superlinear convergence condition on the full Hessian approximation is essentially the application of the Dennis and Moré (1974) characterization of superlinear convergence for unconstrained optimization algorithms to the SQP procedure. The interested reader is referred to that reference and to Dennis and Schnabel (1983) for further information about the convergence of unconstrained optimization algorithms.

CN.3 Existence of feasible QPS solution

Assumption 2.1(5) is needed to ensure that $\left\| \begin{bmatrix} \hat{\mathfrak{Q}}_k \mathcal{G}_k \\ \mathcal{G}_k^{\mathrm{T}} 0 \end{bmatrix}^{-1} \right\| \in (0, \infty)$. In this chapter, we do not explicitly assume that $\hat{\mathfrak{Q}}$ is positive-definite. However, at least positive-semidefinite $\hat{\mathfrak{Q}}$ is required to ensure descent at every iteration insofar as the global convergence of the algorithm is concerned (see Chapter 7). The finiteness restriction on $\hat{\mathfrak{Q}}$ is particularly in view of the penalty parameter in (1.5).

REFERENCES

Arrow, K.J., F.J. Gould and S.M. Howe (1973). General saddle point result for constrained optimization. *Math. Programming* **5**, 225–234.

Biggs, M.C.B. (1978). On the convergence of some constrained minimization algorithms based on recursive quadratic programming. *J. Inst. Math. Applic.* **21**, 67–81.

Boggs, P.T., J.W. Tolle and P. Wang (1982). On the local convergence of quasi-Newton methods for constrained optimization. *SIAM J. Control Optim.* **20**, 161–171.

Chiang, J. and R. Tapia (1989). Convergence rates for the variable, the multiplier, and the pair in SQP methods. Technical Report 89–2, Rice University.

Coleman, T.F. and A.R. Conn (1984). On the local convergence of a quasi-Newton method for the nonlinear programming problem. *SIAM J. Numer. Anal.* **21**, 755–769.

Dennis, J.E. and J.J. Moré (1974). A characterization of Q-superlinear convergence and its application to quasi-Newton methods. *Math. Comput.* **28**, 549–560.

Dennis, J.E. and R.B. Schnabel (1983). *Numerical Methods for Unconstrained Optimization and Nonlinear Equations*. Prentice-Hall, Englwood Cliffs, New Jersey.

Dixon, L.C.W. (1980). On the convergence properties of variable metric recursive quadratic programming methods. Proceedings of Xth International Symposium on Mathematical Programming, Numerical Optimization Centre Report 110, Hatfield Polytechnic.

Fontecilla, R., T. Steihaug and R.A. Tapia (1987). A convergence theory for a class of quasi-Newton methods for constrained optimization. *SIAM J. Numer. Anal.* **24**, 1133–1151.

Glad, S.T. (1979). Properties of updating methods for the multipliers in augmented Lagrangians. *J. Optim. Theory Applic.* **28**, 135–156.

Gabay, D. (1982). Reduced quasi-Newton methods with feasibility improvement for nonlinearly constrained optimization. *Math. Programming Study* **16**, 18–44.
Han, S.-P. (1976). Superlinearly convergent variable metric algorithms for general nonlinear programming problems. *Math. Programming* **11**, 263–282.
Nocedal, J. and M. Overton (1985). Projected Hessian updating algorithms for nonlinearly constrained optimization. *SIAM J. Numer. Anal.* **22**, 821–850.
Ortega, J. M. and W. C. Rheinboldt (1970). *Iterative Solution of Nonlinear Equations in Several Variables*, Academic Press, London.
Powell, M.J.D. (1978). The convergence of variable metric methods for nonlinearly constrained optimization calculations. In *Nonlinear Programming* (J. B. Rosen, O. L. Mangasarian and K. Ritter, eds), pp. 27–63. Academic Press, New York.
Powell, M.J.D. (1983). Variable metric methods for constrained optimization. In *Mathematical Programming: The State of The Art* (A. Bachem, M. Grotschel and B. Korte, eds). Springer-Verlag, Berlin.
Rockafeller, R.T. (1973). A dual approach to solving nonlinear programming problems by unconstrained optimization. *Math. Programming* **5**, 354–373.
Rustem, B. (1984). A class of superlinearly convergent projection algorithms with relaxed stepsizes. *Appl. Math. Optim.* **12**, 29–43.
Rustem, B. (1986). Convergent stepsizes for constrained optimization algorithms. *J. Optim. Theory Applic.* **49**, 135–160.
Rustem, B. (1993). Equality and inequality constrained optimization algorithms with convergent stepsizes. *J. Optim. Theory Applic.* **76**, 429–453.
Rustem, B. (1994). The two-step and three-step Q-superlinear convergence of successive quadratic programming algorithms. *J. Optim. Theory Applic.* **83**, 613–619.
Stoer, J. (1985). The convergence of sequential quadratic programming methods for solving nonlinear programs. Institut für Angewandte Mathematik und Statistik der Universität Würzburg.
Stoer, J. and R.A. Tapia (1986). On the characterization of Q- superlinear convergence of quasi-Newton methods for constrained optimization. Technical Report 84–2, Rice University.
Tapia, R.A. (1977). Diagonalized multiplier methods and quasi-Newton methods for constrained optimization. *J. Optim. Theory Applic.* **22**, 135–194.
Tapia, R.A. (1986). On secant updates for use in general constrained optimization. Technical Report 84–3, Rice University.
UNIDO (1972). *Guidelines for Project Evaluation.* United Nations Publications, UNIDO, Vienna.

CHAPTER 9

Algorithms for Equilibria and Games

Consider the extension of the basic optimal policy problem to situations involving multiple decision makers pursuing their individual distinct objectives, in competition with each other. The policy adopted by one decision maker affects the policies of others. Such competitive situations arise among individuals and firms, as well as nations. The solution to the problem is a Nash equilibrium strategy that ensures that no decision maker will benefit by deviating from this equilibrium.

We introduce the best replay algorithm as the intuitive solution to the computation of equilibria, and discuss relaxed variants with improved convergence properties. We also consider Newton-type algorithms. These aim to satisfy the optimality conditions for each player, based on the framework of Section 1.4 and Chapter 6. An essential ingredient is the computational mapping (1.4.1) that evaluates the endogenous variables of the common system, given the actions of each decision maker. This involves the computational solution of the underlying model.

In order to make their evaluation as efficient as possible, the models are preprocessed by reordering the equations in order to minimize the dimension of the simultaneous block. The model solution algorithms are then applied to this block. The Newton- type algorithms evaluate numerically, as required, selected columns of the Jacobian of the simultaneous part. Provisions also exist for similar systems to be solved, if possible, without actually reinitializing the Jacobian. One of the algorithms also uses the Broyden update to improve the Jacobian. Global convergence is maintained by an Armijo-type stepsize strategy.

The global and local convergence of the quasi-Newton algorithm is discussed. Convergence under relaxed descent directions is established. The achievement of unit stepsizes is related to the accuracy of the Jacobian approximation. Furthermore, a simple derivation of the Dennis-Moré characterization of the Q-superlinear convergence rate is given.

The model equation reordering algorithm is also described. The model is reordered to define heart and loop variables. This is also applied recursively to the subgraph formed by the loop variables to reduce the total number of above-diagonal elements in the Jacobian of the complete system. The extension of the solution algorithms to consistent expectations are discussed.

1 COMPUTATION OF EQUILIBRIA AND SOLUTION OF NONLINEAR EQUATIONS

1.1 Games and equilibria

We extend the discussion in Chapter 1, and the optimal decision problem (1.1.2), to the case of multiple decision makers. For the ith decision maker, $i = 1, \ldots, N$, we are specifically concerned with the competitive situation of a

Nash game in which the ith decision maker seeks to optimize the ith objective with respect to the ith set of decision variables. We thus have the problem

$$\min_{Y,U^i} \{f^i(Y, U^1, U^2, \ldots, U^i, \ldots, U^N) \mid g(Y, U^1, U^2, \ldots, U^i, \ldots, U^N) = 0\}, \quad (1.1)$$

where U^i is the vector of decision variables and f^i is the objective function of the ith decision maker. The equality constraints represent the model of the underlying system, which can be used to evaluate Y, given the values of $U^1, U^2, \ldots, U^i, \ldots, U^N$. Thus, we have a mapping similar to (1.4.1), given by

$$Y = \mathscr{Y}(U^1, U^2, \ldots, U^i, \ldots, U^N). \quad (1.2)$$

This mapping can be used to eliminate the output variables Y from the problem and express (1.1) as

$$\min_{U^i} \{\mathscr{F}^i(U^1, U^2, \ldots, U^i, \ldots, U^N)\}, \quad i = 1, \ldots, N, \quad (1.3)$$

where

$$\mathscr{F}^i(U^1, U^2, \ldots, U^i, \ldots, U^N) = f^i(\mathscr{Y}(U^1, U^2, \ldots, U^i, \ldots, U^N), U^1, U^2, \ldots, U^i, \ldots, U^N).$$

For this problem, the Nash equilibrium is the point from which no decision maker would wish to deviate unilaterally, since such an action would not improve the position of that decision maker. Thus, the Nash equilibrium $U^i_*, i = 1, \ldots, N$, is defined by

$$\mathscr{F}^i(U^1_*, U^2_*, \ldots, U^i_*, \ldots, U^N_*) \leqslant \mathscr{F}^i(U^1_*, U^2_*, \ldots, U^i, \ldots, U^N_*), \quad \forall U^i, \quad i = 1, \ldots, N$$

(see e.g. Ho, 1970; Basar and Olsder, 1982). Hence, $\mathscr{F}^i(U^1_*, U^2_*, \ldots, U^i_*, \ldots, U^N_*)$ is the unconstrained minimum of $\mathscr{F}^i$, with respect to U^i, and this is evaluated simultaneously for all $i = 1, \ldots, N$. According to the necessary conditions of an optimum, discussed in Proposition 1.2.3, this is equivalent to the solution of the simultaneous system of equations

$$\nabla_{U^i} \mathscr{F}^i(U_1, U_2, \ldots, U^i, \ldots, U^N) = 0, \quad i = 1, \ldots, N. \quad (1.4)$$

We note that the second-order sufficiency condition in Proposition 1.2.3 also requires that the Hessian of $\mathscr{F}^i(\cdot)$, with respect to U^i, should be positive definite *at the solution*. By Theorems 1.2.2 and 1.2.3, this is ensured if each $\mathscr{F}^i(\cdot)$ is convex with respect to U^i in a neighbourhood of the solution. The point, $(U^1_*, U^2_*, \ldots, U^i_*, \ldots, U^N_*)$, satisfying (1.4), is a minimum of $\mathscr{F}^i$, with respect to U^i, and thus a Nash solution. Hence, at the Nash equilibrium, the objective function of the ith decision maker satisfies

$$\langle v, \nabla^2_{U^i} \mathscr{F}^i(U^1_*, U^2_*, \ldots, U^i_*, \ldots, U^N_*) v \rangle > 0, \quad \forall v \neq 0, \quad i = 1, \ldots, N. \quad (1.5)$$

Clearly, for an algorithm solving (1.4), this is a difficult condition to ensure. It is therefore desirable that all points satisfying (1.4) also satisfy (1.5)—In other words, that $\mathscr{F}^i(\cdot)$ be globally convex with respect to U^i. As not all objective functions satisfy such a global convexity requirement, there is a need to consider possible reformulations of the basic problem. The spirit of the reformulation is to seek a solution in a neighbourhood in which convexity is assured.

One possible approach is a convexification procedure similar to the one described in Chapter 7. Given constraints on the decision variables of the ith decision maker,

$$g^i(U^i) = 0, \qquad i = 1, \ldots, N,$$

we assign penalty parameter $c^i \geqslant 0$ to these constraints and reformulate the basic problem as

$$\min_{U^i} \{\mathscr{F}^i(U^1, U^2, \ldots, U^i, \ldots, U^N) + \frac{1}{2}c^i\langle g^i(U^i), g^i(U^i)\rangle \Big| g^i(U^i) = 0\}, \qquad i = 1, \ldots, N.$$

The discussion in Section 7.1 applies straightforwardly to ensure the convexity of this problem. We have avoided including other decision variables in $g^i(U^i) = 0$, due to conceptual and practical difficulties (see Rosen, 1965). An alternative convexification may be constructed around a desired level of the decision variables. Suppose that the decision maker is seeking a solution as close as possible to a desired level of decision variables, denoted by U^{id}. Then, for some $c^i \geqslant 0$, (1.3) can be reformulated as

$$\min_{U^i} \{\mathscr{F}^i(U^1, U^2, \ldots, U^i, \ldots, U^N) + \frac{1}{2}c^i\|U^i - U^{id}\|_2^2\}, \qquad i = 1, \ldots, N.$$

$\|U^i - U^{id}\|_2^2$ is not a hard constraint as $g^i(U^i) = 0$. For c^i chosen sufficiently large to ensure (1.5), this does convexify the original problem.

A similar approach involves variable values, instead of fixed U^{id}, in the form a sequence converging to the Nash solution (see Bertsekas, 1979). To illustrate the significance of this choice, consider the following algorithm for the two-person Nash equilibrium.

Algorithm 1.1 (Best replay)
Step 0: Given U_0^1, U_0^2, tol, set $k = 0$.

Step 1: Compute

$$U_{k+1}^1 = \arg\min_{U^1} \{\mathscr{F}^1(U^1, U_k^2)\}.$$

and

$$U_{k+1}^2 = \arg\min_{U^2} \{\mathscr{F}^2(U_k^1, U^2)\}.$$

Step 2: Equilibrium check: if

$$\left\| \begin{bmatrix} U_{k+1}^1 \\ U_{k+1}^2 \end{bmatrix} - \begin{bmatrix} U_k^1 \\ U_k^2 \end{bmatrix} \right\| \leqslant \text{tol},$$

stop. Otherwise set $k = k + 1$, **go to Step 1**.

Here tol is a tolerance value that depends on the accuracy to which each optimization problem can be solved in practice. This is essentially a Jacobi-type algorithm, with each player optimizing given the previous iteration for the other player (see Ortega and Rheinbold, 1970). The convexification due to Rockafeller (1976) and Bertsekas (1979) is introduced to ensure that, while the solutions in **Step 2** aim at achieving the best for each player, they also remain close to the previous iteration. Thus, with $c^1, c^2 \geqslant 0$, the problems in **Step 1** above become

$$\min_{U^1} \{\mathscr{F}^1(U^1, U_k^2) + c^1 \|U^1 - U_k^1\|^2\}$$

and

$$\min_{U^2} \{\mathscr{F}^2(U_k^1, U^2) + c^2 \|U^2 - U_k^2\|^2\}.$$

The application of a Gauss–Seidel-type procedure (Ortega and Rheinbold, 1970) leads to **Step 1** being replaced by

$$U_{k+1}^1 = \arg \min_{U^1} \{\mathscr{F}^1(U^1, U_k^2) + c^1 \|U^1 - U_k^1\|^2\}$$

and

$$U_{k+1}^2 = \arg \min_{U^2} \{\mathscr{F}^2(U_{k+1}^1, U^2) + c^2 \|U^2 - U_k^2\|^2\}.$$

If the above procedures converge then it can be verified that they converge to the solution. However, convergence may be slow, and convergence itself cannot be always assured. In order to overcome this, overrelaxation procedures may be adopted, involving scalars α, ν as relaxation parameters. The relaxation of the Jacobi procedure is given by

$$U_{k+1}^1 = \alpha U_k^1 + (1 - \alpha)[\arg \min_{U^1} \{\mathscr{F}^1(U^1, U_k^2) + c^1 \|U^1 - U_k^1\|^2\}]$$

and

$$U_{k+1}^2 = \nu U_k^2 + (1 - \nu)[\arg \min_{U^2} \{\mathscr{F}^2(U_k^1, U^2) + c^2 \|U^2 - U_k^2\|^2\}].$$

The overrelaxation of the Gauss–Seidel procedure is given by

$$U_{k+1}^1 = \alpha U_k^1 + (1 - \alpha)[\arg \min_{U^1} \{\mathscr{F}^1(U^1, U_k^2) + c^1 \|U^1 - U_k^1\|^2\}]$$

and

$$U_{k+1}^2 = \nu U_k^2 + (1 - \nu)[\arg \min_{U^2} \{\mathscr{F}^2(U_{k+1}^1, U^2) + c^2 \|U^2 - U_k^2\|^2\}].$$

The parameters α and ν are determined so that the algorithms are convergent, with a fast rate of convergence. The relaxation strategies discussed above are based on the work of Basar (1987) and Khachiyan and Erlich (1978). Jacobi- and Gauss–Seidel-type algorithms have the advantage that they can be implemented on parallel computers by assigning the optimization to different processors (see Bertsekas and Tsitsiklis, 1989; Bertsekas and Tseng, 1994).

1.2 Solution of systems of equations

The remaining discussion in this chapter is concerned with the solution of a system of nonlinear equations arising either from (1.4) or from the computation of the mapping (1.2), given the model $g(Y, U^1, U^2, \ldots, U^i, \ldots, U^N) = 0$. The Nash equilibrium solution computed in a dynamic setting is as an *open-loop* strategy in that the decision makers evaluate their best policy over time and implement it (Basar and Olsder, 1982). The same consideration also applies to the algorithms in the previous section. This may cause problems if there is a change in the original conditions prevailing at the time the decision was made. A feedback strategy that relates future decisions to the state of the system at that juncture may be more desirable. One possible way of obtaining such a strategy is by considering a parametrized decision rule of the type

$$U^i = \mathscr{U}^i(Y, \theta^i), \qquad i = 1, \ldots, N$$

where $\mathscr{U}^i(\cdot)$ is the feedback of the ith decision maker and θ^i is the parameter vector that relates Y to U^i. The model and the feedback strategies of all decision makers can be considered as a simultaneous system and a mapping established that relates a given value of $\theta^i, i = 1, \ldots, N$, to $Y, U^i, i = 1, \ldots, N$. The Nash strategy can then be formulated in terms of the optimization of θ^i for the ith decision maker. The details of this parametrised feedback approach are discussed further in Chapter 11 for a single decision maker. The extension to the Nash strategy discussed above is straightforward. The parametrized feedbacks simply entail optimization over θ^i, rather than U^i in open-loop decisions. Therefore, we confine the discussion below to the open-loop case.

We return to the solution of the system (1.4) and the decision models. The models are essentially large systems of nonlinear difference equations, which have to be solved over a given period of time. For large, and in particular linear, systems, the received wisdom from numerical analysis is that a Gauss–Seidel successive overrelaxation (GS-SOR) algorithm is appropriate especially if strict accuracy is not particularly demanded. Systems exceeding 200 simultaneous equations are, in this case, referred to as 'large' (see Ortega, 1972). Although economic models usually exceed this size, Nepomiastchi and Ravelli (1978) have argued that economic systems can be reordered to reduce the simultaneous blocks, during each time period, to well below this limit. This allows efficient implementations of the Newton or Broyden algorithms to solve the simultaneous blocks (Becker and Rustem, 1993).

Consider the following model at time period t:

$$y_t = \Gamma_t(y_t, y_{t-1}, y_{t-2}, \ldots, e_t, \hat{x}_t), \tag{1.6}$$

$$e_t = y_{t+1}, \; e_T = y_T, \; t = 1, \ldots, T, \tag{1.7}$$

where y_t is the endogenous variable vector, e_t is the expectation of y_{t+1}, restricted by (1.7) under perfect foresight, and $\hat{x}_t$ are the fixed exogenous values. The model equations (1.6) can be reordered such that, for each time period, the nonzero elements of the Jacobian matrix take the form characterized by Figure 1.1(b). For simplicity, we omit the time subscript whenever the time structure is clear.

Following Nepomiastchi and Ravelli (1978), Gabay *et al.* (1980) and Henk Don and Gallo (1987) argue that econometric models as in Figure 1.1(a) can be expressed as in Figure 1.1(b), and a Newton-type algorithm can be used to solve the purely simultaneous part of the model, called the *loop block*. In this chapter, we use the same basic approach, with a number of modifications. The algorithm of Levy and Low (1988) is adopted to reorder the model equations. We apply a column partitioning and sequential evaluation of the columns of the Jacobian of (1.6) with respect to y_t, while keeping e_t fixed at $\bar{e}_t$. This simplifies the Newton algorithm, in which the columns of the Jacobian are evaluated sequentially and at certain intervals. We call this the *Newton successive column evaluation* (N-SCE) algorithm.

As an alternative strategy, we use Broyden's secant update to further improve the Jacobian approximation. The previous algorithm is simply augmented by Broyden's rank-one update. We call this the *Newton–Broyden* (N-B-SCE) algorithm.

In both N-SCE and N-B-SCE algorithms, global convergence is ensured by an Armijo-type stepsize strategy, based on an error norm merit function. Another feature of both algorithms is that the Jacobian of (1.6), computed for the block of equations at period t, is kept and reused, as required, in the solution of corresponding blocks. The justification for this is past observations that model Jacobians do not change substantially over different time periods and also across different solutions. This is aimed to avoid the initial evaluation of the Jacobian of a block, and can result in substantial savings.

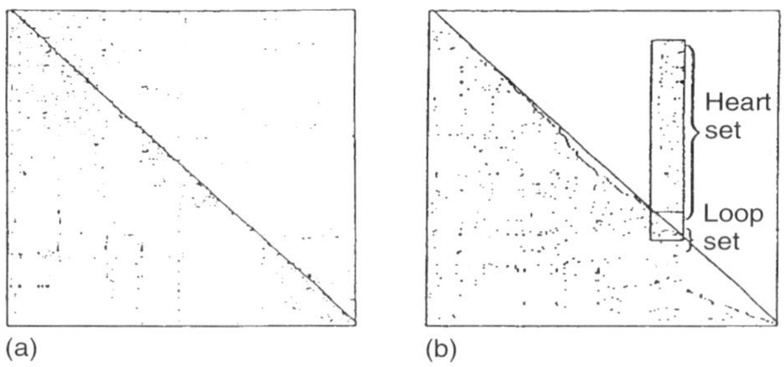

Figure 1.1 (a) The nonzero elements of the Jacobian of the USA model of OECD (1989) in the form (1.6) for a single time period, before reordering – without perfect foresight, with $e_t = \bar{e}_t$. (b) The nonzero elements of the model in (a) after reordering.

The solution algorithm is extended to the rational expectation restriction (1.7) on the system. This requires a forward solution, with which partial derivatives are obtained. The Newton algorithm for the expectations uses this information backwards in time to gradually reduce the error between the expectations estimates and the corresponding endogenous variables.

The purpose of this chapter is to discuss the computational and theoretical properties of efficient Broyden-or Newton-type algorithms. The N-SCE and N-B-SCE algorithms are often contrasted with variants of the Gauss–Seidel successive overrelaxation (GS-SOR) algorithm with optimal relaxation factors. Let $f_t^i(\cdot)$ denote a reduced form of (1.1) from which all variables other than those corresponding to the loop block have been eliminated. We discuss how this may be done in Section 2. The basic GS-SOR algorithm can be described as follows.

Algorithm 1.2 (Gauss–Seidel SOR for solving the system at time t)

Step 0: Given $\alpha, \gamma \in (0, 2), e_t = \bar{e}_t, \text{tol}_1, \text{tol}_2, \text{tol}_3$ and an initial value y_{t0}, set $j = 0$.

Step 1: For equation $i = 1, \ldots, n$, evaluate

$$y^i_{t(j+1/2)} = \alpha f^i_t(y^1_t(j+1/2), \ldots, y^{i-1}_{t(j+1/2)}, y^{i+1}_{t(j)}, \ldots, y^n_{t(j)}) + (1-\alpha) y^i_{t(j)},$$

$$y^i_{t(j+1)} = \gamma y^i_{t(j+1/2)} + (1-\gamma) y^i_{t(j)}.$$

Step 2: If, for all i,

$$|y^i_{t(j)}| > \text{tol}_1 \quad \text{and} \quad \frac{|y^i_{t(j+1)} - y^i_{t(j)}|}{|y^i_{t(j)}|} < \text{tol}_2,$$

or

$$|y^i_{t(j+1)} - y^i_{t(j)}| < \text{tol}_3,$$

we have convergence then **stop**. Else, set $j = j + 1$ and **go to Step 1**.

The values of the relaxation parameters α, γ can be tuned empirically for a given model to improve the convergence of the algorithm. In numerical experiments with the OECD (1989) Interlink macroeconomic models of Spain and the USA discussed in Becker and Rustem (1993), the parameters were set at $\text{tol}_1 = 0.5 \times 10^{-8}$, $\text{tol}_2 = 0.2 \times 10^{-3}$, $\text{tol}_3 = 10^{-8}$, and a search over the relaxation factors to improve the behaviour of the algorithms led to the values $\alpha = \gamma = 1$. However, there is little theoretical justification for this algorithm. Hence, if it fails to converge, it is difficult to establish the reason, and remedy the situation, using theoretical analysis. In addition, the structure of the system being solved is restricted to the form given in **Step 1**. For every equation in the system, we need to define a variable for which that equation is being solved.

This is a restriction for complex systems with no obvious choice of a unique variable, given each equation.

The Newton algorithm, on the other hand, is intended for nonlinear systems, and, as discussed in Section 3 (see also Ortega and Rheinboldt, 1970; Dennis and Schnabel, 1983), it does have theoretical analysis to assist the user. Also, it does treat the equations as a system without a particular structure, and computes the solution for a general system. The Armijo stepsize strategy used to ensure global convergence in the N-SCE and N-N-SCE algorithms is shown to achieve unit stepsizes upon satisfaction of an accuracy condition on the Jacobian approximation. We also give a simple derivation of the Dennis–Moré (1974) characterization of Q-superlinear convergence rates for the N-SCE and N-B-SCE algorithms.

The ease of implementing the GS-SOR variants is, to some extent, responsible for the popularity of this algorithm. Improved software facilities such as modelling languages and differentiation schemes have eliminated this relative advantage of GS-SOR. There are also studies suggesting the superiority of Gauss–Seidel variants in terms of computational efficiency (Norman, Lasdon and Hsin, 1983; Hughes-Hallett, 1984). In the computation of Nash equilibria in game theory, Basar (1987) also provides encouraging results. On the other hand, Gabay *et al.* (1980) and Henk Don and Gallo (1987) have found GS-SOR to be inferior in performance to the Newton methods they have considered. Also, the convergence and accuracy of the solution can be more easily regulated in Newton-type algorithms. This does help to distinguish between the error in the solution and actual disturbances in a noisy decision framework.

A popular Newton method for solving the loop block of the model in Figure 1.1(b) is the *simplified Newton method*, which uses the same initial Jacobian for the loop block, unless forced to reevaluate by lack of progress. Another popular method is the Broyden secant method (see Gabay *et al.*, 1980). Henk Don and Gallo (1987) apply Broyden's method with relatively poor results. This is somewhat contradicted by experience with sparse Broyden updates to the model Jacobian reported in Rustem and Zarrop (1981) and also in Li (1989). The difference in performance may, perhaps, be explained by the different models used in these exercises. Furthermore, the convergence of the simplified Newton algorithm is ensured under restrictive conditions and its rate of convergence is slow, unless started in a close neighbourhood of the solution (Ortega and Rheinboldt, 1970). Thus, the choice of the initial points also may affect the algorithm. Gabay *et al.* (1980) report divergence of the simplified Newton algorithm from bad initial points.

The Broyden update may not always converge elementwise to the Jacobian at the solution. Dennis and Schnabel (1983) give an example in which the update converges to a different matrix. However, this does not alter the desirable theoretical properties of the method, especially when it is coupled with periodic numerical reevaluation of the Jacobian (Moré and Trangenstein, 1976). Furthermore, numerical experience with the method is generally reported to be good. We compare the periodic reevaluation of the columns of the Jacobian (N-SCE) and a method that augments this with the Broyden update (N-B-SCE).

In Section 2, we discuss the N-SCE and N-B-SCE algorithms. In Section 3, we establish the global and local convergence properties of the N-SCE and N-B-SCE algorithms. In Section 4, we discuss the extension to handle the rational expectations conditions. In Section 5, we discuss the Levy and Low (1988) algorithm for ordering the equations.

2 NEWTON-TYPE ALGORITHMS

At time t, with $e_t = \bar{e}_t$, the model (1.6) can be structured in the form of a series of lower-triangular and square sub-blocks (see Section 5 below). The variables corresponding to the former can be solved in one forward sweep. The square blocks require an iterative algorithm for their solution. A square block can be decomposed into disjoint sets of variables called the heart and loop sets. The heart set is distinguished by the fact that if the loop variables are removed, or fixed in value, the heart variables can be evaluated, as a lower-triangular system, in a single sweep. Thus, we need only be concerned with the heart and loop sets in individual square blocks. Let typical heart and loop variables be denoted by $y_{\rm h}$ and y_ℓ respectively. Omitting the time subscript, a typical square block of (1.6) can be written as

$$y_{\rm h} = \Gamma_{\rm h}(y_{\rm h}, y_\ell,), \tag{2.1a}$$

$$\Gamma_\ell(y_{\rm h}, y_\ell) = 0, \tag{2.1b}$$

in which all other known variable values have been substituted. Given the lower-triangular structure of the heart block, (2.1a) can be computationally evaluated as a mapping from the loop to the heart variables,

$$y_{\rm h} = \tilde{\Gamma}_{\rm h}(y_\ell),$$

and can be used to reduce the loop equations (2.1b) to

$$\mathscr{W}(y_\ell) \equiv \Gamma_\ell\big(\tilde{\Gamma}_{\rm h}(y_\ell, y_\ell),\big) = \Gamma_\ell(y_{\rm h}, y_\ell) = 0.$$

Remark

The solution of $\mathscr{W}(y_\ell) = 0$ defines both y_ℓ and $y_{\rm h}$. Therefore it is desirable to minimise the dimension of y_ℓ. In Section 5, we discuss an algorithm for this purpose.

Let $y_\ell \in \mathbb{R}^n$ and $\mathscr{W} : \mathbb{R}^n \to \mathbb{R}^n$. In this section, we consider methods that solve $\mathscr{W}(y_\ell) = 0$. We ignore the subscript ℓ for convenience.

Definition

The system evaluated at point y_j is denoted by $\mathscr{W}_j = \mathscr{W}(y_j)$, and, similarly, the Jacobian $\nabla\mathscr{W}_j = \nabla\mathscr{W}(y_j)$.

Definition

The algorithms usually compute a selected number of columns of $\nabla \mathscr{W}_j$ during an iteration. We use $\nabla \hat{\mathscr{W}}_j$ to denote this and the Broyden update approximation to $\nabla \mathscr{W}_j$. Algorithm 2.2, resulting from the use of the Broyden update, is also called a *quasi-Newton algorithm*.

Algorithm 2.1 (N-SCE) (Newton successive column evaluation)

Step 0: Given

y_0:	the initial value;
δ:	step for numerical evaluation of derivatives in **Step 2** (default: $\sqrt{\text{macheps}}$);
J_{maxit}:	maximum allowed iterations;
$\mu \in [0, \frac{1}{2})$:	the stepsize factor;
μ_1:	the stepsize increase factor (> 1.0; default: 3.0);
$\tilde{\tau} \in (0, 1)$:	the stepsize decrease factor (default: $\frac{1}{2}$);
$k_e \in [1, n]$:	the number of columns to be reevaluated at any one time; $\beta_1 > 0.73, \beta_2 < 0.3$;
ξ:	minimum tolerance level for stepsize (default: macheps);
$s_{\max}$:	maximum number of iterations allowed without any column evaluation.

If there does not exist an available $\nabla \hat{\mathscr{W}}$ then initialize the LQ factors of $[\nabla \hat{\mathscr{W}}] = LQ^{\mathrm{T}}$ as $Q_j = I, L_j = 0$ (Q is orthogonal, $Q^{\mathrm{T}} Q = I$, and L is lower-triangular). Set count of columns to be evaluated, $k = n$; the next column to be evaluated, $i = 1$.

Else use the existing LQ factors of $\nabla \hat{\mathscr{W}}$. Set $k = 0, i = 0$.

Step 1: Start the solution algorithm. Set the iteration count $j = 0, y_j = y^{\mathrm{c}}$. Compute the current error $\mathscr{W}^{\mathrm{c}} = \mathscr{W}(y^{\mathrm{c}})$, the current error norm $\|\mathscr{W}^c\|_2^2$, and set $\mathscr{W}_j = \mathscr{W}^c$. Set the current convergence rate $\rho^{\mathrm{c}} = 0$. Set the columns evaluated at the current point $u = 0$.

Step 2: While

$k > 0$

evaluate column i of the Jacobian using finite differences by computing

$$\Delta\epsilon = \frac{\mathscr{W}(y^{\mathrm{c}} + \delta \mathscr{I}^{\mathrm{i}}) - \mathscr{W}^{\mathrm{c}}}{\delta},$$

and update the LQ factors of $\nabla \hat{\mathscr{W}}$ by removing any existing column i and then updating by the rank-one matrix $\Delta\epsilon \mathscr{I}^{i\mathrm{T}}$ ($\mathscr{I}^i$ is the ith column of the identity matrix).

If $i = n$ then set $i = 1$. Else $i = i + 1$.

Set $k = k - 1, u = u + 1$.

Endwhile.

Set steps without any column evaluation, $s = 0$.
Reinitialization of stepsize after evaluating columns of $\nabla \hat{\mathscr{W}}$: set the stepsize factor $\tau = 1$.

Step 3: If $j \geqslant J_{\text{maxit}}$ then return with failure. Compute d and y^n:

(i) $\nabla \hat{\mathscr{W}}^c d = LQ^T d = -\mathscr{W}$, (2.2)

(ii) $y^n = y^c + \tau d$; (2.3)

take a step, and evaluate $\|\mathscr{W}(y^n)\|_2^2$. If the inequality

$$\|\mathscr{W}^n\|_2^2 - \|\mathscr{W}^c\|_2^2 \leqslant -\tau\mu\|\mathscr{W}^c\|_2^2 \tag{2.4}$$

is not satisfied then the step has failed. **Go to Step 4.**
If $\|\mathscr{W}(y^n)\|_2^2$ is small enough then return success (convergence): $y_* = y^n$.
Set

$$\rho^n = 1 - \frac{\|\mathscr{W}^n\|_2^2}{\|\mathscr{W}^c\|_2^2}, \quad \rho^a = \frac{\rho^n + \rho^c}{2} \quad \text{(the average convergence rate)},$$

$$\|\mathscr{W}^c\|_2^2 = \|\mathscr{W}^n\|_2^2, \quad \rho^c = \rho^n, \quad y^c = y^n,$$

$$j = j + 1,$$

$$y_j = y^c$$

Update count since last move, $u = 0$.
Steps since last update $s = s + 1$.
If $s \geqslant s_{\max}$ or if $\rho^a < \beta_2$ then **go to Step 5.**
Increase the stepsize: if $\rho^a > \beta_1$ then $\tau = \min\{1, \mu_1\tau\}$
Go to Step 3 (i).

Step 4: If $u < n$ then **go to Step 5.**
Reduce stepsize $\tau = \tilde{\tau}\tau$.
If τ is large enough ($\tau > \zeta$) then **go to Step 3 (ii).** Otherwise, line search fails and the algorithm terminates with failure.

Step 5: Number of columns to reevaluate $k = \min\{k_e, n - u\}$. **Go to Step 2.**

The stepsize-increasing procedure in **Step 3** is only activated when the current stepsize is less than unity and the Jacobian is not reevaluated. A new evaluation of the Jacobian, albeit partial, causes the initial trial stepsize to be reset to unity. The reason for not resetting the stepsize to unity in **Step 3** is to reduce the number of expected trials in **Step 3**, (2.4). Furthermore, the stepsize is increased upwards in **Step 3** at a faster rate than the rate of decrease in **Step 4**.

Algorithm 2.2 (Quasi-Newton algorithm (N-B-SCE): combined Newton successive column evaluation—Broyden update)
The difference between this and the above algorithm is in **Step 2:**

Step 2: Unless $k_e = n$ and $s_{\max} = 1$, which would amount to all the columns being reevaluated by numerical differentiation at every iteration (i.e. the *discrete* Newton algorithm (Ortega and Rheinboldt, 1970)), at the end of **Step 2** update the Jacobian matrix using

$$\nabla \hat{\mathscr{W}}_j = \nabla \hat{\mathscr{W}}_{j-1} + \frac{[\mathscr{W}_j - \mathscr{W}_{j-1} - \nabla \hat{\mathscr{W}}_{j-1}(y_j - y_{j-1})](y_j - y_{j-1})^{\mathrm{T}}}{\langle (y_j - y_{j-1}), (y_j - y_{j-1}) \rangle}. \quad (2.5)$$

3 CONVERGENCE OF NEWTON-TYPE ALGORITHMS

In this section, we consider the convergence properties of the basic algorithm. We establish global convergence with relaxed descent directions, and relate the achievement of unit stepsizes to the difference of the Jacobian approximation and the actual Jacobian at the solution. We also discuss the local convergence rate by providing a simple derivation of the Dennis–Moré characterization of Q-superlinear convergence.

The direction d_j, computed in **Step 3,** need not necessarily satisfy (2.2). The global convergence of the algorithm requires that d_j be a descent direction for the sum-of-squares merit function such that

$$\langle [\nabla \mathscr{W}]^{\mathrm{T}} \mathscr{W}_j, d_j \rangle \leqslant -m \|d_j\|^2 \quad (3.1\mathrm{a})$$

for some $m > 0$. This is clearly satisfied by the steepest descent and, with a nonsingular Jacobian, the Newton directions. The quasi-Newton direction satisfies

$$\langle [\nabla \hat{\mathscr{W}}_j]^{\mathrm{T}} \mathscr{W}_j, d_j \rangle \leqslant -m \|d_j\|^2. \quad (3.1,\mathrm{b})$$

The stepsize is adjusted to ensure the inequality

$$\|\mathscr{W}(\bar{y})\|_2^2 - \|\mathscr{W}_j\|_2^2 \leqslant \tau_j \mu \langle [\nabla \mathscr{W}_j]^{\mathrm{T}} \mathscr{W}_j, d_j \rangle, \quad (3.2)$$

where $\langle [\nabla \mathscr{W}_j]^{\mathrm{T}} \mathscr{W}_j, d_j \rangle \leqslant 0$. If, even for small stepsizes, the Newton or the quasi-Newton direction (2.2) does not satisfy (3.2), then the steepest descent direction $d_j = -[\nabla \mathscr{W}_j]^{\mathrm{T}} \mathscr{W}_j$ or even $-d_j$ may be tried (Li, 1989).

In the algorithms of Section 2, an approximate Jacobian is used. If (3.2) is not satisfied for $\tau_j = 1$, with this approximation then the algorithm computes the Jacobian numerically before attempting to reduce the stepsize in order to satisfy the stepsize criterion (3.2).

Assumptions 3.1

(1) There exists a solution y_* to $\mathscr{W}(y) = 0$.

(2) $\mathscr{W} \in \mathbb{C}^1(\mathbb{R}^n)$;

(3) The direction d_j satisfies (3.1).

Assumption 3.1(2) is required to ensure that the algorithm is well defined. We also require that $\nabla\mathcal{W}$ be Lipschitz-continuous to establish the monotonic decrease of the sequence $\{\|\mathcal{W}_j\|_2^2\}$ discussed in Theorem 3.1 below and also to establish global convergence. The monotonic decrease may also be obtained without Lipschitz continuity if it is assumed that $\|\mathcal{W}_j\| = \phi\|d_j\|$, for some $\phi \in (0, \infty)$. The latter is satisfied both by the steepest descent with a nonsingular Jacobian approximation and with the Newton and quasi-Newton directions.

In Theorems 3.3–3.5, we further require $\mathcal{W} \in \mathbb{C}^2(\mathbb{R}^n)$ to establish the achievement of unit stepsizes (i.e. $\tau_j = 1$). The descent condition in Assumption 3 is satisfied by the Newton and steepest descent directions. For the quasi-Newton direction, the algorithm ensures satisfaction by refining the Jacobian approximation. If the descent condition is not satisfied, even after a full numerical evaluation of the Jacobian, with reduced δ, in **Step 2**, then the algorithms fail. In Theorem 3.1, we have allowed for a general descent direction including the Newton, quasi-Newton or steepest descent steps. For example, one reasonable way of establishing the quasi-Newton case is by assuming $\|\mathcal{W}_j\| = \phi\|d_j\|$ as in Theorem 3.1 (ii).

Theorem 3.1 (Monotonic decrease)
Let Assumptions 3.1 be satisfied and, furthermore, let either

(i) $\nabla\mathcal{W}$ *be Lipschitz-continuous*: $\|\nabla\mathcal{W}(y) - \nabla\mathcal{W}(z)\| \leqslant \ell\|y - z\|$, *for* $\ell > 0$; *or*

(ii) d_j *satisfies* $\|\mathcal{W}_j\| = \phi\|d_j\|$, *for some* $\phi \in (0, \infty)$;

(iii) d_j *satisfies (3.1).*

Then, the stepsize computed in the N-SCE and N-B-SCE algorithms is such that $\tau_j \in (0, 1]$ *and hence the sequence* $\{y_j\}$ *generates a corresponding monotonically decreasing sequence* $\{\|\mathcal{W}_j\|_2^2\}$.

Proof
Using the first-order expansion of $\|\mathcal{W}\|_2^2$, we have

$$\|\mathcal{W}_{j+1}\|_2^2 - \|\mathcal{W}_j\|_2^2 = \tau_j\langle[\nabla\mathcal{W}_j]^{\mathrm{T}}\mathcal{W}_j, d_j\rangle$$
$$+\tau_j\int_0^1\langle[\nabla\mathcal{W}(y(t))]^{\mathrm{T}}[\mathcal{W}(y(t)) - \mathcal{W}_j] + \{[\nabla\mathcal{W}(y(t))]^{\mathrm{T}} - [\nabla\mathcal{W}_j]^{\mathrm{T}}\}\mathcal{W}_j, d_j\rangle\,\mathrm{d}t, \tag{3.3}$$

where $y(t) = y_j + t\tau_j d_j$. For (i), given the Lipschitz continuity and (3.1a), we have

$$\|\mathcal{W}_{j+1}\|_2^2 - \|\mathcal{W}_j\|_2^2 = \tau_j\langle[\nabla\mathcal{W}_j]^{\mathrm{T}}\mathcal{W}_j, d_j\rangle\left[1 - \frac{\tau_j}{2m}(\psi^2 + \ell\|\mathcal{W}_j\|)\right], \tag{3.4}$$

where $\|\nabla\mathcal{W}\| \leqslant \psi$. The scalar $\mu \in (0, 1)$ in the stepsize strategy (3.2) clearly determines τ_j such that†

† In Theorems 3.1 and 3.2, the bound $\mu \in (0, 1)$ is sufficient. However, we have adopted the tighter bound $\mu \leqslant \frac{1}{2}$, since the latter is required to establish the convergence to unit stepsizes.

$$\mu \leqslant 1 - \frac{\tau_j}{2m}(\psi^2 + \ell\|\mathscr{W}_j\|) \leqslant \frac{1}{2}.$$

By (3.1), there exists a $\tau_j \in [0, 1]$ satisfying the inequalities (3.2) and (3.3). Suppose τ^0 is the largest $\tau \in [0, 1]$ satisfying these inequalities. Thus, all $\tau \leqslant \tau^0$ also satisfy these conditions, and that the stepsize strategy selects a $\tau_j \in [\bar{\tau}\tau^0, \tau^0]$, where $\bar{\tau} \in (0, 1)$ is input in **Step 0.** By (3.1), it follows that $\{\|\mathscr{W}_j\|_2^2\}$ is a monotonically decreasing sequence. For (ii), we can use (3.3) to derive a relationship similar to (3.4) by invoking $\|\mathscr{W}_j\| = \phi\|d_j\|$. □

We discuss the global convergence of the basic Newton and quasi-Newton algorithms. A review of convergence concepts is given in **CN. 2** in Chapter 6.

Lemma 3.1
Let the assumptions of Theorem 3.1 be satisfied. Then we have

$$\lim_{j\to\infty} \langle [\nabla \mathscr{W}_j]^{\mathrm{T}} \mathscr{W}_j, d_j \rangle = 0. \tag{3.5}$$

Proof
The level set

$$\mathbb{F} = \{y \in \mathbb{R}^n \mid \|\mathscr{W}(y)\|_2^2 \leqslant \|\mathscr{W}(y_{j_0})\|_2^2\}$$

is bounded for some $j_0 \geqslant 0$. We state the proof for the Lipschitz-continous case (i) in Theorem 3.1. Case (ii) can also be similarly established using the compactness of $\mathbb{F}$. Given $\mu \in (0, \frac{1}{2})$, by (3.4), the choice

$$\tau_0 = \min\left\{1, \frac{1-\mu}{\frac{1}{2m}(\psi^2 + \ell\|\mathscr{W}_j\|)}\right\}$$

always satisfies the stepsize strategy (3.2). Clearly, τ_j chosen in the algorithms is in the range $\tau_j \in [\bar{\tau}\tau_0, \tau_0]$, and thereby also satisfies (3.2). As $\mathscr{W}$ is Lipschitz-continuous and $\mathbb{F}$ is compact, there exists a scalar $M < \infty$ and $\|\mathscr{W}_j\| < M$. Thus, as $\ell \in (0, \infty)$ and $m > 0$, we have $\tau_j \geqslant \hat{\tau} > 0, \forall j$, for some $\hat{\tau}$ and stepsize strategy (3.2). The boundedness of $\|\mathscr{W}(y)\|_2^2$ on $\mathbb{F}$ and $\langle [\nabla \mathscr{W}_j]^{\mathrm{T}} \mathscr{W}_j, d_j \rangle \leqslant 0$ imply

$$0 \leqslant \mu \sum_j \tau_j |\langle [\nabla \mathscr{W}_j]^{\mathrm{T}} \mathscr{W}_j, d_j \rangle| \leqslant \sum_j (\|\mathscr{W}_j\|_2^2 - \|\mathscr{W}_{j+1}\|_2^2) < \infty,$$

which yields (3.5). □

Lemma 3.2
Inequality (3.1) and Lemma 3.1 imply $\lim_{j\to\infty} \|d_j\| = 0$.

Proof
The result follows from (3.1) and (3.5). □

Hence, we can show the global convergence for any algorithm satisfying (3.1) as well as $\|\mathscr{W}_j\| = \phi\|d_j\|$, for some $\phi \in (0, \infty)$. As mentioned earlier, this is satisfied by the Newton and quasi-Newton directions and the steepest descent direction with a nonsingular Jacobian.

Theorem 3.2 (Global convergence)
Let Assumptions 3.1 be satisfied and let d_j satisfy $\|\mathscr{W}_j\| = \phi\|d_j\|$, for some $\phi \in (0, \infty)$. Then the algorithms either terminate at a solution of the system $\mathscr{W}(y) = 0$ or generate an infinite sequence $\{y_j\}$ with a subsequence $j \in J \subset \{0, 1, \ldots\}$ such that $\{\|d_j\|\} \to 0$, and thus every accumulation point y_* of the infinite sequence $\{y_j\}$ is a solution of $\mathscr{W}(y) = 0$.

Proof
By Lemmas 3.1 and 3.2, there exists a subsequence $\{y_j\}, j \in J$, such that $\|d_j\| \to 0$. Let there exist y_* such that $\{y_j\} \to y_*$. The existence of such points is ensured, since the algorithms decrease $\|\mathscr{W}(y_j)\|_2^2$ at each iteration, thereby ensuring $y_j \in \mathbb{F}$, with $\mathbb{F}$ compact. The result then follows by letting $j \to \infty, j \in J$, since $\mathscr{W}(y_*) = \lim_{j\to\infty} \mathscr{W}(y_j) = \lim_{j\to\infty} \phi\|d_j\| = 0$. □

We demonstrate convergence to unit stepsizes in terms of a condition on the Jacobian or its approximation. To establish these results, we need to further assume that d_j is a Newton or quasi- (or approximate) Newton direction, and strengthen the assumption on differentiability. We show that the convergence to unit stepsizes depends on the Jacobian. We first establish the result for an exact Jacobian, and in Theorem 3.4 discuss the case for an approximate Jacobian such as a quasi-Newton approximation.

Theorem 3.3 (Convergence to unit stepsizes—exact Jacobian)
Let Assumptions 3.1 be satisfied. Also let $\mathscr{W} \in \mathscr{C}^2(\mathbb{R}^n)$ and

$$m\|v\|^2 \leqslant \langle v, [\nabla\mathscr{W}_j]^{\mathrm{T}}[\nabla\mathscr{W}_j]v\rangle \leqslant \mathscr{M}\|v\|^2, \qquad \mathscr{M} \geqslant m > 0, \quad \forall v \neq 0.$$

Then there is a stage J such that strategy (3.2) is satisfied with $\tau_j = 1, \forall j \geqslant J$.

Proof
Premultiplying the relationship $\nabla\mathscr{W}_j d_j = -\mathscr{W}_j$ by $\mathscr{W}_j$, we obtain

$$\langle([\nabla\mathscr{W}_j]^{\mathrm{T}}\mathscr{W}_j, d_j\rangle = -\langle\mathscr{W}_j, \mathscr{W}_j\rangle = -\langle d_j, [\nabla\mathscr{W}_j]^{\mathrm{T}}[\nabla\mathscr{W}_j]d_j\rangle. \tag{3.6}$$

Consider the second-order expansion of $\|\mathscr{W}\|_2^2$:

$$\begin{aligned}
&\|\mathscr{W}_{j+1}\|_2^2 - \|\mathscr{W}_j\|_2^2 = \langle[\nabla\mathscr{W}_j]^{\mathrm{T}}\mathscr{W}_j, d_j\rangle + \frac{1}{2}\langle d_j, [\nabla\mathscr{W}_j]^{\mathrm{T}}[\nabla\mathscr{W}_j]d_j\rangle \\
&+ \int_0^1 (1-t)\langle d_j, \{\mathfrak{Q}(y_j + td_j) - [\nabla\mathscr{W}_*]^{\mathrm{T}}[\nabla\mathscr{W}_*] + [\nabla\mathscr{W}_*]^{\mathrm{T}}[\nabla\mathscr{W}_*] \\
&- [\nabla\mathscr{W}_j]^{\mathrm{T}}[\nabla\mathscr{W}_j]\}d_j\rangle\,\mathrm{d}t,
\end{aligned} \tag{3.7}$$

where $\mathfrak{Q}(\cdot)$ is the Hessian of $\|\mathscr{W}(\cdot)\|_2^2$, given by $[\nabla\mathscr{W}(\cdot)]^{\mathrm{T}}[\nabla\mathscr{W}(\cdot)]+\Sigma_i\nabla^2\mathscr{W}^i(\cdot)\mathscr{W}^i(\cdot)$, with the last term vanishing at the solution $\mathscr{W}(y_*)=0$. Using (3.6), (3.7) becomes

$$\|\mathscr{W}_{j+1}\|_2^2-\|\mathscr{W}_j\|_2^2\leqslant\langle[\nabla\mathscr{W}_j]^{\mathrm{T}}\mathscr{W}_j,d_j\rangle\left[\frac{1}{2}-(\varsigma_j+\frac{1}{2}\xi_j)\right]$$

where

$$\varsigma_j=\int_0^1(1-t)\|\mathfrak{Q}(y_j+td_j)-[\nabla\mathscr{W}_*]^{\mathrm{T}}[\nabla\mathscr{W}_*]\|\,\mathrm{d}t$$
$$\xi_j=\|[\nabla\mathscr{W}_*]^{\mathrm{T}}[\nabla\mathscr{W}_*]-[\nabla\mathscr{W}_j]^{\mathrm{T}}[\nabla\mathscr{W}_j]\|.$$

Since, by Theorem 3.2, $\{y_j\}\to y_*$, we have $\{\varsigma_j\},\{\xi_j\}\to 0$. For $\tau_j=1$, the scalar $\mu\in[0,\frac{1}{2})$ of the stepsize strategy is bounded by

$$\mu\leqslant\frac{1}{2}-\left(\varsigma_j+\frac{1}{2}\xi_j\right).$$

Thus, when $\frac{1}{2}-\mu\geqslant\varsigma_j+\frac{1}{2}\xi_j$, the stepsize strategy is satisfied with $\tau_j=1$. □

The main difficulty about the use of approximate Jacobians is that d_j does not always satisfy the descent condition $\langle[\nabla\mathscr{W}_j]^{\mathrm{T}}\mathscr{W}_j,d_j\rangle\leqslant 0$, and that $\langle[\nabla\hat{\mathscr{W}}_j]^{\mathrm{T}}\mathscr{W}_j,d_j\rangle\leqslant 0$ does not necessarily imply descent. If descent is not ensured then global convergence cannot be established. A better Jacobian approximation is required, and the algorithms in the previous section aim to achieve this.

Theorem 3.4 (Convergence to unit stepsizes—approximate Jacobian)
Let Assumptions 3.1 be satisfied. Also let $\mathscr{W}\in\mathscr{C}^2(\mathbb{R}^n)$ and

$$m\|v\|^2\leqslant\langle v,[\nabla\hat{\mathscr{W}}_j]^{\mathrm{T}}[\nabla\hat{\mathscr{W}}_j]v\rangle\leqslant\mathscr{M}\|v\|^2,\qquad \mathscr{M}\geqslant 0,\ m>0,\quad \forall v\neq 0,$$

and let d_j that solves

$$\nabla\hat{\mathscr{W}}_jd=-\mathscr{W}_j \tag{3.8}$$

satisfy the descent condition

$$\langle[\nabla\mathscr{W}_j]^{\mathrm{T}}\mathscr{W}_j,d_j\rangle\leqslant-\langle\mathscr{W}_j,\mathscr{W}_j\rangle. \tag{3.9}$$

Then the stepsize strategy is satisfied for some $\tau_j\in(0,1]$, and the monotonic decrease of the sequence $\{\|\mathscr{W}_j\|_2^2\}$ is ensured. Also, there is a number χ such that if

$$\|[\nabla\mathscr{W}_*]^{\mathrm{T}}[\nabla\mathscr{W}_*]-[\nabla\hat{\mathscr{W}}_j]^{\mathrm{T}}[\nabla\hat{\mathscr{W}}_j]\|\leqslant\chi \tag{3.10}$$

then the stepsize strategy is satisfied for $\tau_j=1$.

Proof
Consider the second-order expansion (3.7):

$$\begin{aligned}\|\mathscr{W}_{j+1}\|_2^2 - \|\mathscr{W}_j\|_2^2 \leqslant \tau_j\langle[\nabla\hat{\mathscr{W}}_j]^{\mathrm{T}}\mathscr{W}_j, d_j\rangle + \tau_j\langle[\nabla\mathscr{W}_j - \nabla\hat{\mathscr{W}}_j]^{\mathrm{T}}\mathscr{W}_j, d_j\rangle \\ + \frac{1}{2}(\tau_j)^2\langle d_j, [\nabla\hat{\mathscr{W}}_j]^{\mathrm{T}}[\nabla\hat{\mathscr{W}}_j]d_j\rangle \\ + \int_0^1\left(\tau_j)^2(\zeta_j + \frac{1}{2}\|[\nabla\mathscr{W}_*]^{\mathrm{T}}[\nabla\mathscr{W}_*] - [\nabla\hat{\mathscr{W}}_j]^{\mathrm{T}}[\nabla\hat{\mathscr{W}}_j]\|\right)\|d_j\|^2\mathrm{d}t\end{aligned} \tag{3.11}$$

From (3.8) and (3.9), we note that $\langle[\nabla\mathscr{W}_j - \nabla\hat{\mathscr{W}}_j]^{\mathrm{T}}\mathscr{W}_j, d_j\rangle \leqslant 0$. Also, from (3.8), we have

$$\langle([\nabla\hat{\mathscr{W}}_j]^{\mathrm{T}}\mathscr{W}_j, d_j\rangle = -\langle\mathscr{W}_j, \mathscr{W}_j\rangle = -\langle d_j, [\nabla\hat{\mathscr{W}}_j]^{\mathrm{T}}[\hat{\nabla}\mathscr{W}_j]d_j\rangle. \tag{3.12}$$

Using (3.12), (3.11) becomes

$$\begin{aligned}\|\mathscr{W}_{j+1}\|_2^2 - \|\mathscr{W}_j\|_2^2 \leqslant \tau_j\langle[\nabla\hat{\mathscr{W}}_j]^{\mathrm{T}}\mathscr{W}_j, d_j\rangle \\ \times\left\{1 - \frac{1}{2}\tau_j\left[1 + \frac{1}{m}\left(2\zeta_j + \left\|[\nabla\mathscr{W}_*]^{\mathrm{T}}[\nabla\mathscr{W}_*] - [\nabla\hat{\mathscr{W}}_j]^{\mathrm{T}}[\nabla\hat{\mathscr{W}}_j]\right\|\right)\right]\right\}\end{aligned}$$

Thus, there exists an $\tau_j \in (0, 1]$ such that

$$\mu \leqslant 1 - \frac{1}{2}\tau_j\left[1 + \frac{1}{m}\left(2\zeta_j + \left\|[\nabla\mathscr{W}_*]^{\mathrm{T}}[\nabla\mathscr{W}_*] - [\nabla\hat{\mathscr{W}}_j]^{\mathrm{T}}[\nabla\hat{\mathscr{W}}_j]\right\|\right)\right] < \frac{1}{2}, \tag{3.13}$$

which satisfies the stepsize strategy. The monotonic decrease of the sequence $\{\|\mathscr{W}_j\|_2^2\}$ follows from this property. Global convergence follows from Theorem 3.2. Thus, $\{y_j\} \to y_*$ and $\{\zeta_j\} \to 0$.

We can use (3.13) to conclude that if $\chi > 0$ is such that

$$\frac{1}{2} - \mu \geqslant \frac{1}{m}(2\zeta_j + \chi)$$

(in view of $\{\zeta_j\} \to 0$, this defines the number χ) then the stepsize strategy holds with $\tau_j = 1$. □

The local superlinear convergence rate of the quasi-Newton algorithm is established by the Dennis–Moré (1974) characterisation (see CN.4 of Chapter 6 and CN.2 of Chapter 8). The following is a simple derivation of this characterization that only requires the once continuous differentiability of $\mathscr{W}$ and does not require the Lipschitz continuity of the Jacobian.

Lemma 3.3
Let $\{y^j\} \to y^*$. Then $\{y^j\}$ is Q-superlinearly convergent if and only if $\|d^j\| \leqslant \gamma^j\|d^{j-1}\|, \lim_{j\to\infty}\gamma^j = 0$.

Proof

Suppose $\|d_j\| \leqslant \gamma_j \|d_{j-1}\|, \lim_{j\to\infty} \gamma_j = 0$, holds. We have

$$\begin{aligned}\|y_* - y_j\| &\leqslant \lim_{t\to\infty} \sum_{i=j}^{t-1} \|y_{i+1} - y_i\| \\ &\leqslant \gamma_j \|d_{j-1}\|(1 + \omega + \omega^2 + \omega^3 + \ldots) \\ &\leqslant \frac{\gamma_j}{1-\omega}(\|y_j - y_*\| + \|y_* - y_{j-1}\|),\end{aligned}$$

for some $\omega \in [0, 1)$. As $\{\gamma_j\} \to 0, \omega$ is chosen such that $\gamma_j + \omega < 1, \forall j \geqslant J_0$. J_0 is an integer and is such that $\gamma_j < 1, \forall j \geqslant J_0$. Rearranging the above expression,

$$\begin{aligned}\frac{\|y_j - y_*\|}{\|y_j - y_*\| + \|y_{j-1} - y_*\|} &\leqslant \frac{\gamma_j}{1-\omega}, \\ \|y_* - y_j\| &\leqslant \frac{\gamma_j}{1-\omega-\gamma_j}\|y_* - y_{j-1}\|,\end{aligned} \tag{3.14}$$

which establishes the Q-superlinear convergence of $\{y_j\}$.

Suppose that $\|y_* - y_j\| \leqslant \beta_j \|y_* - y_{j-1}\|, \lim_{j\to\infty} \beta_j = 0$, with $\beta_j < 1$. This yields the inequality

$$\begin{aligned}\|y_* - y_j\| &\leqslant \beta_j \|y_* - y_j\| + \beta_j \|d_{j-1}\| \\ &\leqslant \frac{\beta_j}{1-\beta_j}\|d_{j-1}\|.\end{aligned} \tag{3.15}$$

Next, consider

$$-\nabla \hat{\mathscr{W}}_j d_j = \mathscr{W}_j = \mathscr{W}_* + \int_0^1 \nabla \mathscr{W}(y_j + t(y_* - y_j))(y_* - y_j)\, dt,$$

which yields, for $c_1 \in [0, \infty)$,

$$\|d_j\| \leqslant c_1 \|y_* - y_j\|. \tag{3.16}$$

Using (3.16) in (3.15) leads to the required result:

$$\|d_j\| \leqslant c_1 \frac{\beta_j}{1-\beta_j}\|d_{j-1}\|.$$

□

We give a simple proof of the Dennis–Moré (1974) characterization of the local Q-superlinear convergence rate of the algorithms.

Theorem 3.5

Let $D \subset \mathbb{R}^n$ be an open convex set, let Assumptions 3.1 be satisfied, and let $\{\nabla \hat{\mathscr{W}}_j\}$ be a sequence of nonsingular matrices and suppose that for some $y_0 \in D$ the sequence generated by

$$y_{j+1} = y_j + d_j \quad \text{with} \quad \nabla \hat{\mathscr{W}}_j d_j = -\mathscr{W}_j \tag{3.17}$$

remains in D, *and* $\lim_{j\to\infty} y_j = y_*$. *Then* $\{y_j\}$ *satisfies* $\|d_j\| \leqslant \gamma_j \|d_{j-1}\|$, $\lim_{j\to\infty} \gamma_j = 0$, and thence *converges Q-superlinearly to* y_*, *in some norm* $\|\cdot\|$, *and* $\mathscr{W}(y_*) = 0$ *if and only if*

$$\lim_{j\to\infty} \frac{\|(\nabla\hat{\mathscr{W}}_j - \nabla\mathscr{W}_*)d_j\|}{\|d_j\|} = 0.$$

Proof

In order to establish this result in both directions, we need only consider (3.17) and the first-order expansion of $\mathscr{W}(x)$:

$$-\nabla\hat{\mathscr{W}}_j d_j = \mathscr{W}_j = \mathscr{W}_{j-1} + \nabla\hat{\mathscr{W}}_{j-1} d_{j-1} + \Psi_j + (\nabla\mathscr{W}_* - \nabla\hat{\mathscr{W}}_{j-1})d_{j-1},$$

where

$$\Psi_j = \int_0^1 [\nabla\mathscr{W}(y(t)) - \nabla\mathscr{W}_*]d_{j-1}\,\mathrm{d}t,$$

with $\{\psi_j\} \to 0$ and $\mathscr{W}_{j-1} + \nabla\hat{\mathscr{W}}_{j-1} d_{j-1} = 0$. As $\nabla\hat{\mathscr{W}}_j$ is nonsingular, dividing the above expression by $\|d_{j-1}\|$ yields the required result. □

4 COMPUTATION OF PERFECT FORESIGHT

Having computed the model solution at a given value of e_t, say e_j, it is necessary to ensure the perfect-foresight requirement given by (1.7). Enforcing (1.7) directly destroys the causal (block lower-triangular structure) of the system, since the present expectation of a future variable is being forced to take the actual future (perfect-foresight) value. It is the causal structure that allows the reduction of the system solution, through (2.1), to the solution of the smaller system $\mathscr{W}(y_t) = 0$. We utilize the efficient solution of the causal system to solve the model with perfect foresight. Thus, we relax the perfect-foresight restriction and solve the system along an approximate value of the expectation. This yields a solution along which the Jacobian of the lower-triangular block can be estimated. We use this derivative information to compute an improved expectation by iterating towards the achievement of perfect foresight (taking a quasi-Newton direction to solve (4.2) below). This strategy is in the spirit of most decomposition algorithms. One advantage is that the derivative information required (in (4.4) below) to iterate towards perfect foresight can be evaluated using back-substitution.

The procedure can be seen as enforcing feasibility with respect to (1.7). The problem structure permits the decomposition of the computation of the solution of the system without (1.7) and the enforcement of (1.7). The derivative information required to enforce (1.7) is obtained through this decomposition. Once feasibility is achieved, the corresponding endogenous values will differ from those computed by the algorithm in Section 2. It is then necessary to set e_t at the newly determined value and return to the algorithms of Section 2 to recompute the model solution. The process continues until a fixed point is reached that solves (1.6) and (1.7) simultaneously. The model solution of Section 2 is also necessary to evaluate the derivatives required below.

Consider the problem

$$\min_{e_t;t=1,\ldots,T} \left\{ \sum_{t=1}^{T-1} (y_{t+1} - e_t)^2 + (y_T - e_T)^2 \,\middle|\, y_t = \Gamma_t(y_t, y_{t-1}, y_{t-2}, \ldots, e_t, \hat{x}_t); t = 1, \ldots, T \right\}. \tag{4.1}$$

Consider the stacked variables

$$\mathscr{E} = \begin{bmatrix} e_1 \\ \vdots \\ e_t \\ \vdots \\ e_T \end{bmatrix}, \qquad Y = \begin{bmatrix} y_1 \\ \vdots \\ y_t \\ \vdots \\ y_T \end{bmatrix},$$

which can be used to express the stacked model as $F(Y, \mathscr{E}) = 0$. These are essentially the constraint equations in stacked form. Thus, (4.1) can be rewritten as

$$\min_{\mathscr{E}} \{J(Y, \mathscr{E}) | F(Y, \mathscr{E}) = 0\}, \tag{4.2}$$

where $J(Y, \mathscr{E})$ is the objective function in (4.1). Starting from an arbitrary initial point $\mathscr{E}_0$, a sequence of points $\{\mathscr{E}_{\mathscr{K}}\}$ is generated using the quasi-Newton algorithm,

$$\mathscr{E}_{\mathscr{K}+1} = \mathscr{E}_{\mathscr{K}} - [\hat{\mathfrak{Q}}_{\mathscr{K}}]^{-1} \nabla_{\mathscr{E}} J(Y_{\mathscr{K}}, \mathscr{E}_{\mathscr{K}}) \tag{4.3}$$

where $\hat{\mathfrak{Q}}_{\mathscr{K}}$ is a quasi-Newton approximation to the Hessian $\nabla^2_{\mathscr{E}} J(Y_{\mathscr{K}}, \mathscr{E}_{\mathscr{K}})$, and

$$\nabla_{\mathscr{E}} J(Y_{\mathscr{K}}, \mathscr{E}_{\mathscr{K}}) = \frac{\partial J(Y, \mathscr{E})}{\partial \mathscr{E}} + \left[\frac{\mathrm{d}Y}{\mathrm{d}\mathscr{E}}\right]^{\mathrm{T}} \frac{\partial J(Y, \mathscr{E})}{\partial Y} \bigg|_{Y_{\mathscr{K}}, \mathscr{E}_{\mathscr{K}}} \tag{4.4}$$

We can overcome the difficulty of evaluating $[dY/d\mathscr{E}]$ for large models. From the model equations, we have

$$\frac{\mathrm{d}Y}{\mathrm{d}\mathscr{E}} = -\left[\frac{\partial F}{\partial Y}\right]^{-1} \frac{\partial F}{\partial \mathscr{E}}. \tag{4.5}$$

Hence, (4.3) becomes

$$\nabla_{\mathscr{E}} J(Y_{\mathscr{K}}, \mathscr{E}_{\mathscr{K}}) = \frac{\partial J(Y, \mathscr{E})}{\partial \mathscr{E}} - \left[\frac{\partial F}{\partial \mathscr{E}}\right]^{\mathrm{T}} \Pi \bigg|_{Y_{\mathscr{K}}, \mathscr{E}_{\mathscr{K}}},$$

where Π is the solution of the system

$$\left[\frac{\partial F}{\partial Y}\right]^{\mathrm{T}} \prod = \left[\frac{\partial J(Y, \mathscr{E})}{\partial Y}\right].$$

Since $[\partial F / \partial Y]^{\mathrm{T}}$ is block upper-triangular and $[\partial J(Y, \mathscr{E}) / \partial Y]$ is a vector, $\prod$ can be solved for using block back-substitution.

The overall algorithm for computing the model solution with perfect foresight can be summarized as follows:

Algorithm 4.1 (Perfect foresight)

Step 0: Given an initial point Y_0, set $\mathscr{E}_0 = Y_0$ and set $\mathscr{K} = 0$.

Step 1: Compute $Y_{\mathscr{K}}$ that solves the model $F(Y, \mathscr{E}_{\mathscr{K}}) = 0$ using the algorithms of Section 2.

Step 2: Compute the partial derivatives required in (4.4) and (4.5) evaluated at $Y_{\mathscr{K}}, \mathscr{E}_{\mathscr{K}}$

Step 3: Compute $\mathscr{E}_{\mathscr{K}+1}$ using (4.3).

Step 4: Test for convergence. If not converged, set $\mathscr{K} = \mathscr{K} + 1$; **go to Step 1**.

5 REORDERING SYSTEMS OF EQUATIONS

In order to apply the above algorithm efficiently, we need to reduce the dimension, n of the solution space. This is done by using an algorithm (Levy and Low, 1988) to find a small number of variables whose removal makes the whole model lower-triangular. The removed variables are then those for whose values we must solve in each time period. The algorithm proceeds in several stages. At the outset, we assume that we have a matching of variables and equations, i.e. to each endogenous variable we can associate a single equation. This is normally easy for econometric models, since economists usually introduce equations in order to 'explain' particular variables, but is easy to handle in any case (see e.g. Duff, 1981). In the case where such an assignment of variables to equations is impossible, the equation system is either under- or over-determined, and there is probably something amiss.

Given a matching, we construct an initial incidence digraph $D_i = (V, E)$, where

$$V = \{v \mid v \text{ is a variable}\}$$

and

$$E = \{(u, v) \mid v, u \in V, v \neq u, u_t \text{ appears in the equation for } v_t\}.$$

D contains no self-loops, i.e. edges (v, v), and no parallel edges.

Next, we use the linear time algorithm of Tarjan (1982) to find all the strongly connected components of D; our implementation is based on Duff and Reid (1978). A strongly connected component of D corresponds to a subdigraph $\tilde{D} = (\tilde{V}, \tilde{E})$ such that for any $\tilde{v}, \tilde{u} \in \tilde{V} \exists$ a path $\tilde{v} \to \tilde{u}$ entirely within $\tilde{D}$. Trivial SCCs consist of a single vertex, nontrivial SCCs correspond to sets of variables that mutually depend on each other, i.e. they correspond to sets of equations that cannot be solved as a lower-triangular system.

Given the decomposition of D into SCCs $\tilde{D}_i$, we first order the $\tilde{D}_i$ using the natural ordering (the SCCs of a graph form an acyclic graph), and combine adjacent trivial components to produce a set of subdigraphs $\hat{D}_i, i = 1, \ldots, b$. Some of the $\hat{D}_i$ correspond to blocks of equations that can be solved as a lower-triangular system, and some (which correspond to one of the $\tilde{D}_i$) to square blocks that must be solved iteratively. For the lower-triangular blocks, block ordering is trivial. For the square blocks, we apply the algorithm of Levy and Low (1988) to find a cutset; removal of the cutset leaves a lower-triangular block, and the whole block is then ordered as this lower-triangular block (the so-called 'heart'), followed by the cutset (or loop) variables.

The whole algorithm is applied recursively to the subgraph formed by the loop variables of each square block, as suggested in Hughes Hallet and Fisher (1990). This reduces the total number of above diagonal elements in the Jacobian of the complete system.

Algorithm 5.1 (Levy and Low, 1988)
$G = \hat{D}_i$, where $\hat{D}_i$ is a strongly connected digraph.
$S = \emptyset$, the empty set.
While G is non empty
 While G nonempty and suitable candidates remain apply one of the following contractions:
  If $\exists$ an edge $v \to v$ then contract G by removing v and all edges leaving v. Add v to S.
  If $\exists v$ st indegree $(v) = 0$ then contract G by removing v and all edges leaving v.
  If $\exists v$ st outdegree $(v) = 0$ then contract G by removing v and all edges entering v.
  If $\exists v$ st indegree $(v) = 1$ then contract G by combining v and its ancestor u, and relabelling edges $v \to w$ as $u \to w$, ensuring that parallel edges are deleted.
  If $\exists v$ st outdegree $(v) = 1$ then contract G by combining v and its descendant u, and relabelling edges $w \to v$ as $w \to u$, ensuring that parallel edges are deleted.
 Endwhile
 If G is nonempty then select a vertex v using a heuristic and contract G by removing v and all edges entering and leaving v. Our implementation uses that vertex which has a maximum indegree $\times$ outdegree. Add v to the cutset S.
Endwhile

Levy and Low show that this algorithm can be implemented to run in $O(|E| \log |V|)$ time (E and V being the number of edges and vertices respectively), and it produces a minimal cutset if the heuristic step is never needed. It is also shown that their algorithm produces a minimal cutset on a wider class of digraphs than other efficient (non-exponential-time) algorithms.

Figure 1.1 in Section 1 shows the effect of the above algorithm when it is used to reorder the USA model of the OECD (1989). The algorithm is applied to minimize the dimension of the loop set.

6 CONCLUDING REMARKS

The computation of equilibria can be done using a number of algorithms. The best replay and relaxation-type algorithms are relatively easy to implement, while Newton-type algorithms require some processing of the model to evaluate the Jacobian. The advantage of the N-B-SCR and N-SCR algorithms is that, if required, they coincide with the Newton and discrete Newton algorithms, and thence enjoy the corresponding convergence and superlinear convergence properties discussed in Section 3. It has been shown that the stepsize τ is guaranteed to converge to unity as the Jacobian approximation is sufficiently close to its value at the solution. A novel short proof of Q-superlinear convergence has been given.

The performance of the N-SCR and N-B-SCR algorithms could perhaps be improved by adopting a better way of perturbing the Jacobian columns (see e.g. Li, 1989). However, the imposition of perfect foresight would destroy the block structure and thereby any attempt to exploit that structure in partitioning the columns.

The convergence and accuracy of the solution can be more easily regulated in Newton-type algorithms then in GS-SOR. This does help to distinguish between the error in the solution and actual disturbances in a noisy economy.

The computational advantage of the Newton-type algorithms diminishes as the size of the loop block increases. Yet, most large econometric models are within the range of Newton-type algorithms. The work done, per iteration, in GS-SOR algorithms increases approximately linearly with the size of the loop set. The work done, per iteration, by Newton-type algorithms increases approximately with the square of the same size. It is reasonable to expect that for loop blocks of size 80 or under, Newton-type algorithms will outperform GS-SOR.

One advantage of Newton-type algorithms, regardless of size, is that they solve for the zero of a system $\mathscr{W}(y) = 0$, and it is possible to apply them to models that are not in normal form. It is not necessary to specify the variable for which an equation is to be solved. This arises if a number of equations are reversed so that the variable for which an equation is to be solved is no longer clear. GS-SOR cannot handle such general formulations, whereas Newton-type algorithms can. Unnormalized equations can be added to the loop block with no great loss for Newton-type algorithms.

REFERENCES

Basar, T. (1987) 'Relaxation techniques and asynchronous algorithms for on-line computation of non-cooperative equilibria'. *J. Econ. Dynam. Control* **11**, 531–550.

Basar, T. and G.L. Olsder (1982). *Dynamic Noncooperative Games*. Academic Press, New York.

Becker, R. and B. Rustem (1993). Algorithms for solving nonlinear dynamic decision models. *Ann. Oper. Res.* **44**, 117–142.

Bertsekas, D.P. (1979). Convexification procedures and decomposition methods for non-convex optimisation problems. *J. Optim. Theory Applic.* **29**, 169–197.

Bertsekas, D.P. and P. Tseng (1994). Partial proximal minimization algorithms for convex programming. *SIAM J. Optim.* **4**, 551–572.

Bertsekas, D.P. and J.N. Tsitsiklis (1989). *Parallel and Distributed Computing*. Prentice-Hall, Eaglewood Cliffs, NJ.

Dennis, J.E. and J.J. Moré (1974). A characterisation of superlinear convergence and its application to quasi-Newton methods. *Math. Comput.* **28**, 549–560.

Dennis, J.E. and R.B. Schnabel (1983). *Numerical Methods for Unconstrained Optimization and Nonlinear Equations*, Prentice-Hall Inc, New Jersey.

Duff, I.S. (1981). On algorithms for obtaining a maximum transversal. *ACM Trans. Math. Software* **7**, 315–330, 387–390.

Duff, I.S. and J.K. Reid (1978). An implementation of Tarjan's algorithm for the block triangularisation of a matrix. *ACM Trans. Math. Software* **4**, 137–147, 189–192.

Gabay, D., P. Nepomiastchy, M. Rachdi and A. Ravelli (1980). Numerical methods for the simulation and optimal control of large-scale macroeconomic models. In *Applied Stochastic Control in Econometrics and Management Science* (A. Bensoussan, P. Kleindorfer and C. Tapiero, (eds). North-Holland, Amsterdam.

Henk Don, F.J. and G.M. Gallo (1987). Solving large sparse systems of equations in econometric models. *J. Forecasting* **6**, 167–180.

Ho, Y.C. (1970). Differential games, dynamic optimization and generalized control theory. *J. Optim. Theory Applies.* **6**, 179–209.

Hughes-Hallett, A.J. (1984). Multiparameter iterative techniques for solving econometric models. DP 1, ESRC Macroeconomic Modelling Bureau, University of Warwick.

Hughes-Hallett, A.J. and P.G. Fisher (1987). Should econometricians use Newton methods for model solution? DP 10, ESRC Macroeconomic Modelling Bureau, University of Warwick.

Khachiyan, L.G. and A.I. Erlich (1978). Serial game processes of solution of convex programming problems. *Engng Cyber.* **16**, 147–151

Levy, H. and D.W. Low (1988). A contraction algorithm for finding small cycle cutsets. *J. Algorithms* **9**, 470–493.

Li, G. (1989). 'Succesive column correction algorithms for solving sparse nonlinear systems of equations'. *Math. Programming* **43**, 187–208.

Moré, J.J. and J.A. Trangenstein (1976). On the global convergence of Broyden's method. *Math. Comput.* **30**, 523–540.

Nepomiastchi, P. and A. Ravelli (1978). 'Adapted methods for solving and optimizing quasi-triangular models.' *Ann. Econ. Social Meas.* **6**, 555–562.

Norman, A.L., L.S. Lasdon and J.K. Hsin (1983). A comparison of methods for solving and optimizing a large non-linear macroeconomic model. *J. Econ. Dynam. Control* **6**, 3–24.

OECD (1989). *Interlink Model Manual*. Paris.

Ortega, J. (1972). *Numerical Analysis: A Second Course*. Academic Press, New York.

Ortega, J. and W.C. Rheinboldt (1970). *Iterative Solution of Nonlinear Equations in Several Variables*. Academic Press, New York.

Rockafeller, R.T. (1976). Monotone operators and the proximal point algorithm. *SIAM J Control Optim.* **14**, 877–898.

Rosen, J.B. (1965). Existence and uniqueness of equilibrium points for concave *N*-person games. *Econometrica* **33**, 520–534.

Rustem, B. and . . Zarrop (1981). . . .

Tarjan, R. (1982). 'Depth first search and linear graph algorithms'. *SIAM J. Comput.* **1**, 146–160.

PART IV

Uncertainty

CHAPTER 10

Multiple Objectives under Uncertainty Mean-Variance Optimization for Multicurrency Portfolios

In this chapter, we discuss decision making under uncertainty. Typically, the basic objectives are maximizing expected, or mean, return and minimizing expected risk, or the return variance. These are essentially conflicting objectives, since increased return is normally achieved by adopting increasingly risky strategies. The conflicting objectives entail decision alternatives that need to be carefully analysed before the decision. The starting point of the study is the classical mean-variance approach, widely used in portfolio optimization in finance. It reduces to a quadratic programming formulation, solved by the algorithms discussed in Chapter 2. We introduce this basic framework, and extend it to provide an exact and computable solution to the general quadratic mean-variance problem with the multiplicative effect of exchange rate and return uncertainties. Multiple currencies are considered, with multiple investments in each currency.

In extending the classical mean-variance formulation, we augment it with currency risk. The overall risk is thus the multiplicative effect of return risks, within a currency, and the currency risk. Risk may be interpreted in terms of—though not restricted to—past forecast errors. We develop a framework within a general structure for return and exchange rate forecasts and risks. We show that the quadratic mean-variance structure is maintained, and provide solutions to computational questions in the evaluation of the optimal portfolio. By extending conventional analysis, we arrive at a surprisingly simple solution.

1 INTRODUCTION TO PORTFOLIO OPTIMIZATION

Decision making under uncertainty may be formulated as the simultaneous optimization of the expected total return and the associated risk, measured in terms of the return variance. There is clearly a trade-off in emphasizing the maximization of return and the minimization of risk. The more risk-seeking the decision maker, the greater will the emphasis be on expected return. Similarly, as the decision maker considers more risk-averse, or cautious, policies, the importance of minimizing the variance of the returns will increase.

In this chapter, we consider the classical quadratic mean-variance portfolio optimization problem (Markowitz, 1959; Sharpe, 1970) and its extension to portfolios in different currencies. The multiple objectives we consider are thus expected return and its risk due to the return in a given currency and the currency risk. The importance of international diversification is well recognized in the literature (e.g. Elton and Gruber, 1991; French and Poterba, 1991;

Grauer and Hakanson 1987; Hull and White, 1987; Stulz, 1981; Uppal, 1993). Some of these studies consider currency hedging (e.g. Adler and Jorion, 1992; Kritzman, 1992), whereas others discuss approximate solutions. Solnik (1974) formulates the effect of the nonlinearity arising from joint return and exchange uncertainties, but only considers hedging to take account of currency risk. Makin (1978) discusses the exchange rate as a separate problem. The main issue arises from the product of two uncertain variables, since, as Adler and Dumas (1983) point out, the product of two normal variates is not necessarily normal. Adler and Dumas (1983) consider a log-linear objective and a combination of logarithmic and hedge portfolios.

1.1 Mean-variance optimization within a single currency

Consider the return vector $y_t^i \in \mathbb{R}^{y^i}$ on y^i investments in currency i, at time t:

$$y_t^i = \mathscr{E}[y_t^i] + \rho_t^i,$$

where $\mathscr{E}$ denotes expectation and $\rho_t^i \sim \mathscr{N}(0, \Lambda_{\rho\rho}^i)$, with $\Lambda_{\rho\rho}^i = \mathscr{E}[(\rho_t^i)(\rho_t^i)^{\mathrm{T}}]$. As the models we consider in this chapter are static, we shall omit any reference to time and drop the subscript t. We have multiple currencies $i = 1, \ldots, \mathscr{I}$. There are many ways of estimating $y^i = \mathscr{E}[y^i] + \rho^i$, and we refer the interested reader to the literature on finance (e.g. Elton and Gruber, 1991). To introduce the portfolio problem in a single currency, let $u^i \in \mathbb{R}^{y^i}$ denote the vector of decision variables. Each element of u^i denotes the weight, in terms of the proportion of the total budget, to be attached to each investment. The expected return on the portfolio is then given by

$$\mathscr{E}[\langle u^i, y^i\rangle] = \mathscr{E}[\langle u^i, \mathscr{E}[y^i] + \rho^i\rangle] = \langle \mathscr{E}[y^i], u^i\rangle.$$

The risk of this portfolio can be measured in terms of the statistical variance of the portfolio return, given by

$$\begin{aligned}
\operatorname{var}\langle u^i, y^i\rangle &= \mathscr{E}[\langle u^i, y^i\rangle - \mathscr{E}[\langle u^i, y^i\rangle]]^2 \\
&= \mathscr{E}[\langle u^i, y^i\rangle]^2 - 2\mathscr{E}[\langle u^i, y^i\rangle \mathscr{E}[\langle u^i, y^i\rangle]] + (\mathscr{E}[\langle u^i, y^i\rangle])^2 \\
&= \mathscr{E}[\langle u^i, y^i\rangle]^2 - (\mathscr{E}[\langle u^i, y^i\rangle])^2 \\
&= \mathscr{E}[\langle u^i, \mathscr{E}[y^i] + \rho^i\rangle\langle(\mathscr{E}[y^i] + \rho^i,), u^i\rangle] - [\langle u^i, \mathscr{E}[y^i]\rangle]^2 \\
&= (\langle u^i, \mathscr{E}[y^i]\rangle)^2 + 2\mathscr{E}[\langle u^i, \mathscr{E}[y^i]\rangle\langle\rho^i, u^i\rangle] + \mathscr{E}[\langle u^i, \rho^i\rangle\langle\rho^i, u^i\rangle] - (\langle u^i, \mathscr{E}[y^i]\rangle)^2 \\
&= \langle u^i, \mathscr{E}[\rho^i, \rho^{i\mathrm{T}}]u^i\rangle.
\end{aligned}$$

The decision maker's attitude to risk determines the importance attached to the maximization of the portfolio return versus the minimization of risk, measured by the above variance expression. Let $\beta \in [0, 1]$ be a fixed scalar. An optimal mean-variance portfolio in currency i is given by a combination of the mean of the portfolio return and its variance:

$$\max\{(1-\beta)\langle \mathscr{E}[y^i], u^i\rangle - \beta\langle u^i, \Lambda^i_{\rho\rho} u^i\rangle \mid \langle 1, u^i\rangle = 1, u^i \geqslant 0\},$$

where 1 is the vector with all unit elements. Here, the portfolio return is being maximized simultaneously with the minimization of the risk associated with it. The discussion below does not depend on the nature of the constraints on u^i. We have assumed that the budget is restricted and that the weights attached to each investment, as a proportion of the budget, are nonnegative. These constraints can be altered, as required, to allow for various trading strategies.

We note that, in the above problem, the decision u^i to invest a certain amount in each individual element of y^i given does not affect the expected return. In other words, the case

$$y^i = \mathscr{E}[y^i(u^i)] + \rho^i$$

in which $y^i(u^i)$, the return being a function of the investment, does not occur. The market is essentially assumed to be too large to be affected by the actions of an individual. Such dependence normally leads to nonlinear formulations, discussed in Chapter 11.

1.2 The effect of investments in multiple currencies

The rest of this chapter is essentially devoted to computing the mean and variance of a product of two random variables. Basically, we consider the mean and variance of a quadratic function of the variables. The quadratic formulation is essentially a weighted inner (scalar) product of an augmented vector of the two random variables (Rustem, 1995). An exact, and computable, solution to the full mean-variance problem is provided below.

Consider the exchange rate e^i_t associated with currency i at time t

$$e^i_t = \mathscr{E}[e^i_t] + \epsilon^i_t,$$

where $\epsilon^i_t \sim \mathscr{N}(0, \Lambda^i_{\epsilon\epsilon})$ is the random error between the exchange rate and its forecast. As in the case of the return model, we shall omit the subscript t, since the model we consider is static. Clearly, *the exchange rate corresponding to the base currency is unity with certainty*. This does not alter our discussion, but needs to be taken account of during computation. The return on investments in currency i is converted to the base currency by evaluating $e^i \times y^i$.

Exchange rates are notoriously difficult to model. However, forecasts are available based on macroeconomic analysis. In its simplest form, $\mathscr{E}[e^i]$ can be interpreted as such a forecast. The error between past forecasts and the actual exchange rate is characterized by ϵ^i. The historical value of $\Lambda^i_{\epsilon\epsilon}$ can thus be simply evaluated and, along with the forecast $\mathscr{E}[e^i]$, used in the determination of the mean-variance portfolio.

Let y and ρ denote the $\sum_{i=1}^{\mathscr{I}} y^i$-dimensional return vectors for all investments in all currencies, and the corresponding uncertainties, respectively. Hence, we have

$$y = \mathscr{E}[y] + \rho. \tag{1.1}$$

Let e denote the $\mathscr{J}$-dimensional vector of exchange rates:

$$e = \mathscr{E}[e] + \epsilon, \tag{1.2}$$

with

$$\nu \equiv \begin{bmatrix} \rho \\ \epsilon \end{bmatrix} \sim \mathscr{N}(0, \Lambda) \tag{1.3}$$

and

$$\Lambda = \begin{bmatrix} \Lambda_{\rho\rho} & \Lambda_{\rho\epsilon} \\ \Lambda_{\epsilon\rho} & \Lambda_{\epsilon\epsilon} \end{bmatrix} = \begin{bmatrix} \mathscr{E}(\rho\rho^{\mathrm{T}}) & \mathscr{E}(\rho\epsilon^{\mathrm{T}}) \\ \mathscr{E}(\epsilon\rho^{\mathrm{T}}) & \mathscr{E}(\epsilon\epsilon^{\mathrm{T}}) \end{bmatrix}.$$

Most approaches for modelling returns provide an estimate of $\Lambda_{\rho\rho}$ based on historical data. We have already considered the evaluation of $\Lambda_{\epsilon\epsilon}$ based on the error between past forecasts and historical exchange rates. $\Lambda_{\epsilon\rho}$ can be similarly evaluated based on the error between past forecasts of returns, exchange rates and their respective past actual values.

In Section 2, we extend the traditional mean-variance portfolio optimization problem to investments made in different currencies that take account of currency risk. In Section 3, we discuss the relationships for computational implementation.

2 MULTICURRENCY MEAN-VARIANCE PORTFOLIOS

Each element of y is associated with an exchange rate. To introduce the multicurrency investment portfolio, we define an operation that assigns an exchange rate to a particular investment. Let the matrix

$$\mathscr{O} \in \mathbb{R}^{(\sum_{i=1}^{\mathscr{J}} y^i) \times \mathscr{J}}$$

denote the operator that assigns the exchange rate e^i to an investment in y. Let o_{ij} be the $i$jth element of $\mathscr{O}$; then, we have

$$o_{ij} = \begin{cases} 1 & \text{if ith investment is in jth currency,} \\ 0 & \text{otherwise.} \end{cases}$$

Clearly, $\mathscr{O} = I$ if there is only one investment in each currency. In general, the exchange rate corresponding to each element in the vector y is given by $\mathscr{O}e$. Let

$$\operatorname{diag}[\mathscr{O}e] \in \mathbb{R}^{(\sum_{i=1}^{\mathscr{J}} y^i) \times (\sum_{i=1}^{\mathscr{J}} y^i)}$$

be the diagonal matrix whose jth diagonal element is given by the jth element of $\mathscr{O}e$. Then, the return in the base currency is given by

$$Y = (\operatorname{diag}[\mathscr{O}e])y \in \mathbb{R}^{(\sum_{i=1}^{\mathscr{J}} y^i)}. \tag{2.1}$$

Let $u \in \mathbb{R}^{(\sum_{i=1}^{\mathcal{I}} y^i)}$ denote the portfolio weights for the multicurrency problem. Then, the total return on the portfolio is given by

$$\langle Y, u\rangle = \langle y, (\text{diag}[u])\mathcal{O}e\rangle. \tag{2.2}$$

Lemma 2.1
Let ν_t be the vector-valued random variable given above in (1.3), and let Ω be a symmetric matrix. Then, we have

$$\mathcal{E}[\langle \nu_t, \Omega\nu_t\rangle] = \text{trace}(\Lambda\Omega)$$

(see CN.1).

Proof
Let ν_t^i denote the ith element of ν_t and Ω_{ij} denote the ijth element of Ω. Evaluating the quadratic, we obtain

$$\mathcal{E}[\langle \nu_t, \Omega\nu_t\rangle] = \mathcal{E}\left[\sum_{i,j} \nu_t^i \Omega_{ij} \nu_t^j\right] = \sum_{i,j} \Omega_{ij}\mathcal{E}[\nu_t^i\nu_t^j] = \sum_{i,j} \Omega_{ij}\Lambda_{ij},$$

from which the required result follows. □

Proposition 2.1
Let y and e be given by (1.1) and (1.2). Then, we have

$$\mathcal{E}[\langle Y, u\rangle] = \langle \mathcal{E}[y], (\text{diag}[u])\mathcal{O}\mathcal{E}[e]\rangle + \frac{1}{2}\text{trace}[\Lambda\Omega(u)], \tag{2.3a}$$

for

$$\Omega(u) = \begin{bmatrix} 0 & \text{diag}[u]\mathcal{O} \\ \mathcal{O}^{\mathrm{T}}\text{diag}[u] & 0 \end{bmatrix} \tag{2.3b}$$

Proof
Using (1.1) and (1.2), we can write (2.2) as

$$\begin{aligned} \mathcal{E}[\langle Y, u\rangle] &= \mathcal{E}[\langle \mathcal{E}[y] + \rho, (\text{diag[u]})\mathcal{O}(\mathcal{E}[e] + \epsilon)\rangle] \\ &= \langle \mathcal{E}[y], (\text{diag}[u])\mathcal{O}\mathcal{E}[e]\rangle + \mathcal{E}[\langle \rho, (\text{diag}[u])\mathcal{O}\epsilon\rangle]. \end{aligned}$$

Using Lemma 2.1, the desired result follows. □

To consider the variance term, we define the matrix

$$\mathcal{M} = \begin{bmatrix} \text{diag}[\mathcal{O}\mathcal{E}[e]] \\ \mathcal{O}^{\mathrm{T}}\text{diag}\mathcal{E}[y] \end{bmatrix} \in \mathbb{R}^{(\mathcal{I}+\sum_{i=1}^{\mathcal{I}} y^i)\times(\sum_{i=1}^{\mathcal{I}} y^i)}.$$

The first $(\sum_{i=1}^{\mathcal{I}} y^i)$ rows of $\mathcal{M}$ consist of the diagonal matrix whose diagonal elements are given by $\mathcal{O}\mathcal{E}[e]$. The last $\mathcal{I}$ rows are given by the product of $\mathcal{O}$ with the diagonal matrix whose diagonal elements are given by $\mathcal{E}[y]$.

Lemma 2.2
Let ν_t be the vector-valued random variable given above in (1.3), and let Ω be a symmetric matrix. Then, we have

$$\mathscr{E}[\langle \nu_t, \Omega\nu_t\rangle^2] = [\text{trace}(\Lambda\Omega)]^2 + 2\text{trace}(\Lambda\Omega)^2.$$

Proof
Let the matrix $\Lambda^{1/2}$ be symmetric and $\Lambda^{1/2}\Lambda^{1/2} = \Lambda$, and let $\nu_t = \Lambda^{1/2}b_t$ (see CN.2). Thus, we have

$$\mathscr{E}[b_t b_t^{\mathrm{T}}] = I \quad \text{and} \quad \langle \nu_t, \Omega\nu_t\rangle = \langle b_t, Bb_t\rangle,$$

where $B = \Lambda^{1/2}\Omega\Lambda^{1/2}$ is a symmetric matrix. The components of the vector b_t, with the ith component denoted by b_t^i, are hence uncorrelated normally distributed random variables with

$$\mathscr{E}[(b_t^i)^2] = 1, \quad \mathscr{E}[(b_t^i)^4] = 3, \quad \forall i,$$

where we have used the fact that the fourth moment of a normally distributed random variable with unit variance is 3 (see e.g. Papoulis, 1991, Chap. 5). Consider the transformed expression

$$\mathscr{E}[\langle b_t, Bb_t\rangle^2] = \mathscr{E}\left[\sum_{i,j,k,l} b_t^i b_t^j b_t^k b_t^l B_{ij} B_{kl}\right],$$

where B_{mn} is the mnth element of B. The only nonzero terms arise from equality of all indices or equality in pairs, i.e.

$$\begin{array}{lll} \text{(a)} & i=j, \quad k=l, \quad i\neq k; \\ \text{(b)} & i=k, \quad j=l, \quad j\neq i; \\ \text{(c)} & i=l, \quad j=k, \quad i=j; \\ \text{(d)} & i=j=k=l. \end{array}$$

Thus, we have

$$\begin{aligned} \mathscr{E}[\langle b_t, Bb_t\rangle^2] &= \sum_{\substack{i,k\\ i\neq k}} B_{ii}B_{kk} + \sum_{\substack{i,j\\ i\neq j}} (B_{ij})^2 + \sum_{\substack{i,k\\ i\neq k}} B_{ij}B_{ji} + 3\sum_i (B_{ii})^2 \\ &= \sum_{i,k} B_{ii}B_{kk} + 2\sum_{i,j}(B_{ij})^2 \\ &= (\text{trace}\, B)^2 + 2\,\text{trace}\,(B^2), \end{aligned}$$

where the terms in the first equality on the right correspond to (a), (b), (c), (d) respectively. Noting that for two square matrices $D, F, \text{trace}\,(DF) = \text{trace}\,(FD)$ (see CN. 1), we have

$$\operatorname{trace} B = \operatorname{trace} \Lambda^{1/2}\Omega\Lambda^{1/2} = \operatorname{trace} = \Lambda^{1/2}\Lambda^{1/2}\Omega = \operatorname{trace} \Lambda\Omega,$$
$$\operatorname{trace} B^2 = \operatorname{trace} \Lambda^{1/2}\Omega\Lambda^{1/2}\Lambda^{1/2}\Omega\Lambda^{1/2} = \operatorname{trace} \Lambda\Omega\Lambda\Omega = \operatorname{trace} (\Lambda\Omega)^2. \qquad \square$$

Proposition 2.2
Let y and e be given by (1.1) and (1.2). Then we have

$$\operatorname{var}[\langle Y, u\rangle] = \langle \mathscr{M}u, \Lambda\mathscr{M}u\rangle + \frac{1}{2}\operatorname{trace}[\Lambda\Omega(u)]^2. \tag{2.4}$$

Proof
Using (1.1) and (1.2), we can write (2.3) as

$$\begin{aligned}
\operatorname{var}[\langle Y, u\rangle] &= \mathscr{E}[\langle y, (\operatorname{diag}[u])\mathscr{O}e\rangle - \mathscr{E}[\langle y, (\operatorname{diag}[u])\mathscr{O}e\rangle]]^2 \\
&= \mathscr{E}[\langle y, (\operatorname{diag}[u])\mathscr{O}e\rangle - \langle \mathscr{E}[y], (\operatorname{diag}[u])\mathscr{O}\mathscr{E}[e]\rangle \\
&\quad - \frac{1}{2}\operatorname{trace}[\Lambda\Omega(u)]]^2 \\
&= \mathscr{E}[\langle \mathscr{E}[y] + \rho, (\operatorname{diag}[u])\mathscr{O}(\mathscr{E}[e] + \epsilon)\rangle \\
&\quad - \langle \mathscr{E}[y], (\operatorname{diag}[u])\mathscr{O}\mathscr{E}[e]\rangle - \frac{1}{2}\operatorname{trace}[\Lambda\Omega(u)]]^2 \\
&= \mathscr{E}[\langle \mathscr{E}[y], (\operatorname{diag}[u])\mathscr{O}\varepsilon\rangle + \langle \rho, (\operatorname{diag}[u])\mathscr{O}\mathscr{E}[e]\rangle \\
&\quad + \langle \rho, (\operatorname{diag}[u])\mathscr{O}\varepsilon\rangle - \frac{1}{2}\operatorname{trace}[\Lambda\Omega(u)]]^2 \\
&= \mathscr{E}[\langle \mathscr{M}u, \nu\rangle + \frac{1}{2}\langle \nu, \Omega(u)\nu\rangle - \frac{1}{2}\operatorname{trace}[\Lambda\Omega(u)]]^2 \\
&= \mathscr{E}[\langle \mathscr{M}u, \nu\rangle]^2 + \frac{1}{4}\mathscr{E}[\langle \nu, \Omega(u)\nu\rangle]^2 \\
&\quad - \frac{1}{2}\operatorname{trace}[\Lambda\Omega(u)]\mathscr{E}[\langle \nu, \Omega(u)\nu\rangle] + \frac{1}{4}\{\operatorname{trace}[\Lambda\Omega(u)]\}^2 \\
&= \mathscr{E}[\langle \mathscr{M}u, \nu\rangle]^2 + \frac{1}{2}\operatorname{trace}[\Lambda\Omega(u)]^2,
\end{aligned}$$

where we invoke Lemma 2.2 in the last equality $\square$

As the mean of the portfolio return is given by (2.3) and its variance is given (2.4), the mean-variance optimization problem can now be formulated to take account of currency risk as well. We summarize this result in Theorem 2.1.

Theorem 2.1
Consider the mean-variance portfolio in which the return Y is the product of the return in a local currency and the exchange rate, given by (2.1). Then, the mean-variance portfolio

$$\max\{(1-\beta)\mathscr{E}[\langle Y, u\rangle] - \beta \operatorname{var}[\langle Y, u\rangle] \mid \langle 1, u\rangle = 1, u \geqslant 0\}$$

is given by

$$\max\left\{(1-\beta)\left[\langle\mathscr{E}[y],(\operatorname{diag}[u])\mathscr{O}\mathscr{E}[e]\rangle+\frac{1}{2}\operatorname{trace}[\Lambda\Omega(u)]\right]\right.$$
$$\left.-\beta\left[\langle\mathscr{M}u,\Lambda\mathscr{M}u\rangle+\frac{1}{2}\operatorname{trace}[\Lambda\Omega(u)]^2\right]\bigg|\langle\mathit{1},u\rangle=1,u\geqslant 0\right\}. \tag{2.5}$$

The simpler case of one investment per currency can be obtained by considering $\mathscr{O}$ as an identity matrix. In order to compute optimal portfolios, the differentials of the trace terms in (2.3) and (2.4), with respect to portfolio weights, need to be easily computable. This is discussed in the next section.

3 COMPUTING THE OPTIMAL PORTFOLIO

A computational difficulty in the evaluation of the optimal mean-variance portfolio as a quadratic programming problem (see e.g. Gill, Murray and Wright, 1981; Luenberger, 1984) is the trace terms in (2.3) and (2.4). Although the mean-variance objective, (2.2) and (2.4), is still clearly a quadratic function of the portfolio weights u, the quadratic objective function can be simplified by an explicit evaluation of the differential of the trace terms with respect to u. For (2.4), we consider

$$f(\Omega(u))\equiv\operatorname{trace}[\Lambda\Omega(u)]^2$$

and

$$\frac{\partial f(\Omega(u))}{\partial u_s}=\sum_{k,l}\frac{\partial f(\Omega(u))}{\partial\Omega_{kl}}\quad\frac{\partial\Omega_{kl}}{\partial u_s} \tag{3.1}$$

where u_s is the sth element of u and Ω_{kl} is the kth element of $\Omega(u)$. We invoke a result due to Athans (1968) to yield

$$\frac{\partial f}{\partial\Omega}=2\Lambda\Omega(u)\Lambda. \tag{3.2}$$

Evaluating $\partial\Omega_{kl}/\partial u_s$, it can be verified from the structure of $\Omega(u)$ that the only nonzero elements are

$$\frac{\partial\Omega_{s,n+j(s)}}{\partial u_s}=\frac{\partial\Omega_{n+j(s),s}}{\partial u_s}=1,\qquad j(s)\in\{1,\ldots,\mathscr{J}\},$$

where $j(s)$ is the index of the element of e indicating the exchange rate corresponding to u_s. Using (3.1) and (3.2), we get

$$\begin{aligned}\frac{\partial f(\Omega(u))}{\partial u_s}&=2\{[\Lambda\Omega(u)\Lambda]_{s,n+j(s)}+[\Lambda\Omega(u)\Lambda]_{n+j(s),s}\}\\&=4[\Lambda\Omega(u)\Lambda]_{s,n+j(s)}.\end{aligned}$$

The block containing this element in $\Lambda\Omega(u)\Lambda$ is

$$[\Lambda_{\rho\varepsilon}\mathcal{O}^{\mathrm{T}}(\mathrm{diag}\,[u])\Lambda_{\rho\varepsilon} + \Lambda_{\rho\rho}(\mathrm{diag}\,[u])\mathcal{O}\Lambda_{\varepsilon\varepsilon}].$$

As n can be eliminated from $n + j(s)$ by just considering this block, we have

$$\frac{\partial f(\Omega(u))}{\partial u_s} = 4[\Lambda_{\rho\varepsilon}\mathcal{O}^{\mathrm{T}}(\mathrm{diag}\,[u])\Lambda_{\rho\varepsilon} + \Lambda_{\rho\rho}(\mathrm{diag}\,[u])\mathcal{O}\Lambda_{\epsilon\epsilon}]_{s,j(s)}.$$

Consider the evaluation of the differential

$$\frac{\partial\,\mathrm{trace}\,[\Lambda\Omega(u)]}{\partial u_s}$$

arising from (2.3). This can be computed using the above discussion and the relationship

$$\frac{\partial\,\mathrm{trace}\,[\Lambda\Omega(u)]}{\partial\Omega} = \Lambda,$$

which is also due to Athans (1968).

4 CONCLUDING REMARKS

Mean-variance optimization is the classical tool for decision making under uncertainty—in particular, portfolio optimization. We have introduced the basic approach, and have extended it to construct the multiple currency version of quadratic mean-variance portfolio optimization. The resulting portfolio takes account of the return risks in investments, in a currency, and the currency risk, with respect to a base currency. The mean-variance portfolio, given by (2.5), ensures a risk averse policy with respect to both risks. Thus, the formulation provides an exact solution to the problem of uncertainty in investments in multiple currencies.

5 COMMENTS AND NOTES

CN.1 Trace of a matrix

For a square $(n \times n)$ matrix A, with its elements denoted by a_{ij},

$$\mathrm{trace}\,(A) = \sum_{i=1}^{n} a_{ii}.$$

For $(n \times n)$ matrices D, F, we have the property

$$\mathrm{trace}(DF) = \sum_{i=1}^{n}(DF)_{ii} = \sum_{i=1}^{n}\sum_{j=1}^{n} D_{ij}F_{ji} = \sum_{j=1}^{n}\sum_{i=1}^{n} F_{ji}D_{ij} = \sum_{j=1}^{n}(FD)_{jj} = \mathrm{trace}\,(FD).$$

CN.2 Decomposition of a matrix A

The decomposition of a symmetric positive-semidefinite $(n \times n)$ matrix A, given by

$$A = U[\text{diag}\,(\lambda^1, \ldots, \lambda^n)]\,U^{\mathrm{T}}, \qquad U^{\mathrm{T}}U = I,$$

where λ^i is the ith eigenvalue of A, can be used to establish

$$A^{1/2} = U[\text{diag}\,(\lambda^1, \ldots, \lambda^n)]^{1/2}U^{\mathrm{T}}$$

(see e.g. Horn and Johnson, 1993, Chap. 7).

REFERENCES

Adler, M. and B. Dumas (1983). International portfolio choice and corporate finance. *J. Finance* **38**, 925–984.

Adler, M. and P. Jorion (1992). Universal currency hedges for global portfolios. *J. Portfolio Management* **18** (4), 28–35.

Athans, M. (1968). The matrix minimum principle. *Information and Control* **11**, 592–606.

Elton, E.J. and M.J. Gruber (1991). *Modern Portfolio Theory and Investment Analysis*. Wiley, New York.

French K. and J.M. Poterba (1991). Investor disunification and international equity markets. *Am. Econ. Rev.* **81**(2), 222–226.

Gill, P., W. Murray and M. Wright (1981). *Practical Optimization*. Academic Press, London.

Grauer, F. and N. Hakanson (1987). Gains from international diversification: 1968–85 returns to portfolios of stocks and bonds. *J. Finance* **42**, 721–738.

Horn, R.A. and C.R. Johnson (1993). *Matrix Analysis*. Cambridge University Press.

Hull, J. and A. White (1987). Hedging the risks from writing foreign currency options. *J. Int. Money Finance* **6**, 131–152.

Kritzman, M. (1992). The optimal currency hedging policy with biased forward rates. *J. Portfolio Management* **19** (4), 94–100.

Luenberger, D.G. (1984). *Linear and Nonlinear Programming*. Addison-Wesley, Reading, MA.

Makin, J. (1978). Portfolio theory and the problem of exchange risk. *J. Finance* **33**, 517–534.

Markowitz, H. (1959). *Portfolio Selection: Efficient Diversification of Investments*. Wiley, New York.

Papoulis, A. (1991). *Probability, Random Variables and Stochastic Processes*, 3rd edn. McGraw-Hill, New York.

Rustem, B. (1995). Computing optimal mean-variance portfolios. *J. Econ. Dynam. Control* **19**, 901–908.

Sharpe, W. (1970). *Portfolio Theory and Capital Markets*. McGraw-Hill, New York.

Solnik, B.H. (1974). Equilibrium model of the international market. *J. Econ. Theory* **8**, 500–524.

Stulz, R.M. (1981). On the effects of barriers to international investment. *J. Finance* **36**, 923–934.

Uppal, R. (1993). A general equilibrium model of international portfolio choice. *J. Finance*, **48**, 529–553.

CHAPTER 11

Mean-Variance Optimization: The Nonlinear Case

In this chapter, we extend the mean-variance approach of Chapter 10 to robust policies with nonlinear systems. In the linear case, the approach in Chapter 10 leads to the minimization of a convex combination of the mean and variance of the uncertain problem. The actions of the decision maker do not affect the underlying system. In this chapter, we consider the case when the decision maker can affect the system, as investment in a company might affect its performance and, in turn, influence the return on the investment.

The method computes robust strategies. These can be in the form of *simple* decision rules, as parametrized feedback strategies. The parameters are determined to minimize a convex combination of the objective function and the sensitivity, or variance, of the problem to uncertainty. The sensitivity, or variance, is defined either in terms of the original objective function or the individual endogenous variables.

The effect of uncertainty on nonlinear systems can be evaluated using Monte Carlo simulations. This effect may be biased due to the nonlinearity of the system. We discuss a method for computing expectations of nonlinear systems based on evaluating the bias from certainty equivalence using Monte Carlo simulations.

1 INTRODUCTION TO THE NONLINEAR PROBLEM

In this chapter, we consider a robust optimal policy formulation to decision making under uncertainty for nonlinear systems. As a simple example, consider the return in Chapter 10, given by $y = \mathcal{G}[y] + \rho$. If this return is dependent on the decision variable u,

$$y(u) = \mathcal{G}[y(u)] + \rho,$$

then the resulting mean-variance decision becomes

$$\max\{(1-\beta)\langle\mathcal{G}[y(u)], u\rangle - \beta\,\mathrm{var}\,[\langle y(u), u\rangle] | \langle \mathit{1}, u\rangle = 1; u \geqslant 0\},$$

for $\beta \in [0, 1]$, where the mean is no longer linear in u, nor is the variance necessarily a quadratic.

In the case of a linear model (e.g. where $y(u)$ is a linear function of u) and a quadratic objective function, the problem has been extensively studied (see e.g. Hughes-Hallett, 1984; Karp, 1985; Mitchell, 1979; Stöppler; 1979; Whittle, 1982, 1990), but little attention seems to have been devoted to the nonlinear case. We use as a starting point an approach for nonlinear systems originally discussed in Karakitsos, Rustem and Zarrop (1980), and consider its generalization to parametrized feedback decision rules, and robustness of output or endogenous variables, and we refine the evaluation of expectations in nonlinear systems using Monte Carlo simulations. Consider the policy optimization problem

$$\min_{Y,U,\theta} \mathscr{E}\{f(Y,U) \mid g(Y,U,\epsilon)=0; U=\mathscr{U}(Y,\theta)\}, \tag{1.1}$$

where $\mathscr{E}$ denotes expectation, ϵ is the vector of random uncertainty in model parameters, forecasts or residual variables. Formulation (1.1) is essentially a *discrete-time dynamic optimization* problem: as in Chapter 1, U, Y and ϵ are given by (1.1.1b). Y and U are respectively the endogenous, or output, variables and decision variables, or controls, of the system. f is the policy objective function and g is the underlying model. The relationship

$$U = \mathscr{U}(Y,\theta)$$

represents simple parametrized (feedback) decision rules, with the time structure written in stacked form. $\theta \in \mathbb{R}^\theta$ is the vector of simple feedback decision rule parameters, which are to be determined optimally. An example of $\mathscr{U}$ that does not depend on Y is the polynomial

$$u_t = \theta_0 + \theta_1 t + \theta_2 t^2.$$

A practical use of a polynomial is discussed in Section 6. An example of $\mathscr{U}$ that depends on Y is discussed in detail in Karakitsos and Rustem (1984, 1986) in connection with a large nonlinear macroeconometric model of the UK economy. The interest rate, like the decision variable u_t, is determined with a feedback from endogenous money supply and exchange rate.

The feedback may be ignored if simple decision rules are not required, and the optimal policy is then computed over Y, U only. Due to their *comprehensibility*, simple parametrized feedbacks are sometimes used as *simple rules* that relate the decision to be made to the current state of the system. Without simple feedback rules, the basic approach has been used to formulate robust capacity expansion under demand uncertainty (Paraskevopoulos, Karakitsos and Rustem, 1991) to consider macroeconomic policy (Becker, Hall and Rustem, 1994) and to design chemical engineering processes under uncertainty (Darlington *et al.* 1997).

As in Chapter 1, the framework in (1.1) is a *static* transcription of the dynamic optimization problem (Cannon, Cullum and Polak, 1970; Polak, 1971). The reason for this transcription is partly for simplicity of terminology and partly due to the nonlinear nature of the problem. The static formulation allows the application of nonlinear optimization algorithms for the iterative solution of optimal policy problem, and avoids the difficulties of a general nonlinear dynamic programming framework. The dynamic structure of the problem, as well as the structure of the feedback law, is exploited in computation.

1.1 The stochastic problem

In its general nonlinear form, problem (1.1) seems to have received little attention. It can be solved using stochastic approximation techniques (see e.g. Ermoliev, 1983; Gaivoronski, 1988; Marti, 1990; Kushner and Clark, 1978; Polyak, 1978; Syski, 1988). For linear stochastic systems, Kendrick (1981) discusses the adaptive control strategy. These algorithms present a real computational burden for large systems. Stochastic programming, on the other hand, also provides a computationally demanding solution, but with a limited scope for the formulation of probability densities, which we try to improve upon in this chapter (see e.g. Ben-Tal and Teboulle, 1986; Ben-Tal and Ben-Israel, 1991; Dembo, 1990; Dupacova, 1990; Escudero *et al.* 1993; Kall, 1976; Kall and Wallace, 1994).

To develop a feeling as to how stochastic approximation techniques work, and what is meant by their convergence, we summarize one of the well-known algorithms and the associated convergence result. First, we use the feedback $U = \mathscr{U}(Y, \theta)$ to eliminate U from (1.1), and consider

$$\min_{Y,\theta} \mathscr{E}\{f(Y, \mathscr{U}(Y, \theta) \mid g(Y, \mathscr{U}(Y, \theta), \epsilon) = 0\}.$$

Applying the Lagrange-multiplier method (Polyak, 1978) to the above, we have,

$$\begin{bmatrix} Y_{k+1} \\ \theta_{k+1} \end{bmatrix} = \begin{bmatrix} Y_k \\ \theta_k \end{bmatrix} - \gamma_k \left[\nabla_{\left[\begin{smallmatrix} Y \\ \theta \end{smallmatrix}\right]} f(Y_k, \mathscr{U}(Y_k, \theta_k)) + \nabla g^{\mathrm{T}}(Y_k, \mathscr{U}(Y_k, \theta_k), \epsilon)\lambda_k \right],$$
$$\lambda_{k+1} = \lambda_k + \gamma_k g(Y_k, \mathscr{U}(Y_k, \theta_k), \epsilon)$$

where $\nabla_{\left[\begin{smallmatrix} Y \\ \theta \end{smallmatrix}\right]} f$ is the gradient of f with respect to Y and θ, λ_k is the vector of Lagrange multipliers or shadow prices, and $g(Y_k, \mathscr{U}(Y_k, \theta_k), \epsilon)$ is the observed value of $g(Y_k, \mathscr{U}(Y_k, \theta_k), 0)$, which is corrupted by noise. Similarly, $\nabla g(Y_k, \mathscr{U}(Y_k, \theta_k), \epsilon)$ is the noise-corrupted version of

$$\nabla_{\left[\begin{smallmatrix} Y \\ \theta \end{smallmatrix}\right]} g(Y_k, \mathscr{U}(Y_k, \theta_k), 0).$$

The parameter γ_k satisfies the condition

$$\sum_{k=0}^{\infty} \gamma_k = \infty, \qquad \sum_{k=0}^{\infty} \gamma_k^2 < \infty.$$

This is a well-known condition for the convergence of stochastic approximation methods. Under specific conditions, discussed in detail in Polyak (1978), ensuring the twice-differentiability of the deterministic part of the problem, the existence of a solution and the positive-definiteness of the Hessian of the Lagrangian, the independence of the random disturbances with zero mean and bounded variances in the neighbourhood of the solution, we are assured that

$$\mathscr{P}\left[\begin{bmatrix} Y_k \\ \theta_k \end{bmatrix} \to \begin{bmatrix} Y_* \\ \theta_* \end{bmatrix}\right] \geqslant 1 - \delta,$$

$$\delta = c_1 \mathscr{E}\left[\left\| \begin{matrix} Y_0 - Y_* \\ \theta_0 - \theta_* \end{matrix} \right\|^2 + \|\lambda_0 - \lambda_*\|^2\right] + c_2\sigma^2 \sum_{k=0}^{\infty} \gamma_k^2.$$

The scalars c_1 and c_2 are constants that depend on the deterministic part of the problem. $\mathscr{P}$ denotes probability and σ denotes the bound on the variance of the disturbances.

In the deterministic case, algorithms for solving nonconvex nonlinear programs converge to a local solution. In the presence of random disturbances, there is a generally nonzero probability that the algorithm will leave the region of convergence. (A region of convergence signifies an open domain. If the initial Y_0, U_0 are chosen within this domain, the algorithm converges to the usually isolated local optimum in the domain.) Consequently, there is no convergence in the usual sense of probability one. The convergence result above is thus stated with probability $1 - \delta$, where δ decreases when the performance of the initial approximation is better, the variance bound on the disturbances lower and the step size γ_k shorter.

The above method converges at a rate $O(1/k)$. With $\gamma_k = \gamma/(k + a)$, for any $\delta_0 > 0$, one can choose γ, a, ρ, c so that, with

$$\mathscr{E}\left[\left\| \begin{matrix} Y_0 - Y_* \\ \theta_0 - \theta_* \end{matrix} \right\|^2 + \|\lambda_0 - \lambda_*\|^2\right] \leqslant \rho,$$

we shall have

$$\mathscr{P}\left[\left\| \begin{matrix} Y_0 - Y_* \\ \theta_0 - \theta_* \end{matrix} \right\|^2 + \|\lambda_0 - \lambda_*\|^2 \leqslant \frac{c}{k + a}\right] \geqslant 1 - \delta_0.$$

An algorithm such as the one discussed above presents a real computational burden for large models. Even though its convergence may be assured, the rate of convergence is slow and the possibility of its being stopped before coming close to the solution cannot be discounted. There are, however, other conceptual reasons why it need not be used for economic policy design.

1.2 The decision making process under uncertainty

Policy design is a sequential process. Optimal policy is determined in order to anticipate future observations of endogenous variables and take account of impending random effects. In a closed-loop, feedback, setting of the dynamic problem, certainty equivalence generally applies for the linear model, with known parameters/quadratic objective function/Gaussian disturbance case (see CN.1). As certainty equivalence does not necessarily apply to nonlinear problems, the optimal feedback may have to be redesigned once the value of the disturbance is known. The latter stage may be viewed as a relinearization along the new path due to the disturbance and the reevaluation of the feedback along this path. Thus, once the random effects are actually measured, it may be necessary to adjust and correct the policy to account for the disturbances.

In a dynamic setting, we are seeking a robust policy—possibly feedback—that anticipates the possibility of a disturbance. The robust policy is determined for the whole decision period. Once the disturbance has occurred and is measured, the policy for future time periods may need to be recomputed as required as part of a receding-horizon control or sequential design process. In order to define a robust policy that reduces this need, we introduce the concept of sensitivity. In special cases such as for linear models and quadratic objective functions, this also has a stochastic mean-variance interpretation. The concept is introduced in Section 2, and we discuss its applications to the robustness of the objective function and the robustness of the endogenous variables in Sections 3 and 5. In Section 4, we refine the calculation of expected values, by Monte Carlo simulations, for control exercises with nonlinear models.

The sensitivity approach in this chapter is applicable to all nonlinear models that are differentiable at least once. With this assumption, it is applied to the capacity expansion problem in Paraskevopoulos, Karakitsos and Rustem (1991) and to chemical engineering processes described in Section 7, and Darlington *et al.* (1997). Twice continuous differentiability is further required, in Section 3, to ensure the continuity of the optimal policy as a function of the uncertainty and sensitivity. This simplifies the evaluation of derivatives during a robust policy optimization calculation. On the other hand, the approach can be assigned a mean-variance interpretation for models of the following special structure:

$$\begin{bmatrix} Y \\ U \end{bmatrix} = \begin{bmatrix} \mathcal{Y}_y(\theta) + \hat{N}_y(\theta)\epsilon \\ \mathcal{Y}_u(\theta) + \hat{N}_u(\theta)\epsilon \end{bmatrix}, \tag{1.2}$$

where

$$\hat{N}_y(\theta) \in \mathbb{R}^{(y\times\mathcal{T})\times(\epsilon\times\mathcal{T})} \quad \text{and} \quad \hat{N}_u(\theta) \in \mathbb{R}^{(u\times\mathcal{T})\times(\epsilon\times\mathcal{T})}.$$

Without the feedback restriction in (1.1), this reduces to

$$Y = \mathcal{Y}(U) + \hat{N}(U)\epsilon,$$

which is further refined in Section 4 to take account of model nonlinearities when computing expectations.

For linear models, the idea of mean-variance control and robust control of the objective function is discussed in Stöppler (1979), Mitchell (1979) and Whittle (1982). Also for linear models, Engwerda and Otter (1989) discuss the choice of the objective function for stable minimum-variance control.

We consider the computation of robust policies with nonlinear models. The general discussion is in the context of parametrized simple-feedback decision rules. The method simplifies to open-loop or full-feedback strategies by the straightforward elimination of the restriction of parametrized feedbacks. By minimizing the appropriate criteria, optimally chosen strategy and feedback parameters can yield policies that are robust for either the endogenous variables or the policy objective function.

The disadvantage of the robust approach discussed in this chapter is that the perturbation results on which it is based restrict it to a neighbourhood of the robust optimal solution. The local nature of robustness is well known (Pratt, 1964). Nevertheless, it is the robust solution of the nonlinear problem, and, as shown in the examples in Section 3, *this neighbourhood can be sufficiently large*. The advantages are that the approach is implementable for nonlinear systems; it is a deterministic approach that also has a stochastic, mean-variance, interpretation in the linear case; and, it has desirable properties for nonlinear systems. It is also relatively easy to compute, and does not involve Monte Carlo simulations in the computation of the optimal robust policy.

2 THE SENSITIVITY APPROACH TO ROBUST POLICY

Consider the policy optimization problem

$$\min_{Y,U,\theta} \{\hat{f}(Y,U,\beta) \mid g(Y,U,\epsilon)=0;\ U=\mathscr{U}(Y,\theta)\}. \tag{2.1}$$

$\beta \in [0,1]$ is the parameter that emphasizes the degree of robustness of the policy, where

$$\hat{f}(Y,U,0)=f(Y,U).$$

The robust strategies are not restricted to simple parametrized feedback rules. Full-feedback or open-loop strategies can also be computed with the same approach.

Given the values of ϵ and θ, the model solution program simultaneously computes the model equations $g(Y,U,\epsilon)=0$ and determines the value of the feedback $U=\mathscr{U}(Y,\theta)$ to yield the corresponding values of $[Y,U]$. Thus, we can write

$$\begin{bmatrix} Y \\ U \end{bmatrix} = \begin{bmatrix} \mathscr{Y}_y(\theta,\epsilon) \\ \mathscr{Y}_u(\theta,\epsilon) \end{bmatrix}. \tag{2.2}$$

This is a mapping provided by the model solution evaluating $[Y,U]$ from the model and feedback law. Using (2.2) to eliminate $[Y,U]$ from (2.1) yields an unconstrained optimization problem in θ.

$$\min_{\theta} \{\hat{\mathscr{F}}(\theta,\hat{\epsilon})\}, \tag{2.3}$$

where

$$\hat{\epsilon} = \begin{bmatrix} \beta \\ \epsilon \end{bmatrix}$$

and

$$\hat{\mathscr{F}}(\theta, \hat{\epsilon}) = \hat{f}\big(\mathscr{Y}_y(\theta, \epsilon), \mathscr{Y}_u(\theta, \epsilon), \beta\big), \tag{2.4a}$$

which, when $\beta = 0$, becomes

$$\mathscr{F}(\theta, \epsilon) = \hat{\mathscr{F}}\left(\theta, \begin{bmatrix} 0 \\ \epsilon \end{bmatrix}\right) = \hat{f}\big(\mathscr{Y}_y(\theta, \epsilon), \mathscr{Y}_u(\theta, \epsilon), 0\big) = f\big(\mathscr{Y}_y(\theta, \epsilon), \mathscr{Y}_u(\theta, \epsilon)\big). \tag{2.4b}$$

We consider the behaviour of a local solution $\theta(\hat{\epsilon}_0)$ of (2.3) when the base value of $\hat{\epsilon}$, denoted by $\hat{\epsilon}_0$, is subject to perturbation. For simplicity of notation, without loss of generality, we shall assume that $\hat{\epsilon}_0 = 0$. For $\hat{\epsilon}_0 = 0$, the solution of (2.3) yields the optimum deterministic solution to the feedback optimization problem

$$\min_{\theta} \{\mathscr{F}(\theta, 0)\}. \tag{2.5}$$

We apply the results of Fiacco (1976) to establish the basic sensitivity of the solution of (2.3) at $\hat{\epsilon}_0 = 0$.

Proposition 2.1
Let θ_* denote a local isolated solution of (2.5) satisfying the second-order sufficient conditions for a local optimum. If $\hat{\mathscr{F}}$ in the problem

$$\min_{\theta} \{\hat{\mathscr{F}}(\theta, \hat{\epsilon})\} \tag{2.6}$$

is twice continuously differentiable in $\theta, \hat{\epsilon}$ in a neighbourhood of $[\theta_*, 0]$ then for $\hat{\epsilon}$ in a neighbourhood of 0, there exists a unique once continuously differentiable vector function $\theta(\hat{\epsilon})$ satisfying the second-order sufficient conditions for a local minimum of (2.6) such that $\theta(0) = \theta_*$, and, hence, $\theta(\hat{\epsilon})$ is a locally unique minimum of (2.6).

The proof of the above result follows from Fiacco (1976, Theorem 2.1). Corollary 2.1 below relates the perturbations to a first-order approximation of the perturbed solution in the neighbourhood of $\hat{\epsilon}_0 = 0$, and establishes the derivative $d\theta(\hat{\epsilon})/d\hat{\epsilon}$ (Fiacco, 1976, Corollaries 2.1, 3.1).

Corollary 2.1
Let $\nabla_\theta \hat{\mathscr{F}}$ and $\nabla_\theta^2 \hat{\mathscr{F}}$ denote respectively the (column) gradient vector and the second-derivative matrix of $\hat{\mathscr{F}}$, with respect to θ. Under the assumptions of Proposition 2.1, a first-order approximation of $\theta(\hat{\epsilon})$ in a neighbourhood of $\hat{\epsilon} = \hat{\epsilon}_0 = 0$ is given by

$$\theta(\hat{\epsilon}) = \theta_* - [\nabla_\theta^2 \hat{\mathscr{F}}(\theta_*, 0)]^{-1} \frac{\partial[\nabla_\theta \hat{\mathscr{F}}(\theta_*, 0)]}{\partial \hat{\epsilon}} \hat{\epsilon} + O(\|\hat{\epsilon}\|) \tag{2.7a}$$

and the derivative of θ with respect to $\hat{\epsilon}$ can be approximated by

$$\frac{d\theta(\hat{\epsilon})}{d\hat{\epsilon}} = -[\nabla_\theta^2 \hat{\mathscr{F}}(\theta_*, \hat{\epsilon})]^{-1} \frac{\partial[\nabla_\theta \hat{\mathscr{F}}(\theta_*, \hat{\epsilon})]}{\partial \hat{\epsilon}}. \tag{2.7b}$$

In the context of linear continuous-time dynamical systems, the derivative of the state trajectory with respect to perturbed parameters of the system is discussed in Becker, Heunis and Mayne (1979) and Wuu, Becker and Polak (1986).

In subsequent sections, we formulate the policy optimization problem such that the sensitivity of the solution to perturbations is minimized simultaneously with the original policy objective function. The sensitivities are characterized as

$$\frac{\partial f(Y,U)}{\partial \epsilon} \quad \text{and} \quad \frac{\partial Y}{\partial \epsilon}.$$

The former is discussed in Section 3 and the latter in Section 5. The relations of these sensitivities to the variances of the associated stochastic problems are also discussed.

From the point of view of computational implementation, the Hessian required in (2.7b) for the evaluation of

$$\frac{\mathrm{d}\theta(0)}{\mathrm{d}\hat{\epsilon}}$$

will already be available if a Newton-type algorithm is used to compute the point θ_*. An estimate of this matrix will be available if a quasi-Newton algorithm is used. Thus, much of the information required to calculate the sensitivity will already be available. The matrix

$$\frac{\partial \nabla_\theta \hat{\mathscr{F}}(\theta,\hat{\epsilon})}{\partial \hat{\epsilon}}$$

is known once the problem is specified, and need only be evaluated at $[\theta_*, 0]$.

3 ROBUSTNESS WITH RESPECT TO THE POLICY OBJECTIVE FUNCTION

3.1 Sensitivity and robust optimal policy

Consider the deterministic version of the problem

$$\min_{Y,U,\theta} \{ f(Y,U) \mid g(Y,U,0) = 0; U = \mathscr{U}(Y,\theta) \}, \tag{3.1}$$

and the *objective function sensitivity* term

$$f_\epsilon = \left. \frac{\partial f(Y,U)}{\partial \epsilon} \right|_{\epsilon=0} = \left. \frac{\partial Y}{\partial \epsilon}\frac{\partial f(Y,U)}{\partial Y} + \frac{\partial U}{\partial \epsilon}\frac{\partial f(Y,U)}{\partial U} \right|_{\epsilon=0}, \tag{3.2a}$$

where

$$\frac{\partial Y}{\partial \epsilon} = \frac{\partial \mathscr{Y}_y(\theta,\epsilon)}{\partial \epsilon} \tag{3.2b}$$

and

$$\frac{\partial U}{\partial \epsilon} = \frac{\partial \mathscr{Y}_u(\theta, \epsilon)}{\partial \epsilon}, \tag{3.2c}$$

or, equivalently, using the model $g(Y, U, \epsilon)$ and the feedback law $U = \mathscr{U}(Y, \theta)$ directly,

$$\frac{\partial g}{\partial \epsilon} = \frac{\partial g}{\partial Y}\frac{\partial Y}{\partial \epsilon} + \frac{\partial g}{\partial U}\frac{\partial U}{\partial \epsilon}, \tag{3.2d}$$

$$\frac{\partial U}{\partial \epsilon} = \frac{\partial U}{\partial Y}\frac{\partial Y}{\partial \epsilon}, \tag{3.2e}$$

$$\frac{\partial Y}{\partial \epsilon} = \left(\frac{\partial g}{\partial Y} + \frac{\partial g}{\partial U}\frac{\partial U}{\partial Y}\right)^{-1}\frac{\partial g}{\partial \epsilon}. \tag{3.2f}$$

To generate a *robust* policy, an additional penalty term can be augmented to the objective function, for minimizing the policy objective sensitivity, as measured by the column vector (3.2). Consider, therefore, the deterministic problem

$$\min_{Y,U,\theta} \{\hat{f}(Y, U, \beta) \mid g(Y, U, 0) = 0; U = \mathscr{U}(Y, \theta)\} \tag{3.3a}$$

where $0 \leqslant \beta \leqslant 1$ and

$$\hat{f}(Y, U, \beta) = (1 - \beta) f(Y, U) + \beta\langle f_\epsilon, \Sigma f_\epsilon\rangle \tag{3.3b}$$

which is a convex combination of the original objective function and the cost sensitivity $\langle f_\epsilon, \Sigma f_\epsilon\rangle$ for a specified value of β. The matrix Σ is a positive-semidefinite weighting matrix. It is shown below that there is a natural choice for Σ when the problem is reexamined from the risk-aversion point of view. One rationale for adopting $\langle f_\epsilon, \Sigma f_\epsilon\rangle$ is that it reduces to the mean-variance approach in the linear model, quadratic $f(Y, U)$ case. For example, if the model is linearized around the solution path of (3.1), the mean-variance solution to this problem coincides with the solution of (3.1). For $\beta > 0$, problem (3.3) has a solution at which the objective function $f(Y, U)$ is less sensitive to departures from zero perturbation, $\epsilon = 0$, than in the case $\beta = 0$. Using (2.2), $[Y, U]$ can be eliminated from the policy-optimization problem to yield

$$\mathscr{F}(\theta, 0) = f(\mathscr{Y}_y(\theta, 0), \mathscr{Y}_u(\theta, 0)),$$
$$\hat{\mathscr{F}}\left(\theta, \begin{bmatrix}\beta\\0\end{bmatrix}\right) = (1 - \beta)\mathscr{F}(\theta, 0) + \beta\langle f_\epsilon, \Sigma f_\epsilon\rangle.$$

Thus, (3.3a) can be expressed as the unconstrained optimization problem

$$\min_{\theta}\left\{\hat{\mathscr{F}}\left(\theta, \begin{bmatrix}\beta\\0\end{bmatrix}\right)\right\}.$$

This approach generates policies that minimize the objective-function sensitivity to departures of ϵ from the base value $\epsilon = 0$, around a neighbourhood of the solution. Thus, in that neighbourhood, the policy is robust for $\epsilon \neq 0$. In the case of model parameter uncertainty, with the estimated value of the parameter denoted by a, the uncertain parameter is $a + \epsilon$. The system and the derivatives are evaluated at $\epsilon = 0$. The sensitivity of the system to departures of ϵ from its zero mean value is minimized. The local nature of risk-averse policies is well understood (see e.g. Pratt, 1964). The following example illustrates this and considers the usefulness of the *sensitivity approach for large values of* ϵ.

Example 3.1

Consider that, for $u \geqslant 0$, the true model is given by

$$y = (1 + \epsilon)\sqrt{u} \tag{3.4}$$

and the estimated model is given by

$$y = \sqrt{u}. \tag{3.5}$$

The inclusion of ϵ in (3.4) indicates that the true model has a coefficient that differs from the estimated unit value. The objective function is given by

$$f(y, u) = y^2 + (u - 1)^2.$$

Eliminating y from the objective function, we have

$$f(u, \epsilon) = (1 + \epsilon)^2 u + (u - 1)^2 \quad \text{and} \quad f_\epsilon = 2u, \tag{3.6}$$

and, choosing $\Sigma = \frac{1}{4}$, the robust objective function becomes

$$\hat{f}(y(u), u, \beta) = (1 - \beta)[u + (u - 1)^2] + \beta u^2. \tag{3.7}$$

Minimizing (3.7) yields

$$u^*(\beta) = \frac{1}{2}(1 - \beta),$$

which satisfies $u^*(\beta) \geqslant 0$, and, for a completely cautious policy, $\lim_{\beta \to 1} u^*(\beta) = 0$. Evaluating this optimal $u^*(\beta)$ with (3.6) yields

$$f(u^*(\beta), \epsilon) = \frac{1}{2}(1 + \epsilon)^2(1 - \beta) + \frac{1}{4}(1 + \beta)^2.$$

It can be shown that for $\epsilon = 0$ and $\beta > 0$, we have the inequality $f(u^*(\beta), 0) > f(u^*(0), 0)$. This implies that using a robust strategy, with $\beta \neq 0$, may deteriorate the objective function when $\epsilon = 0$. If, however, $\epsilon \neq 0$ then there exist β for which this inequality is reversed for a given ϵ. Equivalently, for given β, and ϵ sufficiently large, this inequality can again be reversed. In particular, if the inequality $(1 + \epsilon)^2 > 1 + \frac{1}{2}\beta$ is satisfied then we have

$$f(u^*(\beta), 0) < f(u^*(0), 0).$$

Thus, the element of robustness, with $\beta \neq 0$, guards against the deterioration of the objective function value for large deviations of the estimated model (3.5) from the true model (3.4). □

Consider the following quadratic objective function, which is widely used in policy optimization:

$$f(Y, U) = \frac{1}{2}\langle Y - Y^{\mathrm{d}}, \mathcal{O}_y(Y - Y^{\mathrm{d}})\rangle + \frac{1}{2}\langle U - U^{\mathrm{d}}, \mathcal{O}_u(U - U^{\mathrm{d}})\rangle, \tag{3.8}$$

where $\mathcal{O}_y$ is a symmetric positive-semidefinite matrix and $\mathcal{O}_u$ is a symmetric positive-definite matrix. The superscript 'd' denotes the desired, or bliss, values of the variables. In this case, the sensitivity term (3.2) is given by

$$f_\epsilon = \left[\frac{\partial \mathcal{Y}_y(\theta, \epsilon)}{\partial \epsilon}\right]^{\mathrm{T}} \mathcal{O}_y(Y - Y^{\mathrm{d}}) + \left[\frac{\partial \mathcal{Y}_u(\theta, \epsilon)}{\partial \epsilon}\right]^{\mathrm{T}} \mathcal{O}_u(U - U^{\mathrm{d}})\Bigg|_{\epsilon=0}. \tag{3.9}$$

In general, the matrices $\partial \mathcal{Y}_y(\theta, \epsilon)/\partial \epsilon$ and $\partial \mathcal{Y}_u(\theta, \epsilon)/\partial \epsilon$ are not constant, and therefore the corresponding $\hat{f}(Y, U, \beta)$ is not a quadratic function. Apart from a purely linear model, one special case that subsumes the linear framework is when the model (2.2) takes the special form (1.2), where

$$\hat{N}_y(\theta) = \frac{\partial \mathcal{Y}_y(\theta, \epsilon)}{\partial \epsilon}\Bigg|_{\epsilon=0} \quad \text{and} \quad \hat{N}_u(\theta) = \frac{\partial \mathcal{Y}_u(\theta, \epsilon)}{\partial \epsilon}\Bigg|_{\epsilon=0}$$

are constant matrices. It is not necessary for $\hat{N}_y$ and $\hat{N}_u$ to be constants. As shown in Paraskevopoulos, Karakitsos and Rustem (1991), the method can be successfully applied without this assumption. However, assuming constant $\hat{N}_y, \hat{N}_u$ for small changes in β from one optimization to the next considerably simplifies the computation of the optimal policy. The algorithm below achieves this by invoking Proposition 2.1. By solving the robust policy problem (3.3) for $\beta = 0$ initially and then solving the problem for gradually increasing values of β, it is possible to regard $\hat{N}_y, \hat{N}_u$ as constant during each optimization. The solution of each optimization can be made to lie arbitrarily close to the previous solution, and, by Proposition 2.1, this closeness is ensured for sufficiently small changes in β from one optimization to the next. Thus, $\hat{N}_y, \hat{N}_u$ can be taken as constants for small variations in β.

Algorithm 3.1. (Robust policies with constant $\hat{N}_y$ and $\hat{N}_u$ during each optimization)

Step 0: Given a small number $\Delta\beta \in (0, 1)$, set $k = 0$, $\beta = \beta_0 = 0$, and solve the optimization problem (3.3a).

Step 1: If $\beta_k = 1$, **stop**. Else set $\beta_{k+1} = \beta_k + \Delta\beta$ and $k = k + 1$.

Step 2: Compute $\hat{N}_y$ and $\hat{N}_u$ along the current optimal solution.

Step 3: Given $\beta = \beta_k$, solve (3.3a) to compute the new optimal solution with constant $\hat{N}_y$ and $\hat{N}_u$ in f_ϵ given by (3.9).

Step 4: If two successive optimal θ values satisfy prescribed convergence criteria, **go to Step 1**. Otherwise, **go to Step 2**.

Proposition 2.1 and Corollary 2.1 ensure that the optimal solution is differentiable *with respect to* β. Thus, by choosing sufficiently small increases to β, we can in turn ensure that successive solutions of (3.3a) are close to each other such that, in this interval, $\hat{N}_y$ and $\hat{N}_u$ can be regarded as constants (see Becker, Hall and Rustem, 1994).

As mentioned earlier, the sensitivity-robustness approach does not depend on the constancy of $\hat{N}_y$ and $\hat{N}_u$. The local-constancy assumption on $\hat{N}_y$ and $\hat{N}_u$ is only a means to simplify the computation. Indeed, in Paraskevopoulos, Karakitsos and Rustem (1991) and Darlington *et al.* (1997), the algorithm is applied without this constancy assumption.

3.2 Mean-variance formulation

The risk-aversion or mean-variance interpretation of the above robust strategy can be discussed using the model (1.2) and the discussion in Chapter 10. For linear models, Stöppler (1979) and Mitchell (1979) have also taken a similar risk-aversion approach. Consider ϵ as a Gaussian random vector $\epsilon \sim \mathcal{N}(0, \Lambda)$. Using the model (1.2), and the same arguments as in Chapter 10, the mean value of the quadratic cost function is given by

$$\begin{aligned}\mathscr{E}[f(Y,U)] = {} & \mathscr{E}[f(\mathscr{Y}_y(\theta), \mathscr{Y}_u(\theta)) \\ & + \langle \hat{N}_y\epsilon, \mathcal{O}_y[\mathscr{Y}_y(\theta) - Y^{\mathrm{d}}]\rangle + \frac{1}{2}\langle \hat{N}_y\epsilon, \mathcal{O}_y\hat{N}_y\epsilon\rangle \\ & + \langle \hat{N}_u\epsilon, \mathcal{O}_u[\mathscr{Y}_u(\theta) - U^{\mathrm{d}}]\rangle + \frac{1}{2}\langle \hat{N}_u\epsilon, \mathcal{O}_u\hat{N}_u\epsilon\rangle] \\ = {} & f(\mathscr{Y}_y(\theta), \mathscr{Y}_u(\theta)) + \frac{1}{2}\,\mathrm{trace}\,[\Lambda(\hat{N}_y^{\mathrm{T}}\mathcal{O}_y\hat{N}_y + \hat{N}_u^{\mathrm{T}}\mathcal{O}_u\hat{N}_u)]. \end{aligned} \tag{3.10}$$

Similarly, the variance of $f(Y, U)$ is given by

$$\begin{aligned}\mathrm{var}\,[f(Y,U)] = {} & \mathscr{E}[f(Y,U) - \mathscr{E}[f(Y,U)]]^2 \\ = {} & \mathscr{E}\Bigg[\frac{1}{2}\langle \epsilon, (\hat{N}_y^{\mathrm{T}}\mathcal{O}_y\hat{N}_y + \hat{N}_y^{\mathrm{T}}\mathcal{O}_u\hat{N}_u)\epsilon\rangle + \langle \epsilon, \hat{N}_y^{\mathrm{T}}\mathcal{O}_y[\mathscr{Y}_y(\theta) - Y^{\mathrm{d}}] \\ & + \hat{N}_u^{\mathrm{T}}\mathcal{O}_u[\mathscr{Y}_u(\theta) - U^{\mathrm{d}}]\rangle - \frac{1}{2}\,\mathrm{trace}\,[\Lambda(\hat{N}_y^{\mathrm{T}}\mathcal{O}_y\hat{N}_y + \hat{N}_u^{\mathrm{T}}\mathcal{O}_u\hat{N}_u)]\Bigg]^2\end{aligned}$$

$$
\begin{aligned}
&= \langle [\mathscr{Y}_y(\theta) - Y^{\mathrm{d}}], \mathcal{O}_y \hat{N}_y \Lambda \hat{N}_y^{\mathrm{T}} \mathcal{O}_y [\mathscr{Y}_y(\theta) - Y^{\mathrm{d}}] \rangle \\
&\quad + \langle [\mathscr{Y}_u(\theta) - U^{\mathrm{d}}], \mathcal{O}_u \hat{N}_u \Lambda \hat{N}_u^{\mathrm{T}} \mathcal{O}_u [\mathscr{Y}_u(\theta) - U^{\mathrm{d}}] \rangle \\
&\quad + 2\langle [\mathscr{Y}_y(\theta) - Y^{\mathrm{d}}], \mathcal{O}_y \hat{N}_y \Lambda \hat{N}_u^{\mathrm{T}} \mathcal{O}_u [\mathscr{Y}_u(\theta) - U^{\mathrm{d}}] \rangle \\
&\quad + \frac{1}{2} \operatorname{trace} [\Lambda(\hat{N}_y^{\mathrm{T}} \mathcal{O}_y \hat{N}_y + \hat{N}_u^{\mathrm{T}} \mathcal{O}_u \hat{N}_u)]^2.
\end{aligned} \tag{3.11}
$$

Thus, the mean-variance objective function

$$(1 - \beta)\mathscr{E}[f(Y, U)] + \beta \operatorname{var} [f(Y, U)]$$

is the same as the robust policy objective function (3.3b) for quadratic objectives, with f_ϵ defined by (3.9), when the trace terms in (3.10) and (3.11) can be treated as constants. This is clearly true for constant $\hat{N}_y, \hat{N}_u$. Obviously, for linear models, this constancy is trivially satisfied. The equivalence of the mean-variance and robust (3.3b) objectives follows if Σ in (3.3b) is identified with the covariance matrix Λ.

The following example illustrates the differences between mean-variance and sensitivity-robustness, and exhibits properties of the latter for nonlinear models.

Example 3.2
A number of points concerning the differences between robustness and mean-variance for nonlinear models can be made using the following static scalar process:

$$\min \mathscr{E}\{f(y) \mid y = \mathscr{Y}(u) + n(u)\epsilon\},$$

where

$$f(y) = y^2 \quad \text{and} \quad \epsilon \sim \mathscr{N}(0, \sigma^2).$$

The robust policy objective function (3.3b) for this problem is given by

$$
\begin{aligned}
\hat{f}(y, \beta) &= (1 - \beta)f(y) + 4\beta\sigma^2 \left(\frac{\partial f}{\partial \epsilon}\bigg|_{\epsilon=0} \right)^2 \\
&= (1 - \beta)[\mathscr{Y}(u)]^2 + 4\beta\sigma^2[\mathscr{Y}(u)]^2[n(u)]^2
\end{aligned} \tag{3.13}
$$

and

$$
\begin{aligned}
\mathscr{E}(f) &= [\mathscr{Y}(u)]^2 + \sigma^2[n(u)]^2, \\
\operatorname{var}(f) &= 2\sigma^4[n(u)]^4 + 4\sigma^2[\mathscr{Y}(u)]^2[n(u)]^2.
\end{aligned} \tag{3.14}
$$

The first term in (3.14) is missing in the objective sensitivity term in (3.13). Therefore, it cannot be expected that a monotonic increase in β necessarily leads to a monotonic decrease in $\operatorname{var}(f)$ for all choices of $\mathscr{Y}(u)$. Numerical examples of monotonic and nonmonotonic behaviour are discussed in Rustem (1994).

Consider the choice

$$f(u) = \sin u, \qquad \mathcal{Y}(u) = \cot u, \qquad u \in [0, \tfrac{1}{2}\pi].$$

When $\epsilon = 0$, we have

$$f(u) = y^2 = \sin^2 u, \quad \frac{\mathrm{d}f}{\mathrm{d}u} = 2\cos u \sin u, \quad \frac{\mathrm{d}^2 f}{\mathrm{d}u^2} = 2(\cos^2 u - \sin^2 u).$$

Thus, for $u = 0, \mathrm{d}f/\mathrm{d}u = 0$, $\mathrm{d}^2 f/\mathrm{d}u^2 > 0$. Hence, f is minimized by $u = y = 0$. However, these values give rise to infinite $\mathcal{E}(f)$ and $\mathrm{var}\,(f)$, as can be confirmed by

$$\begin{aligned}\mathcal{E}(f) &= \sin^2 u + \sigma^2 \cot^2 u,\\ \mathrm{var}\,(f) &= 2\sigma^4 \cot^4 u + 4\sigma^2 \cos^2 u.\end{aligned}$$

On the other hand, given the robust policy objective function

$$\hat{f}(y, \beta) = (1 - \beta)\sin^2 u + 4\beta\sigma^2 \cos^2 u,$$

we see that

$$\begin{aligned}\frac{\mathrm{d}\hat{f}}{\mathrm{d}u} &= 2[(1-\beta) - 4\beta\sigma^2]\cos u \sin u,\\ \frac{\mathrm{d}^2\hat{f}}{\mathrm{d}u^2} &= 2[(1-\beta) - 4\beta\sigma^2]\cos^2 u - \sin^2 u).\end{aligned}$$

Therefore, $\hat{f}$ is minimized by

$$u = \begin{cases} 0 & \text{for } 0 \leqslant \beta \leqslant \dfrac{1}{1+4\sigma^2}, \\ \dfrac{\pi}{2} & \text{for } \dfrac{1}{1+4\sigma^2} < \beta \leqslant 1. \end{cases}$$

Thus, for sufficiently small values of β, the robust strategy leads to $u = 0$ and infinite values of $\mathcal{E}(f)$ and $\mathrm{var}\,(f)$, which are unacceptable. However, if the importance of the sensitivity term is properly emphasised by a high enough value of β, the sensitivity approach leads to the acceptable solution $u = \frac{1}{2}\pi$.

An interesting aspect of the control strategy is its behaviour as the variance of the disturbance increases. Minimizing $\mathcal{E}(f)$ yields the first- and second-order necessary optimality conditions

$$\begin{aligned}\frac{\mathrm{d}f}{\mathrm{d}u} &= \sin^4 u \cos u - \sigma^2 \cos u = 0,\\ \frac{\mathrm{d}^2 f}{\mathrm{d}u^2} &= 4\sin^3 u \cos^2 u - \sin^5 u + \sigma^2 \sin u > 0,\end{aligned}$$

both of which are satisfied by $u = \frac{1}{2}\pi$ for all $\sigma^2 > 1$. The value $u = 0$, often associated with inaction due to infinite variance, does not even solve thc first-order conditions. As $\sigma^2 \to \infty$, the above cautious policy yields $u = \frac{1}{2}\pi$ for $0 < \beta \leqslant 1$, and $u = 0$ for the nonrobust case $\beta = 0$. □

4 COMPUTING EXPECTATIONS USING MONTE CARLO SIMULATIONS

The optimal controls of nonlinear systems are usually computed on the basis that certainty-equivalent controls are sufficiently accurate. In general, the deterministic value of the objective function and its mean value will not be the same for a nonlinear model. There are many studies of nonlinearity that have demonstrated that the divergence between the two can be numerically important (see e.g. Fisher and Salmon, 1983; Fair, 1984; Hall and Henry, 1988; Bianchi and Calzolari, 1982; for a survey, see Chow and Corsi 1982). This section outlines a technique that evaluates, using Monte Carlo simulations, any expectations bias that nonlinearities may introduce. The divergence is subsequently used in the calculation of the optimal solution by augmenting the robust approach discussed in Section 3 (see Becker, Hall and Rustem, 1994).

When the model is linear and the objective function is quadratic, (1.1) can be solved using the stochastic and adaptive control techniques described in Kendrick (1981). In the nonlinear stochastic framework, we can refine the computation of $\mathscr{E}[f(Y, U)]$ by using the approach of Hall and Stephenson (1990). The technique proceeds by computing the expected value of Y, and thence the corresponding bias from certainty-equivalent values, by using Monte Carlo simulations. Let

$$\mathscr{E}[Y] = \hat{Y} + \mathscr{B}. \tag{4.1}$$

Thus, $\mathscr{E}[Y]$ is given by the deterministic solution $\hat{Y}$ plus $\mathscr{B}$, the expected deviation of this deterministic value from $\mathscr{E}[Y]$. This deviation, or bias, is purely due to the nonlinearity of the system, and would vanish for a linear model with $\epsilon \sim \mathscr{N}(0, \Lambda)$. The bias, once computed around a given trajectory, can be included as a shift in the desired values of the endogenous variables. Hence, the essential structure of the quadratic objective function (3.8) can be preserved. Consider the expectation

$$\mathscr{E}[Y - Y^{\mathrm{d}}] = \hat{Y} - Y^{\mathrm{d}} + \mathscr{B} = \hat{Y} - \hat{Y}^{\mathrm{d}}, \tag{4.2a}$$

where

$$\hat{Y}^{\mathrm{d}} = \mathscr{B} - Y^{\mathrm{d}} \tag{4.2b}$$

and

$$\begin{aligned} \mathscr{E}[\langle Y - Y^{\mathrm{d}}, \mathfrak{Q}_y(Y - Y^{\mathrm{d}})\rangle] &= \langle \mathscr{E}[Y - Y^{\mathrm{d}}], \mathfrak{Q}_y \mathscr{E}[Y - Y^{\mathrm{d}}]\rangle \\ &\quad + \mathscr{E}(\langle (Y - Y^{\mathrm{d}}) - \mathscr{E}(Y - Y^{\mathrm{d}}), \mathfrak{Q}_y[(Y - Y^{\mathrm{d}}) - \mathscr{E}(Y - Y^{\mathrm{d}})]\rangle) \\ &= \langle \hat{Y} - \hat{Y}^{\mathrm{d}}, \mathfrak{Q}_y(\hat{Y} - \hat{Y}^{\mathrm{d}})\rangle \\ &\quad + \mathscr{E}(\langle (Y - Y^{\mathrm{d}}) - \mathscr{E}(Y - Y^{\mathrm{d}}), \mathfrak{Q}_y[(Y - Y^{\mathrm{d}}) - \mathscr{E}(Y - Y^{\mathrm{d}})]\rangle), \end{aligned} \tag{4.3}$$

where the first equality can be verified as an identity. If the model

$$Y(U) = \hat{Y}(U) + \mathscr{B}(U) + \hat{N}_y(U)\epsilon \tag{4.4}$$

is adopted then the second term in (4.3) can be simplified by considering the relationship

$$\mathscr{E}[\langle \hat{N}_y\epsilon, \mathfrak{Q}_y\hat{N}_y\epsilon\rangle] = \mathrm{trace}\,(\Lambda\hat{N}_y^{\mathrm{T}}\mathfrak{Q}_y\hat{N}_y), \tag{4.5}$$

due to Lemma 10.2.1, and, hence, ignoring the argument U, we have

$$\mathscr{E}[\langle Y - Y^{\mathrm{d}}, \mathfrak{Q}_y(Y - Y^{\mathrm{d}})\rangle] = \langle \mathscr{E}[Y - Y^{\mathrm{d}}], \mathfrak{Q}_y\mathscr{E}[Y - Y^{\mathrm{d}}]\rangle + \mathrm{trace}\,(\Lambda\hat{N}_y^{\mathrm{T}}\mathfrak{Q}_y\hat{N}_y). \tag{4.6}$$

Thus, we can justify using only the required shift if we assume that the last term in (4.6) is constant. In the algorithms discussed below, we assume that the trace term can be treated as constant and consequently eliminated from the computation. The expected value of the original decision making problem becomes

$$\begin{aligned}\min_{Y,U}\ \mathscr{E}\{f(Y,U) \mid g(Y,U,\epsilon)=0\} &= \min_{Y,U}\ \mathscr{E}\Big\{\frac{1}{2}\langle Y - Y^{\mathrm{d}}, \mathfrak{Q}_y(Y - Y^{\mathrm{d}})\rangle \\ &\quad + \frac{1}{2}\langle U - U^{\mathrm{d}}, \mathfrak{Q}_u(U - U^{\mathrm{d}})\rangle \mid g(Y,U,\epsilon) = 0\Big\} \\ &= \min_{Y,U}\Big\{\frac{1}{2}\langle \hat{Y} - \hat{Y}^{\mathrm{d}}, \mathfrak{Q}_y(\hat{Y} - \hat{Y}^{\mathrm{d}})\rangle \\ &\quad + \frac{1}{2}\langle U - U^{\mathrm{d}}, \mathfrak{Q}_u(U - U^{\mathrm{d}})\rangle \\ &\quad + \mathrm{trace}\,(\Lambda\hat{N}_y^{\mathrm{T}}\mathfrak{Q}_y\hat{N}_y)\Big| g(\hat{Y}, U, 0) = 0\Big\}.\end{aligned} \tag{4.7}$$

Algorithm 4.1 (Certainty-equivalent optimal decisions with nonlinear bias compensation)

Step 1: Compute the initial $\hat{Y}$ and U by computing the optimal deterministic solution of (1.1), with $\epsilon \equiv 0$.

Step 2: Perform a set of Monte Carlo simulations of the model around the base defined by the deterministic optimal solution U, and compute $\mathscr{E}[\mathrm{Y}]$. Thus, compute $\mathscr{B} = \mathscr{E}[Y] - \hat{Y}$.

Step 3: Use $\mathscr{B}$ to formulate and solve the problem

$$\begin{aligned}\min_{Y,U}\Big\{&\frac{1}{2}\langle (Y - \hat{Y}^{\mathrm{d}}), \mathfrak{Q}_y(Y - \hat{Y}^{\mathrm{d}})\rangle \\ &+ \frac{1}{2}\langle U - U^{\mathrm{d}}, \mathfrak{Q}_u(U - U^{\mathrm{d}})\rangle\Big| g(Y, U, 0) = 0\Big\},\end{aligned} \tag{4.8}$$

and return to **Step 1**.

The convergence criterion for the algorithm is that two successive solutions computed in **Step 3** should be sufficiently close to each other. The computed $\mathscr{B}$ in the above algorithm is clearly the value that corresponds to the optimal solution $[Y_*, U_*]$ of (4.8), and hence is a local estimate of the bias. As $[Y_*, U_*]$ is dependent on the model and objective-function specification, the value of $\mathscr{B}$ thereby computed is also dependent on the objective function.

The trace term and $\mathscr{B}$ are assumed to be constant in **Step 3** and (4.8), whereas they may be functions of the control strategy. This limits the algorithm to a first-order estimate of the nonlinear effects of the stochastic disturbances. However, in the case of $\mathscr{B}$, numerical evidence and some intuitive analysis are provided in Becker, Hall and Rustem (1994) for the constancy of this value, even for nonlinear models. $\mathscr{B}$ does not appear to change appreciably after the first iteration of the algorithm. This is despite the fact that the deterministic Newton-type nonlinear optimization algorithm, solving (4.8), itself requires many iterations to attain convergence.

The algorithm does not have a stepsize strategy to ensure convergence in general. It is also not certain whether the algorithm would be computationally viable if it had such a strategy. In practice, this has not been a problem, given numerical experience with $\mathscr{B}$ treated as constant during the nonlinear optimization calculation in **Step 3** (see Becker, Hall and Rustem, 1994).

The robust strategy incorporating the bias $\mathscr{B}$ is formulated using (4.4). The mean of the objective function in the presence of $\mathscr{B}$ has already been discussed in (4.6). Evaluating the variance of the objective function yields

$$\begin{aligned}
\operatorname{var}[f(Y,U)] &= \mathscr{E}[f(Y,U) - \mathscr{E}[f(Y,U)]]^2 \\
&= \mathscr{E}\left[\frac{1}{2}\langle \hat{N}_y\epsilon, \mathfrak{Q}_y\hat{N}_y\epsilon\rangle + \langle \hat{N}_y\epsilon, \mathfrak{Q}_y(\hat{Y} + \mathscr{B} - Y^{\mathrm{d}})\rangle - \frac{1}{2}\operatorname{trace}(\Lambda\hat{N}_y^{\mathrm{T}}\mathfrak{Q}_y\hat{N}_y)\right]^2 \\
&= \langle(\hat{Y} - \hat{Y}^{\mathrm{d}}), \mathfrak{Q}_y\hat{N}_y\Lambda\hat{N}_y^{\mathrm{T}}\mathfrak{Q}_y(\hat{Y} - \hat{Y}^{\mathrm{d}})\rangle + \frac{1}{2}\operatorname{trace}(\Lambda\hat{N}_y^{\mathrm{T}}\mathfrak{Q}_y\hat{N}_y)^2. \qquad (4.9)
\end{aligned}$$

Hence, the bias is again incorporated in the variance by a shift in the desired values. Expression (4.9) indicates that the variance (or sensitivity) term in the objective has to be adjusted with the computed bias. The shift in the desired values, due to $\mathscr{B}$, implies in the robust optimal control case that the algorithm computes the mean of the objective function and minimizes the sensitivity around this mean value. The algorithm evaluates $\mathscr{B}$ and $\hat{N}_y$ sequentially until an optimal solution is achieved and these values reach a fixed point. Computing robust optimal strategies while compensating for any bias due in the endogenous variables to the nonlinearity of the model can be summarized as follows.

Algorithm 4.2 (Robust optimal decisions with nonlinear bias compensation)

Step 0: Given $\sum = \Lambda$, a positive integer j, divide the interval [0, 1] such that $\Delta\beta = 1/j \in (0, 1)$; set $k = 0$, $\beta = \beta_0 = 0$, and solve the deterministic optimization problem.

Step 1: If $\beta_k = 1$, **stop**. Else set $\beta_{k+1} = \beta_k + \Delta\beta$ and $k = k + 1$.

Step 2: Use Monte Carlo simulations to compute $\mathscr{B}$, and set $\hat{Y}^{\mathrm{d}} = Y^{\mathrm{d}} + \mathscr{B}$.

Step 3: Compute $\hat{N}_y|_{\epsilon=0}$ along the current optimal trajectory.

Step 4: Given $\beta = \beta_k$ and $\mathscr{B}$, solve the robust policy problem with nonlinear bias compensation to compute the current optimal trajectory with constant $\hat{N}_y$ in f_ϵ:

$$\min_{\hat{Y},U}\left\{(1-\beta)\frac{1}{2}\left[\langle\hat{Y}-\hat{Y}^{\mathrm{d}},\mathfrak{Q}_y(\hat{Y}-\hat{Y}^{\mathrm{d}})\rangle+\frac{1}{2}\langle U-U^{\mathrm{d}},\mathfrak{Q}_u(U-U^{\mathrm{d}})\rangle\right]\right.$$
$$\left.+\beta\langle(\hat{Y}-\hat{Y}^{\mathrm{d}}),\mathfrak{Q}_y\hat{N}_y\Lambda\hat{N}_y^{\mathrm{T}}\mathfrak{Q}_y(\hat{Y}-\hat{Y}^{\mathrm{d}})\rangle\,\middle|\,g(\hat{Y},U,0)=0\right\}. \tag{4.10}$$

Step 5: Convergence test for the subalgorithm of **Steps 3 and 4**. If two successive optimal solutions (trajectories) generated by **Steps 3 and 4** satisfy prescribed convergence criteria, **go to Step 6**. Otherwise, **go to Step 3**.

Step 6: Convergence test for the overall algorithm. If two successive optimal solutions (trajectories) generated by **Steps 2–5** satisfy prescribed convergence criteria, **go to Step 1**. Otherwise, **go to Step 2**.

The fixed point generated by the algorithm is such that, given the bias, the robust optimum is computed, and the bias reevaluated at this optimum does not alter. **Step 5** tests the convergence of the iterative process due to the constancy assumption of $\hat{N}_y$ in (4.10). **Step 6** tests the convergence of the algorithm with respect to the reevaluated bias. As mentioned in Section 3, the numerical results discussed in Becker, Hall and Rustem (1994) indicate that the convergence of the overall algorithm in **Step 6** is assisted by the insensitivity of $\mathscr{B}$ to changes in policy instrument values. Accounting for the bias ensures that the robust policy framework can be extended to the case when Y is a nonzero-mean random process while ϵ remains zero-mean. The deterministic interpretation of the approach relies on $\mathscr{B}$ being evaluated by Monte Carlo simulations and used to transform Y so that the sensitivity term can be considered for departures of ϵ from zero. The robust policy is computed by incorporating the affine transformation of Y into the sensitivity framework discussed in Section 2.

5 ROBUSTNESS WITH RESPECT TO ENDOGENOUS VARIABLES

In areas such as finance, the main concern is portfolio return and the objective is to maximize its expected value while risk is measured in terms of its variance. In other areas such as engineering design, while the objective function may be a cost criterion, risk may be associated with another design aspect or product variability, in view of uncertainties. In the following formulation, we consider the robustness of the optimal endogenous or output values to perturbations in the system. The parametrized feedback law can be used in minimizing the sensitivity. To illustrate this point, we begin with a linear model and a parametrized feedback law. Consider the system and feedback law

$$Y = NU + \hat{N}_y \epsilon \quad \text{and} \quad U = \Theta(\theta) Y, \tag{5.1}$$

where $N \in \mathbb{R}^{(y \times \mathscr{T}) \times (u \times \mathscr{T})}$ is a given matrix and $\Theta \in \mathbb{R}^{(u \times \mathscr{T}) \times (y \times \mathscr{T})}$ is a matrix describing the feedback parametrized by the vector θ. Using the feedback, we have

$$[I - N\Theta(\theta)] Y = \hat{N}_y \epsilon. \tag{5.2}$$

With $\epsilon \sim \mathcal{N}(0, \Lambda)$, then, assuming that $I - N\Theta(\theta)$ is nonsingular (see CN.2), the variance of Y can be expressed as

$$\mathscr{E}[Y(\theta) Y^{\mathrm{T}}(\theta)] = [I - N\Theta(\theta)]^{-1} \hat{N}_y \Lambda \hat{N}_y^{\mathrm{T}} [I - N\Theta(\theta)]^{-\mathrm{T}}. \tag{5.3}$$

Thus, the parameters of the feedback law do affect the variance of the endogenous variables, and this can be utilized to choose θ such that the variance is minimized in some norm. Consider the problem

$$\min_{Y,U,\theta} \mathscr{E}\left\{(1-\beta) f(Y, U) + \beta \| \mathscr{E}[YY^{\mathrm{T}}] \|_{\mathrm{F}}^2 \middle| Y = NU + \hat{N}_y \epsilon; U = \Theta(\theta) Y\right\}, \tag{5.4}$$

where $\beta \in [0, 1]$ and $\|A\|_{\mathrm{F}} = \sqrt{\sum_i \sum_j |a_{ij}|^2}$ is the Frobenius norm. One advantage of this norm over other matrix norms is that, in the above formulation, it is differentiable as a function of the elements of the matrix. This is an important property for a Newton-type algorithm solving (5.4). If $\beta > 0$, the parameter vector θ of the feedback law is chosen to minimize the norm of the variance of Y as well as the original objective function. Problem (5.4) is another mean-variance formulation where the mean of the original policy objective and the variance of the endogenous variables are minimized simultaneously. This formulation is essentially deterministic, since both the mean and the variance can be evaluated prior to the optimization to yield

$$\min_{Y,U,\theta} \left\{(1-\beta) f(Y, U) + \beta \| \mathscr{E}[YY^{\mathrm{T}}] \|_{\mathrm{F}}^2 \middle| Y = NU; U = \Theta(\theta) Y\right\}. \tag{5.5}$$

For the sensitivity analysis of (5.1), ϵ is interpreted as a vector of uncertain variables or parameters or forecast values that are subject to perturbation. As in Section 3, without loss of generality, we take the base value to be $\epsilon = 0$. Assuming that the matrix on the right of (5.2) is nonsingular, we have

$$\frac{\partial Y}{\partial \epsilon} = \hat{N}_y^{\mathrm{T}} [I - N\Theta(\theta)]^{-\mathrm{T}} \Big|_{\epsilon=0}, \tag{5.6}$$

and define the sensitivity term as

$$\left[\frac{\partial Y}{\partial \epsilon}\right]^{\mathrm{T}} \sum \left[\frac{\partial Y}{\partial \epsilon}\right] = [I - N\Theta(\theta)]^{-1}\hat{N}_y \sum \hat{N}_y^{\mathrm{T}}[I - N\Theta(\theta)]^{-\mathrm{T}}, \tag{5.7}$$

where $\sum$ is a symmetric positive-semidefinite weighting matrix. The right sides of (5.7) and (5.3) are clearly identical when $\sum = \Lambda$. Thus, the sensitivity and mean-variance terms are similar, and the robust policy is defined as the solution of

$$\min_{Y,U,\theta}\left\{(1-\beta)f(Y,U) + \beta\left\|\left[\frac{\partial Y}{\partial \epsilon}\right]^{\mathrm{T}} \sum \left[\frac{\partial Y}{\partial \epsilon}\right]\right\|_{\mathrm{F}}^{2} \,\middle|\, Y = NU;\, U = \Theta(\theta)\,Y\right\}. \tag{5.8}$$

θ can be chosen to minimize the sensitivity term simultaneously with the original policy objective. For $\beta > 0$, the solution of (5.8) ensures that, at the optimal solution, departures of ϵ from its base value have little effect on the optimal value of Y.

For nonlinear systems, the above robust policy approach has an obvious extension. Consider problem (3.1) with the added requirement that the sensitivity of the endogenous variable values at the solution should also be simultaneously minimized:

$$\min_{Y,U,\theta}\{\hat{f}(Y,U,\beta) \mid g(Y,U,0) = 0;\, U = \mathscr{U}(Y,\theta)\}, \tag{5.9a}$$

where

$$\hat{f}(Y,U,\beta) = (1-\beta)f(Y,U) + \beta\left\|\left[\frac{\partial Y}{\partial \epsilon}\right]^{\mathrm{T}} \sum \left[\frac{\partial Y}{\partial \epsilon}\right]\right\|_{\mathrm{F}}^{2}, \tag{5.9b}$$

where $\partial Y/\partial \epsilon$ is given by (3.2b, c) or (3.2e, f). Clearly, the feedback parameters θ that solve (5.9) also address the minimization of the sensitivity.

6 AN EXAMPLE

Consider a chemical engineering process that uses ingredient A to produce B and C. The concentrations of each of these are denoted by C_{A}, C_{B} and C_{C}. The control variable is the reactor temperature $T(t)$ at t. This is a *continuous-time* process. The endogenous variables are the concentrations C_{A}, C_{B} and C_{C}, and the objective is to *maximize* C_{B} at the end of a time horizon of $\mathscr{T} = 150$ minutes. The problem is described in detail in Darlington *et al.* (1997) and Terwiesch, Agrawal and Rippin (1994). As we have a continuous-time process, we need to consider the infinite-dimensional maximization

$$\max_{T(t)} C_B(T(\mathscr{T}), \epsilon_1, \epsilon_2), \qquad t \in [0, \mathscr{T}],$$

where ϵ_1, ϵ_2 are the uncertainties associated with the problem, and the underlying process is described by the differential–algebraic system

$$\frac{\mathrm{d}C_\mathrm{A}}{\mathrm{d}t} + 2k_1(T)C_\mathrm{A}^2 = 0, \tag{6.1a}$$

$$\frac{\mathrm{d}C_\mathrm{B}}{\mathrm{d}t} - k_1(T)C_\mathrm{A}^2 + k_2(T)C_\mathrm{B} = 0, \tag{6.1b}$$

$$\frac{\mathrm{d}C_\mathrm{C}}{\mathrm{d}t} - k_2(T)C_\mathrm{B} = 0, \tag{6.1c}$$

$$k_1(T) - p_1\mathrm{e}^{-2500/T} = 0, \qquad k_2(T) - p_2\mathrm{e}^{-5000/T} = 0, \tag{6.1d}$$

$$p_1 = 0.0444 + \epsilon_1, \qquad \epsilon_1 \sim \mathcal{N}(0, 0.0111),$$

$$p_2 = 6889.0 + \epsilon_2, \qquad \epsilon_2 \sim \mathcal{N}(0, 48.223).$$

The initial conditions are

$$C_\mathrm{A}(0) = 2000; \qquad C_\mathrm{B}(0) = C_\mathrm{C}(0) = 0.$$

In addition, the temperature $T(t)$ is restricted by the bounds

$$298 \leqslant T(t) \leqslant 398.$$

The uncertainty is essentially due to the rates of chemical reactions. The values p_1, p_2 are determined from experimental measurements that may be subject to error. For food and pharmaceutical industries, where the production process is licensed by the authorities, it is important that the optimal operating policies should be robust to variations in the parameters p_1, p_2.

The above dynamic infinite-dimensional optimization problem can be approximated in finite dimensions by considering a polynomial approximation to $T(t)$. Expressing $T(t)$ as a weighted sum of Lagrangian polynomials, given by

$$T(t) = \sum_{i=1}^{N} \theta_i \prod_{j \neq i} \left(\frac{t - t_j}{t_i - t_j} \right), \tag{6.2}$$

$$\theta = [\theta_1, \cdots, \theta_i, \cdots, \theta_N]^\mathrm{T},$$

where $\theta_i, i = 1, \cdots, N$, are the parameters to be determined optimally and t_i are given constants satisfying $t_1 = 0, t_N = \mathcal{T}$, with $t_i < t_{i+1}, i = 1, \cdots, N$. This is a decision-rule parametrization, similar to (1.1). The differential–algebraic system (6.1), (6.2) is thus solved to yield $C_\mathrm{B}(\mathcal{T})$ for given values of θ (Jarvis and Pantelides, 1992).

The above objective function coincides with one of the endogenous variables. Hence, we note that the bias $\mathcal{B}$, discussed in Section 4, cannot be taken into account in the optimization, since $\mathcal{B}$ would simply be an additive constant in the objective. Furthermore, the mean-variance framework presumes that the objective can be expressed as a *quadratic in the uncertain variables*. Therefore, we need to consider the quadratic expansion of the objective function in the neighbourhood of the mean values of the uncertainties. Let

$$\epsilon = \begin{bmatrix} \epsilon_1 \\ \epsilon_2 \end{bmatrix}, \quad \mathscr{E}(\epsilon) = \begin{bmatrix} 0 \\ 0 \end{bmatrix}, \quad \mathscr{E}(\epsilon\epsilon^{\mathrm{T}}) \equiv \Lambda \equiv \begin{bmatrix} 0.0111 & 0 \\ 0 & 48.223 \end{bmatrix},$$

$$C_{\mathrm{B}\epsilon}(\theta) = \nabla_\epsilon C_{\mathrm{B}}(T(\theta), \epsilon)\big|_{\epsilon=\mathscr{E}(\epsilon)}, \qquad \mathscr{C}_\epsilon(\theta) = \nabla_\epsilon^2 C_{\mathrm{B}}(T(\theta), \epsilon)\big|_{\epsilon=\mathscr{E}(\epsilon)}.$$

The quadratic approximation is given by

$$\bar{C}_{\mathrm{B}}(T(\theta), \epsilon) = C_{\mathrm{B}}(T(\theta), \mathscr{E}(\epsilon)) + \langle \epsilon - \mathscr{E}(\epsilon), C_{\mathrm{B}\epsilon}(\theta) \rangle + \frac{1}{2}\langle \epsilon - \mathscr{E}(\epsilon), \mathscr{C}_\epsilon(\theta)[\epsilon - \mathscr{E}(\epsilon)] \rangle. \tag{6.3}$$

We compare the evaluation of

$$\max_\theta \{(1-\beta)\mathscr{E}[\bar{C}_{\mathrm{B}}(T(\theta), \epsilon)] - \beta\mathscr{E}[\bar{C}_{\mathrm{B}}(T(\theta), \epsilon) - C_{\mathrm{B}}(T(\theta), \mathscr{E}(\epsilon))]^2\},$$

subject to the constraints given above, when the mean and variance terms are given by the *first-order estimates*, which omit the quadratic term in (6.3),

$$\mathscr{E}_1[C_{\mathrm{B}}(T(\theta), \epsilon)] = C_{\mathrm{B}}(T(\theta), \mathscr{E}(\epsilon)),$$

$$\mathscr{E}_1[C_{\mathrm{B}}(T(\theta), \epsilon)] - \mathscr{E}_2[C_{\mathrm{B}}(T(\theta), \epsilon)]^2 = \langle C_{\mathrm{B}\epsilon}(\theta), \Lambda C_{\mathrm{B}\epsilon}(\theta) \rangle,$$

and the *second-order estimates*, which utilize all the terms in (6.3),

$$\mathscr{E}_2[C_{\mathrm{B}}(T(\theta), \epsilon)] = C_{\mathrm{B}}(T(\theta), \mathscr{E}(\epsilon)) + \frac{1}{2}\mathrm{trace}[\Lambda\mathscr{C}(\theta)],$$

$$\mathscr{E}_2[C_{\mathrm{B}}(T(\theta), \epsilon)] - \mathscr{E}_2[C_{\mathrm{B}}(T(\theta), \epsilon)]^2 = \langle C_{\mathrm{B}\epsilon}(\theta), \Lambda C_{\mathrm{B}\epsilon}(\theta) \rangle + \frac{1}{2}\ \mathrm{trace}\,[\Lambda\mathscr{C}(\theta)]^2.$$

The mean-variance efficient frontiers for both approximations are given in Figure 6.1. The bias $\mathscr{B}$, which is not directly influenced by the decision, indicates in Figure 6.2 that the second-order approximation performs considerably better. Furthermore, in either case, since the solution is required to be

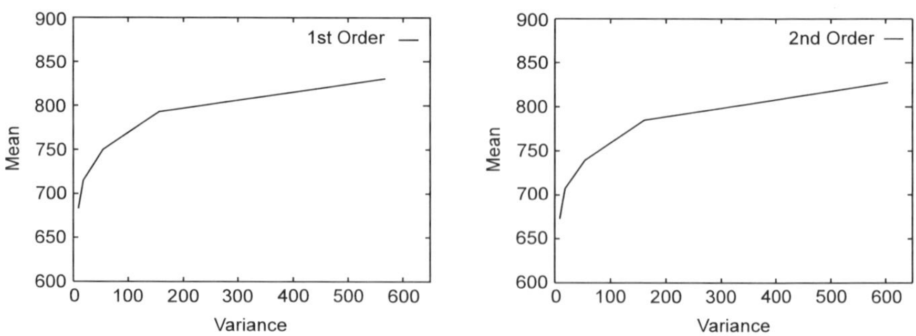

Figure 6.1 Efficient frontier: mean us variance for first and second order approximations

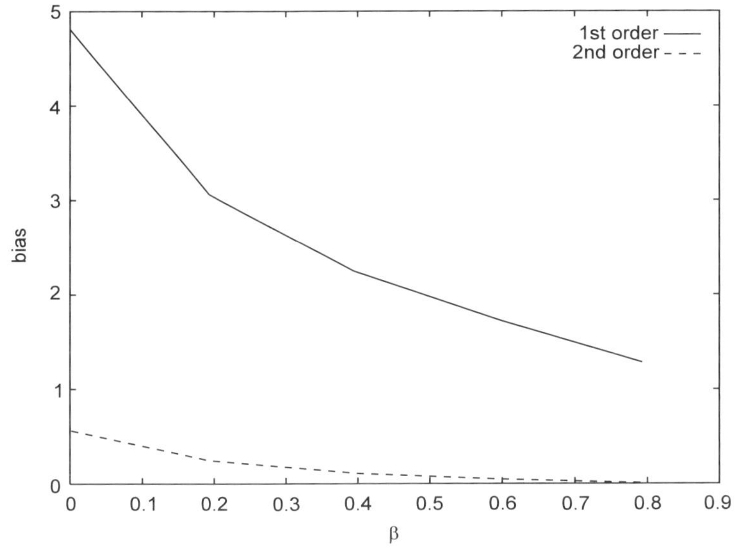

Figure 6.2 Accuracy of first and second order robust formulations

more robust, with greater emphasis on the variance term, a reduction in $\mathscr{B}$ is observed. This is essentially because a reduction in variance, measured around the mean $\mathscr{E}(\epsilon)$, implies that deviations from this mean will have a decreasing effect on the performance of the system. Thus, increased caution ensures certainty-equivalent performance in that the system will operate as predicted by the mean values of the uncertainties.

7 CONCLUDING REMARKS

The robust strategies discussed ensure that either the objective-function value or selected endogenous variables at the optimum are relatively insensitive to perturbations. Both robust strategies are motivated by the sensitivity analysis of nonlinear problems. On the other hand, the mean-variance interpretations of these approaches are based on a linear model. The question, therefore, arises whether the mean-variance interpretation applies for nonlinear models. This may be resolved by considering the linearized model around the nonlinear robust optimum. Clearly, all the robust properties associated with the strategies are valid for the linearized model. Thus, *robust control is at the very least the correct mean-variance strategy for first-order effects*. We also know that if the linearization were done at an arbitrary point (e.g. the base trajectory) and not the robust optimum of the nonlinear problem, the robust objective function could be further reduced by moving to the robust optimal solution of the nonlinear model. Thus, the robust optimal solution of the nonlinear problem is central for the above first-order test. We note that the approach also admits second-order effects, and the example in Section 6 illustrates this flexibility.

8 COMMENTS AND NOTES

CN.1 Separation and certainty equivalence

The *separation principle* is an important result in stochastic control. A decision maker, or controller, of a stochastic system processes the observations, or outputs, of the system, and converts these into a decision, or control, value. The principle establishes that, under certain conditions, this operation separates into two stages: the estimation of the effect of the random disturbance on the system and the computation of the optimal decision, as a function of the estimated effect. *Certainty equivalence* is a more special result, which essentially states that the optimum decision for a stochastic system is the same as that for the related deterministic system where the random variables are replaced by their expected values (see Aoki, 1967; Davis 1977; Kwakernaak and Sivan, 1972).

CN.2 Existence of $[I–N\,\Theta(\theta)]^{-1}$

The assumption that $I - N\Theta(\theta)$ is nonsingular may sometimes be relaxed. For a numerical algorithm computing the solution of an optimization problem such as (5.4), a protection scheme that generates an arbitrarily small perturbation to θ may remedy any singularity during the progress of the algorithm. This is because a rank-deficient matrix is arbitrarily close to a matrix of full rank.

REFERENCES

Aoki, M. (1967). *Optimization of Stochastic Systems*. Academic Press, New York.

Becker, R., S. Hall and B. Rustem (1994). Robust optimal control of stochastic nonlinear economic system. *J. Econ. Dynam. Control* **18,** 125–148.

Becker, R., A. Heunis and D. Mayne (1979). Computer aided design of control systems via optimization. *Proc. IEE* **126,** 573–578.

Ben-Tal, A. and A. Ben-Israel (1991). A recourse certainty equivalent for decisions under uncertainty. *Ann. Oper. Res.* **30,** 3–34.

Ben-Tal, A. and M. Teboulle (1986). Expected utility, penalty functions and duality in stochastic nonlinear programming *Management Sci.* **32**, 1445–1466.

Bianchi, C. and Calzolari, G. (1982). Evaluating forecast uncertainty due to errors in estimated coefficients: empirical comparisons of alternative methods. In *Evaluating the Reliability of Macroeconomic Models* (G. Chow and P. Corsi, eds). Wiley, New York.

Cannon, M., C. Cullum and E. Polak (1970). *Theory of Optimal Control and Mathematical Programming*. McGraw-Hill, New York.

Chow, G. and Corsi, P., eds (1982). *Evaluating the Reliability of Macroeconomic Models*, Wiley, New York.

Darlington, J., C.C. Pantelides, B. Rustem and B.A. Tanyi (1997). An efficient approximate algorithm for optimal robust decisions under uncertainty. Department of Computing, Imperial College.

Davis, M.H.A. (1977). *Linear Estimation and Stochastic Control*. Chapman and Hall, London.

Dembo, R.S. (1990). Scenario Optimization. *Ann. Oper. Res.* **30,** 63–80.
Dupacova, J. (1990). Stability and sensitivity analysis for stochastic programming. *Ann. Oper. Res.* **27,** 115–142.
Engwerda, J. and P. Otter (1989). On the choice of weighting matrices in the minimum variance controller. *Automatica* **25,** 279–285.
Ermoliev, Y. (1983). Stochastic quasigradient methods and their application to systems optimization. *Stochastics* **9,** 1–36.
Escudero, L.F., P.V. Kamesam, A. King and R.J.-B Wetts (1993). Production planning via scenario modelling. *Ann. Oper. Res.* **43,** 311–335.
Fair, R.C. (1984). *Specification Estimation and Analysis of Macroeconometric Models.* Harvard Press, Cambridge, MA.
Fiacco, A.V. (1976). Sensitivity analysis for nonlinear programming using penalty methods. *Math. Programming* **10,** 267–311.
Fisher, P. and Salmon, M. (1983). On evaluating the importance of non-linearities in large macro-econometric models. ESRC Macroeconomic Modelling Bureau, Discussion Paper No. 2, University of Warwick.
Gaivoronski, A. (1988). Implementation of stochastic quasigradient methods. In *Numerical Techniques for Stochastic Optimization* (Y. Ermoliev and R. Wets, eds). Springer-Verlag, Heidelberg.
Hall, S.G. and S.G.B. Henry (1988). *Macroeconomic Modelling.* North-Holland, Amsterdam.
Hall, S. and M.J. Stephenson (1990). Optimal control of stochastic non-linear models. In *Dynamic Modelling and Control of National Economies* (N. Christodoulakis, ed.). Pergamon Press, Oxford.
Hughes-Hallett, A.J. (1984). Optimal stockpiling in a high risk commodity market: the case for copper. *J. Econ. Dynam. Control* **8,** 211–236.
Jarvis, R.B. and C.C. Pantelides (1992). DASOLV—Differential-Algebraic Equation Solver. Technical Report, Centre for Process Systems Engineering, Imperial College, London.
Kall, P. (1976). *Stochastic Linear Programming.* Springer-Verlag, Berlin.
Kall, P. and S.W. Wallace (1994). *Stochastic Programming*, Wiley, Chichester.
Karakitsos, E. and B. Rustem (1984). Optimally derived fixed rules and indicators. *J. Econ. Dynami. Control* **6,** 341–350.
Karakitsos, E. and B. Rustem (1986). Optimal fixed rules and simple feedback laws for nonlinear econometric models. *Automatica* **21,** 169–180.
Karakitsos, E., B. Rustem and M. Zarrop (1980). PROPE Discussion Paper 38, Robust economic policy formulation under uncertainty Imperial College. Presented at the Society of Economic Dynamics and Control Conference, Princeton.
Karp, L.S. (1985). Higher moments in the linear quadratic Gaussian problem. *J. Econ. Dynam. Control* **9,** 41–54.
Kendrick, D.A. (1981). *Stochastic Control for Econometric Models.* McGraw-Hill, New York.
Kushner, H.J. and D.S. Clark (1978). *Stochastic Approximation Methods for Constrained and Unconstrained Systems.* Springer-Verlag, Berlin.
Kwakernaak, H. and R. Sivan (1972). *Linear Optimal Control Systems.* Wiley, New York.
Marti, K. (1990). Stochastic programming: numerical solution techniques by semi-stochastic approximation methods. In *Stochastic versus Fuzzy Approaches to Multi-Objective Mathematical Programming under Uncertainty*, (R. Slowinski and J. Teghem, eds). pp. 23–43. Kluwer, Dordrecht.
Mitchell, D.W. (1979). Risk aversion and macro policy. *Econ. J.* **89,** 913–918.
National Institute (1979). Listing of the Interim Model IV. NIESR Discussion Paper 28, London.

Paraskevopoulos, D., E. Karakitsos and B. Rustem (1991). Robust capacity planning Under uncertainty. *Management Sci.* **37,** 787–800.
Polak, E. (1971). *Computational Methods in Optimization*. Academic Press, New York.
Polyak, B.T. (1978). Nonlinear programming methods in the presence of noise. *Math. Programming* **14,** 87–98.
Pratt, J.W. (1964). Risk aversion in the small and in the large. *Econometrica* **32,** 122–136.
Rustem, B. (1994). Stochastic and robust control of nonlinear economic systems. *Eur. J. Oper. Res.* **73,** 304–318.
Stöppler, S. (1979). Risk minimization by linear feedback. *Kybernetes* **8,** 171–184.
Syski, W. (1988). A method of stochastic subgradients with complete feedback stepsize rule for convex stochastic approximation problems. *J. Optim. Theory Applie.* **59,** 487–504.
Terwiesch, P., M. Agrawal and D.W.T. Rippin (1994). Batch unit optimization with imperfect modelling: a survey. *J. Process Control* **4,** 238–258.
Whittle, P. (1982). *Optimization over Time*, Vol. 1. Wiley, New York.
Whittle, P. (1990). *Risk-Sensitive Optimal Control*. Wiley, New York.
Wuu, T., R. Becker and E. Polak (1986). On the computation of sensitivity functions of linear time-invariant systems responses via diagonalization. *IEEE Trans. Autom. Control* **AC- 31,** 1141–1143.

CHAPTER 12

A Discrete Min–Max Algorithm for Multiple Objectives and Uncertainty

In this chapter, we consider the min–max approach to multiple objectives. This permits the decision maker to focus on the objective with worst value. Multiple-objective decision making can also be regarded as an approach to uncertainty, as in Chapters 10–11. In the present chapter, each objective may arise from a rival description of the uncertain system. Each description, or model, is a system of discrete-time nonlinear dynamic equations. Observed data of the real system does not, in general, provide sufficient information to invalidate all but one of the rival models. In this case, there is uncertainty about which decision model to use. Prudent policy design would suggest that the decision should take into account of all objectives arising from the rival models. This is achieved as a pooling of the objectives. The pooling that is robust to objective function choice is formulated as a constrained min–max problem. The minimization is over the decision variables and the maximization is over the pooling. Only equality constraints are considered.

We use some of the nonlinear programming techniques of Chapter 7, to design an algorithm for nonlinear minimax for equality-constrained problems. A successive quadratic programming algorithm, similar in spirit to the one in Chapters 7 and 8, is considered for the solution of the min–max problem. The stepsize strategy is based on a differentiable penalty function for the constraints. Two alternative quadratic subproblem formulations are proposed. One is a quadratic min–max and the other a quadratic programming problem. The objective function of either subproblem includes a linear term that is dependent on the penalty function. The penalty parameter is determined at every iteration, using a strategy that ensures descent as well as boundedness of the penalty, which needs to be increased a finite number of times.

The global and local convergence of the algorithm is studied. Conditions, involving projected Hessian approximations, are established under which the algorithm achieves unit stepsizes and subsequently Q-superlinear convergence.

1 THE MIN–MAX APPROACH TO MULTIPLE OBJECTIVES

Decision making with multiple objectives sometimes focuses on the worst objective. A particular instance of the optimization of the worst objective is given by the Chebyshev, or min–max, norm (see e.g. Bowman, 1976; Pinto, 1995; Pinto and Rustem, 1996a, b). Decisions are often complicated by the presence of rival models purporting to represent the same underlying system. The models are large systems of nonlinear continuous or discrete-time dynamic equations. Generally, statistical observations and analysis of the real system are not sufficient to rule out all but one of these models. Economic forecasting with rival models is usually resolved by some form of forecast pooling (see e.g. Fuhrer and Haltmaier, 1986; Granger and Newbold, 1974, Lawrence, Edmunson and O'Connor, 1986; Makridakis and Winkler, 1983).

1.1 The decision problem with rival objectives

A similar approach in economic decision making involves the pooling of multiple objective functions derived from rival decision models (see Rustem, 1987). Using an input–output mapping such as (1.4.1), which generically exists for these models, the model equations are used to eliminate the output (or endogenous or state) variables. Alternatively, the model can be treated as part of the equality constraints. Hence, the problem can be formulated in general as the constrained optimization of the pooled objective functions subject to non-linear constraints. Consider, therefore, the problem

$$\min_{x} \{\langle \vartheta, f(x)\rangle \mid g(x) = 0\} \tag{1.1}$$

where $\vartheta \in \mathbb{R}^{\varsigma}, x \in \mathbb{R}^{n}, f : \mathbb{R}^{n} \to \mathbb{R}^{\varsigma}, g : \mathbb{R}^{n} \to \mathbb{R}^{e}$, and f and g are twice continuously differentiable functions. In (1.1) the ith element of f, denoted by f^{i}, represents the ith objective function corresponding to the ith rival objective. The restrictions $g(x) = 0$ are the equality constraints imposed on the decision problem. In the above formulation, the vector ϑ is fixed and defines the pooling weights. Furthermore,

$$\vartheta \in \mathbb{E}_{+}^{\varsigma} \equiv \{\vartheta \in \mathbb{R}^{\varsigma} \mid \vartheta \geqslant 0; \langle \mathit{1}, \vartheta\rangle = 1\}, \tag{1.2}$$

where $\mathit{1} \in \mathbb{R}^{\varsigma}$ denotes the vector with every component unity. It is also generally the case that the number of objectives is much less than the number of decision variables, i.e. $n >> \varsigma$. The solution of (1.1), for fixed ϑ, is discussed in chapter 7.

Although (1.1) provides a means of incorporating all existing rival objectives in the formulation of an optimal policy, it is not necessarily a robust insurance against overreliance on the wrong objective. Chow (1979) initially formulated a robust policy approach for two rival objectives. This approach obtains the optimal policy based only on one objective, and evaluates the effect of this policy if the second objective turns out to actually represent the true system. A 'payoff matrix' is constructed, and the strategy chosen is the optimal strategy based on one objective that inflicts the lesser damage when implemented on the rival objective. We have a discrete choice set of policy strategies. Each member of the set is an optimal policy derived using only one of the objectives as the true representation of the system.

An extension of the above approach is given by considering the worst case:

$$\min_{x} \max_{\vartheta} \{\langle f(x), \vartheta\rangle \mid g(x) = 0, \vartheta \in \mathbb{E}_{+}^{\varsigma}\}, \tag{1.3}$$

where the worst-case objective function is computed *simultaneously* with the minimization over x. Thus, the decision choice is no longer restricted to be discrete, in the sense that the policy is not required to be the optimal based on one of the objectives. It is shown in Rustem (1987, Proposition 7.6) that the solution of (1.3) has a robust character and that the min–max strategy has a noninferior value in view of all the objective functions. The solution of (1.3) yields a value of ϑ corresponding to the robust min–max strategy. If the insurance provided by the robust and cautious policy has too high a cost in policy terms then the decision maker could base the optimal policy on (1.1), with ϑ chosen in the neighbourhood of the min–max value such that the policy can be made 'acceptable' and at the same time as robust as possible.

In Becker *et al.* (1986), two objectives are considered, arising from two large nonlinear models of the UK economy. These are the HM Treasury model and the National Institute of Economic and Social Research Model. The former model is about 800 and the latter is 200 discrete-time nonlinear dynamic equations. The policy optimization problem is taken as a 20-quarter (i.e. 5-year) decision problem. Hence, systems that are 20 times the size of each model need to be solved. Invoking the time structure of each system allows the computation of the endogenous (or output) variables given the decision (or control) variables. The output variables of the problem can thus be eliminated. Alternatively, the model equations can be appended to the equality constraints, $g(x) = 0$. For various practical reasons, the former alternative is usually adopted. This also applies to the experiment reported in Becker *et al.* (1986), where, by evaluating (1.1) for various values of ϑ, the values of ϑ and x that solve the min–max problem (1.3) are computed. The algorithm discussed below computes the min–max solution directly by utilizing a simple but well-defined stepsize strategy as well as a simple quadratic subproblem.

1.2 An example

In order to motivate the min–max formulation, consider the multiple-objective problem (1.1), with f given by

$$f(x) \equiv \begin{bmatrix} f^1(x) \\ f^2(x) \end{bmatrix} = \begin{bmatrix} (ax-b)^2 \\ (cx-d)^2 \end{bmatrix},$$

where $a, b, c, d \neq 0$. $f^1(x)$ and $f^2(x)$ are minimized by $x_{1_*} = b/a$ and $x_{2_*} = d/c$ respectively and

$$f^1(x_{1_*}) = f^2(x_{2_*}) = 0.$$

The value of x that minimizes (1.1) is given by

$$x(\vartheta) = \frac{\vartheta ab + (1-\vartheta)cd}{\vartheta a^2 + (1-\vartheta)c^2}, \tag{1.4}$$

where $\vartheta \in [0, 1]$. The optimality condition of (1.3) for this problem is given by

$$\frac{\partial[\vartheta f^1(x) + (1-\vartheta)f^2(x)]}{\partial x} = 2\vartheta a(ax-b) + 2(1-\vartheta)c(cx-d) = 0. \tag{1.5}$$

This is essentially the same as the optimality condition for (1.1), which yields $x(\vartheta)$ above, for given ϑ. In addition, (1.3) entails conditions for determining ϑ. Assuming, for simplicity, that $\vartheta \in (0, 1)$, we only need to consider

$$\frac{\partial[\vartheta f^1(x) + (1-\vartheta)f^2(x)]}{\partial\vartheta} = (ax-b)^2 - (cx-d)^2 = 0. \tag{1.6}$$

The other conditions relate to the constraints on ϑ:

$$\vartheta \geqslant 0, \quad 1-\vartheta \geqslant 0, \quad \mu^1\vartheta = \mu^2(1-\vartheta) = 0, \quad \mu^1 \geqslant 0, \quad \mu^2 \geqslant 0, \tag{1.7}$$

and, by appropriate choice of the problem parameters, these may be satisfied and thus ignored as far as this example is concerned. Nevertheless, (1.5)–(1.7)

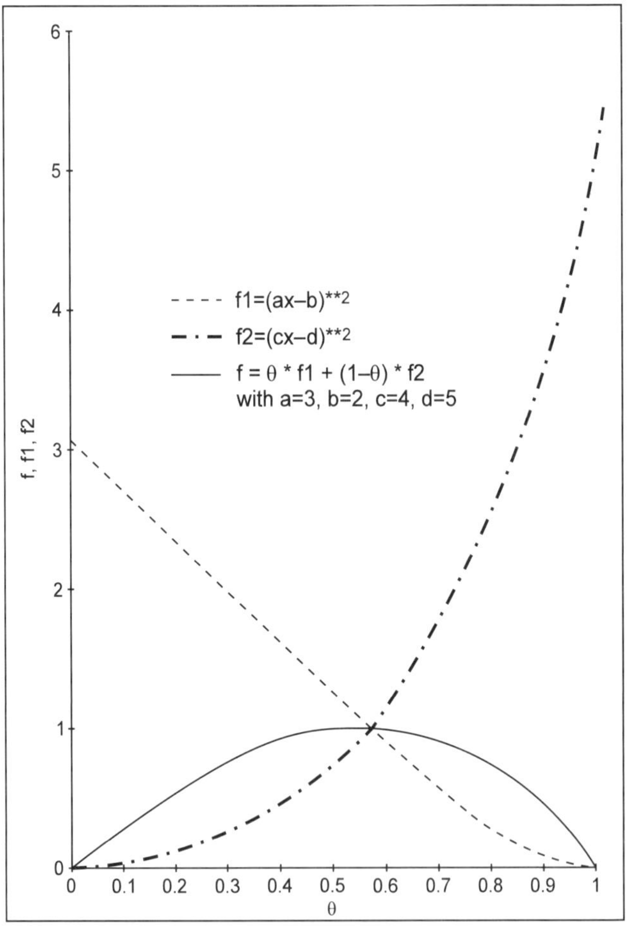

Figure 12.1 Optimal values of f with respect to x, for θ varying between 0 and 1; the corresponding values of f_1, f_2.

are deduced from the optimality conditions discussed in Chapter 1, and are specified in Theorem 2.1 below for the general min–max problem (1.3). The significance of (1.6) is that, when $\vartheta \in (0, 1)$, the min–max solution aims to make both objectives equal. Hence, the solution implied by (1.6) is

$$x_* = \frac{b \pm d}{a \pm c}. \tag{1.8}$$

The value of ϑ implied by this in (1.4) is given by the choice

$$\vartheta = \frac{\pm c}{a \pm c}. \tag{1.9}$$

In other words, for ϑ given by (1.9), (1.4) reduces to (1.8), with the min–max objective function values

$$f^1(x_*) = f^2(x_*) = \frac{(ad - bc)^2}{(a \pm c)^2}. \tag{1.10}$$

In Figure 1.1, the values of $f^1(x(\vartheta))$, $f^2(x(\vartheta))$ and $\vartheta f^1(x(\vartheta)) + (1 - \vartheta) f^2(x(\vartheta))$ are plotted for a given set of the parameters a, b, c, d. The satisfaction of (1.6) and (1.10) is illustrated at the min–max point.

2 PRELIMINARY CONCEPTS AND RESULTS

We introduce certain concepts and results before discussing the algorithm in Section 3. These are intended to provide a framework for the definition of the quadratic subproblem and the role of the penalty parameter.

2.1 The discrete min–max formulation

It is possible to reformulate (1.3) as a discrete min–max problem to be used in the specification of a relatively straightforward quadratic subproblem for the algorithm.

Lemma 2.1

The min–max problem (1.3) is equivalent to

$$\min_x \max_{i \in \{1,2,\ldots,\varsigma\}} \{f^i(x) \mid g(x) = 0\}. \tag{2.1a}$$

Proof

This result, initially proved by Medanic and Andjelic (1971, 1972) and also by Cohen (1981), follows from the fact that the maximum of ς numbers is equal to the maximum of their convex combination. □

It should also be noted that (2.1a) can be solved by the nonlinear programming problem

$$\min_{x,v} \{v \mid f(x) \leqslant Iv; g(x) = 0\} \tag{2.1b}$$

where $v \in \mathbb{R}^1$. The above lemma is used to represent (1.3) with (2.1 a,b) whenever this becomes necessary.

Algorithms for solving (1.3) or (2.1) have been considered by a number of authors, including Charalambous and Conn (1978), Coleman (1978), Conn (1979), Conn and Li (1988), Demyanov and Malomezov (1974), Demyanov and Pevnyi (1972), Dutta and Vidyasagar (1977), Hald and Madsen (1981), Han (1978, 1981), Murray and Overton (1980), Polak, Mayne and Higgins (1988) and Womersley and Fletcher (1986). In the constrained case, discussed in some of these studies, convergence results to unit steplengths, global convergence and local convergence rates have not been established (e.g. Coleman, 1978; Dutta and Vidyagasar, 1977). In this chapter, the dual approach (1.3), formulated originally by Medanic and Andjelic (1971, 1972) and Cohen (1981), is initially adopted. Subsequently, both (1.3) and (2.1) are utilized to formulate the algorithm.

In the algorithm discussed below, a stepsize strategy is described that directly aims at measuring progress towards the min–max solution. The algorithm defines the direction of progress as a quasi-Newton step obtained from a quadratic subproblem. An augmented Lagrangian function is defined, and a procedure is formulated for determining the penalty parameter. The growth in the penalty parameter is required only to ensure a descent property. It is shown that this penalty parameter does not grow indefinitely. Convergence of the overall algorithm, convergence to unit steplengths and Q-superlinear convergence rates are discussed.

In the unconstrained case, the algorithm below is similar to that of Han (1978, 1981). One difference is the stepsize strategy, (3.10a) below. In both the unconstrained and the constrained cases, the conditions for the attainment of unit stepsizes are established below. These are related to the accuracy of the projection of the Hessian approximation used by the quadratic subproblem. Various characterizations of the Q-superlinear convergence of the algorithm are discussed in Section 5. The attainment of unit stepsizes are illustrated with numerical examples in Rustem (1994). These results highlight the properties of the algorithm related to the penalty parameter and the unit stepsize achievement.

2.2 Convexification

The characterization of the solution of a constrained min–max problem such as (1.3) as a saddle point of the Lagrangian function is known to be heavily dependent on the convexity properties of the underlying problem (see Demyanov and Malomezov, 1974; Arrow, Gould and Howe, 1973; Rockafeller, 1973). Motivated by the discussions in Arrow and Hurwicz (1956) and Arrow and Solow (1958), it has been shown that, in the case of the nonlinear programming problem (i.e. just the min case of the min–max problem below), these convexity assumptions can be relaxed via a modified Lagrangian approach (see Arrow, Gould and Howe, 1973; Rockafeller, 1973). This modified Lagrangian approach is essentially a convexification procedure for more general problems. In this section, we invoke these results to characterize the solution of the above min–max problem as a saddle-point. The saddle-point formulation and the convexification are subsequently used by the algorithm discussed in the next section.

Let the Lagrangian function associated with (1.3) be given by

$$L(x, \vartheta, \lambda, \mu, \eta) = \langle f(x), \vartheta \rangle + \langle g(x), \lambda \rangle - \langle \vartheta, \mu \rangle + (\langle 1, \vartheta \rangle - 1)\eta \tag{2.2}$$

where $\lambda \in \mathbb{R}^e$,

$$\mu \in \mathbb{R}^\varsigma_+ = \{\mu \in \mathbb{R}^\varsigma \mid \mu \geqslant 0\}$$

and $\eta \in \mathbb{R}^1$ are the multipliers associated with $g(x) = 0, \vartheta \geqslant 0$ and $\langle 1, \vartheta \rangle = 1$ respectively. The characterization of the min–max solution of (1.3) as a saddle point requires the relaxation of convexity assumptions (see Demyanov and Malomezov, 1974; Cohen, 1981). In order to achieve this characterization, we modify (2.2) by augmenting it with a penalty function. Hence, we define the augmented Lagrangian by

$$L^{\mathrm{a}}(x, \vartheta, \lambda, \mu, \eta, c) = L(x, \vartheta, \lambda, \mu, \eta) + \frac{1}{2} c \langle g(x), g(x) \rangle, \tag{2.3}$$

where the scalar $c \geqslant 0$ is the penalty parameter.

In nonlinear programming algorithms, the penalty parameter c is taken as a constant, is increased by a prefixed rate or is adapted as the algorithm progresses. Biggs (1974), Polak and Mayne (1981) and Polak and Tits (1981) are specific examples of the latter type of strategy. We also adopt an adaptive strategy below. The departure of the strategy below from other works is mainly in the relationship of c and the descent property of the direction of search, discussed in Lemmas 3.2 and 3.4 below. In particular, c is only adjusted to ensure that the direction of search is a descent direction for the penalty function that regulates the stepsize strategy (3.10) below. This approach, discussed originally in Rustem (1992), is an extension of a strategy for nonlinear programming in Chapter 7 and Rustem (1986, 1993).

We let $\mathrm{H}(\cdot)$ and $\mathfrak{Q}(\cdot)$ denote the Hessians of L and L^{a}, with respect to x, evaluated at $(\cdot)$. We also denote by $\mathscr{G}(\cdot)$ the matrix

$$\mathscr{G}(x) = [\nabla_x g^1(x) \vdots \ldots \vdots \nabla_x g^e(x)]. \tag{2.4}$$

Sometimes, $\mathscr{G}(x)$ evaluated at x_k will be denoted by $\mathscr{G}_k$, and $g(x_k)$ will be denoted by g_k. Thus, a local linearization of $g(x)$ at x_k can be written as

$$g(x) = g_k + \mathscr{G}_k^{\mathrm{T}}(x - x_k).$$

Assumption 2.1
The columns of $\mathcal{G}_k$ are assumed to be linearly independent.

This assumption is used to simplify the quadratic subproblem used in the algorithm in Section 3 for solving (1.3). This assumption can be relaxed by increasing the complexity of the quadratic subproblem.

Theorem 2.1
Suppose that $(x_*, \vartheta_*, \lambda_*, \mu_*, \eta_*)$ satisfy the second-order sufficiency conditions for x_*, ϑ_* to be an isolated local solution of the min–max problem (1.3). That is, assume

$$\nabla_{x,\vartheta} L(x_*, \vartheta_*, \lambda_*, \mu_*, \eta_*) = 0, \tag{2.5}$$

$$g(x) = 0, \tag{2.6}$$

$$\langle v, H(x_*, \vartheta_*, \lambda_*, \mu_*, \eta_*)v\rangle > 0, \quad \forall v \neq 0 \in \mathbb{R}^n, \quad \mathcal{G}(x_*)^{\mathrm{T}} v = 0, \tag{2.7}$$

$$\vartheta_* \geqslant 0, \quad \mu_* \geqslant 0, \quad \langle \mu_*, \vartheta_* \rangle = 0, \tag{2.8}$$

$$\langle 1, \vartheta_* \rangle = 1. \tag{2.9}$$

Then, if c is sufficiently large,

(i) the function $L^a(x_*, \vartheta_*, \lambda_*, \mu_*, \eta_*, c)$ has an unconstrained local minimum with respect to x and maximum with respect to ϑ;

(ii) $\forall \lambda \in \mathbb{R}^e$ sufficiently close to λ_*, $L^a(x, \vartheta, \lambda, \mu, \eta, c)$ is strictly convex in x close to x_* and concave in λ, μ, η and ϑ.

Remark
The equivalent result for the min–max formulation (2.1b) follows directly from Arrow, Gould and Howe (1973) and Rockafeller (1973).

Proof
The proof follows directly from Theorem 7.1.1 for x, λ, and from Demyanov and Malomezov (1974, Theorem 5.1) for x, ϑ, μ, η.

We now state the saddle-point property of $L^a(\cdot)$ in the neighbourhood of the min–max solution.

Theorem 2.2
If c in (2.3) is sufficiently large then, under the conditions of Theorem 2.1

$$L^{\mathrm{a}}(x_* \vartheta, \lambda, \mu, \eta, c) \leqslant L^{\mathrm{a}}(x_*, \vartheta_*, \lambda_*, \mu_*, \eta_*, c) \leqslant L^{\mathrm{a}}(x, \vartheta_*, \lambda_*, \mu_*, \eta_*, c),$$

for every x in some neighbourhood of x_* and for every $\vartheta, \lambda, \mu, \eta$.

Remark
The equivalent result for the min–max formulation (2.1b) follows directly from Arrow, Gould and Howe (1973) and Rockafeller (1973).

Proof
The proof follows directly from Theorem 7.1.2 for the nonlinear programming part involving x, λ, and from Demyanov and Malomezov (1974, Theorem 5.2) for x, ϑ, μ, η. □

3 THE ALGORITHM

Consider the objective function

$$\mathscr{F}(x, \vartheta) = \langle \vartheta, f(x) \rangle \tag{3.1}$$

and its linear approximations, with respect to x, at a point x_k,

$$\mathscr{F}_k(x, \vartheta) = \langle \vartheta, f(x_k) + \nabla f(x_k)^{\mathrm{T}}(x - x_k) \rangle, \tag{3.2a}$$

where

$$\nabla f(x) = [\nabla f^1(x) \vdots \ldots \vdots \nabla f^{\varsigma}(x)].$$

We shall sometimes denote $f(x)$ and $\nabla f(x)$, evaluated at x_k, by f_k and ∇f_k respectively. Thus, for $d = x - x_k$, (3.2a) can be written as

$$\mathscr{F}_k(x_k + d, \vartheta) = \langle \vartheta, f_k + \nabla f_k^{\mathrm{T}} d \rangle. \tag{3.2b}$$

The quadratic objective function used to compute the direction of progress is given by

$$q_k(x, \vartheta, c) = \mathscr{F}_k(x, \vartheta) + c\langle \mathscr{G}_k g_k, (x - x_k) \rangle + \frac{1}{2}\langle x - x_k, \hat{\mathfrak{Q}}_k(x - x_k) \rangle, \tag{3.3a}$$

or, alternatively,

$$q_k(d, \vartheta, c) = \mathscr{F}_k(x_k + d, \vartheta) + c\langle \mathscr{G}_k g_k, d \rangle + \frac{1}{2}\langle d, \hat{\mathfrak{Q}}_k d \rangle. \tag{3.3b}$$

The matrix $\hat{\mathfrak{Q}}_k$ is a positive-semidefinite approximation to the Hessian†

$$\mathfrak{Q}_k = \sum_{i=1}^{\varsigma} \vartheta_k^i \nabla^2 f^i(x_k) + \sum_{j=1}^{e} \lambda_k^j \nabla^2 g^j(x_k) + c \mathscr{G}_k \mathscr{G}_k^{\mathrm{T}}. \tag{3.4}$$

It should be noted that the second derivatives due to the penalty term in the augmented Lagrangian (i.e. $c \sum_{j=1}^{e} \nabla^2 g^j(x_k) g^j(x_k)$) are not included in (3.4). The reason for this is discussed in Tapia (1986). Furthermore, since $g^j(x_*) = 0$ at the solution x_*, ignoring this term does not affect the asymptotic properties of the algorithm. The values ϑ_k and λ_k are given by the solution of the

† The assumption of positive-semidefiniteness will be tightened in the later stages of the discussion.

quadratic subproblem in the previous iteration. The direction of progress at each iteration of the algorithm is determined by the quadratic subproblem

$$\min_{d} \max_{\vartheta} \{q_k(d, \vartheta, c_k) \mid \mathcal{G}_k^{\mathrm{T}} d + g_k = 0; \vartheta \in \mathbb{E}_+^{\varsigma}\}. \tag{3.5a}$$

Since the min–max subproblem is more complex, we also consider the quadratic programming subproblem

$$\min_{d,v} \left\{ v + c_k \langle \mathcal{G}_k g_k, d \rangle + \frac{1}{2} \langle d, \hat{\mathfrak{Q}}_k d \rangle \middle| \mathcal{G}_k^{\mathrm{T}} d + g_k = 0; \nabla f_k^{\mathrm{T}} d + f_k \leqslant \mathbf{1} v \right\}. \tag{3.5b}$$

The two subproblems are equivalent. Also, (3.5b) involves fewer variables. It is shown below that the multipliers associated with the inequalities are the values ϑ and that the solution of either subproblem satisfies common convergence properties.

Let the values of d, ϑ, v solving (3.5) be denoted by $d_k, \vartheta_{k+1}, v_{k+1}$. The stepsize along d_k is defined using the equivalent min–max formulation (2.1). Thus, consider the function

$$\psi(x) = \max_{i \in \{1,2,\ldots\varsigma\}} \{f^i(x)\} \tag{3.6}$$

and

$$\psi_k(x) = \max_{i \in \{1,2,\ldots\varsigma\}} \{f^i(x_k) + \langle \nabla f^i(x_k), x - x_k \rangle\}. \tag{3.7}$$

Let $\psi_k(x_k + d_k)$ be given by

$$\psi_k(x_k + d_k) = \max_{i \in \{1,2,\ldots\varsigma\}} \{f^i(x_k) + \langle \nabla f^i(x_k), d_k \rangle\}. \tag{3.8}$$

The stepsize strategy determines τ_k as the largest value of $\tau = (\bar{\tau})^j$, $j = 0, 1, 2, \ldots$, such that x_{k+1} given by

$$x_{k+1} = x_k + \tau_k d_k \tag{3.9}$$

satisfies the inequality

$$\psi(x_{k+1}) + \frac{1}{2} c_{k+1} \langle g_{k+1}, g_{k+1} \rangle - \psi(x_k) - \frac{1}{2} c_{k+1} \langle g_k, g_k \rangle \leqslant \rho \tau_k \Phi(d_k, c_{k+1}), \tag{3.10a}$$

where $\rho \in (0, 1)$ is a given scalar and

$$\Phi(d_k, c_{k+1}) = \psi_k(x_k + d_k) - \psi(x_k) + c_{k+1} \langle g_k, \mathcal{G}_k^{\mathrm{T}} d_k \rangle + \frac{1}{2} \langle d_k, \hat{\mathfrak{Q}}_k d_k \rangle. \tag{3.10b}$$

The stepsize τ_k determined by (3.10) basically ensures that x_{k+1} simultaneously reduces the main objective and maintains or improves the feasibility with respect to the constraints. The penalty term used to measure this feasibility is a quadratic and is consistent with the augmented Lagrangian (2.3). It is shown in Theorem 4.1 below that (3.10) can always be fulfilled by the algorithm.

The determination of the penalty parameter c is an important aspect of the algorithm. This is discussed in the following description.

Algorithm 3.1 (Discrete min–max with nonlinear equality constraints)

Step 0: Given $x_0, c_0 \in [0, \infty)$, and small positive numbers $\delta, \rho, \epsilon, \bar{\tau}$ such that $\delta \in (0, \infty), \rho \in (0, 1), \epsilon \in (0, \frac{1}{2}], \bar{\tau} \in (0, 1), \hat{\mathfrak{Q}}_0$, set $k = 0$.

Step 1: Compute ∇f_k and $\mathscr{G}_k$. Solve the quadratic subproblem (3.5) (choosing (3.5a) or (3.5b) defines a particular algorithm) to obtain d_k, ϑ_{k+1}, and the associated multiplier vector λ_{k+1}. In (3.5a), we also compute μ_{k+1}, η_{k+1}; and in (3.5b), we also compute v_{k+1}.

Step 2: Test for optimality. If optimality is achieved, **stop**. Else **go to Step 3**.

Step 3: If

$$\psi_k(x_k + d_k) - \psi(x_k) - c_k\langle g_k, g_k\rangle + (\epsilon + \frac{1}{2})\langle d_k, \hat{\mathfrak{Q}}_k, d_k\rangle \leqslant 0, \tag{3.11}$$

then $c_{k+1} = c_k$. Else set

$$c_{k+1} = \max\left\{\frac{\psi_k(x_k + d_k) - \psi(x_k) + (\epsilon + \frac{1}{2})\langle d_k, \hat{\mathfrak{Q}}_k d_k\rangle}{\langle g_k, g_k\rangle}, c_k + \delta\right\}. \tag{3.12}$$

Step 4: Find the smallest nonnegative integer j_k with $\tau_k = \bar{\tau}^{j_k}$ in (3.9) such that inequality (3.10) is satisfied.

Step 5: Update $\hat{\mathfrak{Q}}_k$ to compute $\hat{\mathfrak{Q}}_{k+1}$, set $k = k + 1$ and **go to Step 1**.

In **Step 2**, optimality is checked by testing the satisfaction of the first-order optimality conditions of (2.1b). Detailed termination criteria for constrained optimization algorithms are discussed in Gill, Murray and Wright (1981, Chap. 8). In **Step 3**, the penalty parameter c_{k+1} is adjusted to ensure that progress towards feasibility is maintained. In particular, c_{k+1} is chosen to make the direction d_k computed by the quadratic subproblem a descent direction for the penalty function $\psi(x_k) + \frac{1}{2}c_{k+1}\langle g_k, g_k\rangle$. We now summarize the optimality conditions of the quadratic subprobrem (3.5). Subsequently, we establish the descent property of d_k and that c_k determined by (3.12) is not increased indefinitely.

Lemma 3.1

The Lagrangian function for the quadratic subproblem (3.5a) is given by

$$\begin{aligned} L_k(x, \vartheta, \lambda, \mu, \eta, c_k) &= \mathscr{F}_k(x, \vartheta) + c_k\langle g_k, \mathscr{G}_k^{\mathrm{T}}(x - x_k)\rangle \\ &\quad + \frac{1}{2}\langle x - x_k, \hat{\mathfrak{Q}}_k(x - x_k)\rangle - \langle \vartheta, \mu\rangle + (\langle \mathit{1}, \vartheta\rangle - 1)\eta. \end{aligned} \tag{3.13a}$$

The first order necessary conditions of optimality for problem (3.5a) are

$$\nabla f_k \vartheta_{k+1} + \mathcal{G}_k(c_k g_k + \lambda_{k+1}) + \hat{\mathfrak{Q}}_k d_k = 0, \tag{3.13b}$$

$$f_k + \nabla f_k^{\mathrm{T}} d_k - I\mu_{k+1} + I\eta_{k+1} = 0, \tag{3.13c}$$

$$\vartheta_{k+1} \in E_+^\varsigma, \qquad \mu_{k+1} \geqslant 0, \qquad \vartheta_{k+1}^i \mu_{k+1}^i = 0, \qquad \forall i \in \{1, \ldots, \varsigma\}, \tag{3.13d}$$

$$\mathcal{G}_k^{\mathrm{T}} d_k + g_k = 0, \tag{3.13e}$$

and for the quadratic programming subproblem (3.5b) the necessary conditions are given by (3.13b), (3.13e) and

$$f_k + \nabla f_k^{\mathrm{T}} d_k \leqslant I v_{k+1}, \quad \vartheta_{k+1}^i (f_k^i + \langle \nabla f_k^i, d_k \rangle - v_{k+1}) = 0, \quad i = 1, \ldots, \varsigma, \quad \vartheta_{k+1} \in \mathbb{E}_+^\varsigma. \tag{3.13f}$$

Lemma 3.2 (The descent property of d_k)
Let $g_k \neq 0$ (the case when $g_k = 0$ is discussed in Lemma 3.4) and let $\hat{\mathfrak{Q}}_k$ be positive-semidefinite. The direction d_k satisfying the linearized equality constraints (3.13e) of the quadratic subproblem (3.5) and c_{k+1} chosen as in ***Step 3*** *of the above algorithm ensure the inequality*

$$\Phi(d_k, c_{k+1}) \leqslant -\epsilon \langle d_k, \hat{\mathfrak{Q}}_k, d_k \rangle. \tag{3.14}$$

Proof
Using (3.13e), (3.10b) can be expressed as

$$\Phi(d_k, c_{k+1}) = \psi_k(x_k + d_k) - \psi(x_k) - c_{k+1}\langle g_k, g_k \rangle + \frac{1}{2}\langle d_k, \hat{\mathfrak{Q}}_k d_k \rangle \tag{3.15a}$$

$$\leqslant -\epsilon \langle d_k, \hat{\mathfrak{Q}}_k d_k \rangle. \tag{3.15b}$$

c_{k+1} chosen in **Step 3** ensures that (3.15b) is nonpositive. Thereby (3.14) is established. □

Lemma 3.3
Let d_k and ϑ_{k+1} be the solutions of the quadratic subproblem (3.5). Then

$$\psi_k(x_k + d_k) = \max_{\vartheta \in \mathbb{E}_+^\varsigma} \mathcal{F}_k(x_k + d_k, \vartheta) = \mathcal{F}_k(d_k, \vartheta_{k+1}). \tag{3.16}$$

Proof
The first equality follows, as in Lemma 2.1, from the fact that $\psi_k(x_k + d_k)$ is the maximum of ς numbers (evaluated at $x_k + d_k$) and this is equal to the maximum of the convex combination of these numbers.

It can be verified using (3.13a), or its equivalent for subproblem (3.5b), that for $d_k, \lambda_{k+1}, \mu_{k+1}, \eta_{k+1}, v_{k+1}$ solving (3.5), we have

$$\begin{aligned} \vartheta_{k+1} &= \arg\max_{\vartheta \in \mathbb{E}_+^\varsigma} L_k(x_k + d_k, \vartheta, \lambda_{k+1}, \mu_{k+1}, \eta_{k+1}, c_k) \\ &= \arg\max_{\vartheta \in \mathbb{E}_+^\varsigma} \mathcal{F}_k(x_k + d_k, \vartheta). \end{aligned} \tag{3.17}$$

The implication underlying (3.17) is that, since at $d_k, \lambda_{k+1}, \mu_{k+1}, \eta_{k+1}$ the maximization of L_k is independent of terms not included in $\mathscr{F}_k, \vartheta_{k+1}$ maximizes both functions. This establishes the second equality. The same result also follows for subproblem (3.5b), since ϑ is the vector of multipliers of the inequality constraints. □

Lemma 3.4 (The finiteness of the penalty parameter)
Let f and g be once-differentiable, let $\hat{\mathfrak{Q}}_k$ be positive-semidefinite, and let $(d_k, \vartheta_{k+1}, \lambda_{k+1}, \mu_{k+1}, \eta_{k+1})$ generated by (3.5a) and $(d_k, v_{k+1}, \vartheta_{k+1}, \lambda_{k+1})$ computed by (3.5b) be bounded. Then, the following hold.

- *(i) If for some or all k the nonlinear constraints of (1.3) are satisfied, i.e. $g(x_k) = 0$, then the descent property (3.14) of Lemma 3.2 is satisfied for any choice of $c_k \in [0, \infty)$.*
- *(ii) Let c_k be bounded and $g(x_k) = 0$, for some k. Then* ***Step 3*** *of the algorithm chooses $c_{k+1} = c_k$. For this choice of c_{k+1}, the direction d_k ensures the descent property (3.14).†*
- *(iii) $\forall k, \exists c_{k+1} \in [0, \infty)$ satisfying* ***Step 3*** *of the algorithm.*
- *(iv) Let the sequence $\{x_k\}$ generated by the algorithm be bounded. Then, c_k is increased finitely often, and $\exists$ an integer $k_* \geqslant 0$ and $\exists c_* \in [0, \infty)$ such that the algorithm chooses $c_k = c_*, \forall k \geqslant k_*$.*

Proof
To show (i), consider the optimality conditions (3.13b, e) for $g_k = 0$:

$$\begin{aligned} 0 &= \langle d_k, [\nabla f_k \vartheta_{k+1} + \mathscr{G}_k(c_k g_k + \lambda_{k+1}) + \hat{\mathfrak{Q}}_k d_k] \rangle \\ &= \langle d_k, \nabla f_k \vartheta_{k+1} \rangle - c_k \langle g_k, g_k \rangle - \langle g_k, \lambda_{k+1} \rangle + \langle d_k, \hat{\mathfrak{Q}}_k d_k \rangle. \end{aligned}$$

Hence, for $g_k = 0$, we have

$$\langle d_k, \nabla f_k \vartheta_{k+1} \rangle = -\langle d_k, \hat{\mathfrak{Q}}_k d_k \rangle. \tag{3.18}$$

Consider (3.15a) for $g_k = 0$:

$$\begin{aligned} &\Phi(d_k, c_{k+1}) + \epsilon \langle d_k, \hat{\mathfrak{Q}}_k d_k \rangle \\ &\quad = \psi_k(x_k + d_k) - \psi(x_k) - c_{k+1} \langle g_k, g_k \rangle + \left(\frac{1}{2} + \epsilon \right) \langle d_k, \hat{\mathfrak{Q}}_k d_k \rangle \\ &\quad = \psi_k(x_k + d_k) - \psi(x_k) + \frac{1}{2} + \epsilon \langle d_k, \hat{\mathfrak{Q}}_k d_k \rangle \\ &\quad = \max_{\vartheta \in \mathbb{E}_+^{\varsigma}} \langle \vartheta, f_k + \nabla f_k^T d_k \rangle \\ &\qquad - \max_{\vartheta \in \mathbb{E}_+^{\varsigma}} \langle \vartheta, f_k \rangle + \left(\frac{1}{2} + \epsilon \right) \langle d_k, \hat{\mathfrak{Q}}_k d_k \rangle \end{aligned} \tag{3.19}$$

† c_k is the value used in the quadratic subproblems when computing d_k, ϑ_{k+1}, λ_{k+1}, v_{k+1}.

$$
\begin{aligned}
&= \langle \vartheta_{k+1}, f_k + \nabla f_k^{\mathrm{T}} d_k \rangle - \max_{\vartheta \in \mathbb{E}_+^{\varsigma}} \langle \vartheta, f_k \rangle + \left(\frac{1}{2} + \epsilon\right) \langle d_k, \hat{\mathfrak{Q}}_k d_k \rangle \\
&\leqslant (\frac{1}{2} + \epsilon - 1) \langle d_k, \hat{\mathfrak{Q}}_k d_k \rangle \\
&\leqslant 0.
\end{aligned}
$$

Equality (3.19) follows from the definition of $\psi_k(x_k + d_k)$ and that $\psi(x_k)$ is the maximum of the convex combination of ς numbers $f^1(x_k), \ldots, f^{\varsigma}(x_k)$. The subsequent equality follows from Lemma 3.3. The last two inequalities follow from (3.18) and the fact that ϑ_{k+1} does not necessarily optimize $\max_{\vartheta \in \mathbb{E}_+^{\varsigma}} \langle \vartheta, f_k \rangle$. Thus, the descent condition (3.14) is satisfied for $g_k = 0$.

To show (ii), consider the condition (3.11), which is satisfied $\forall c_k \in [0, \infty)$, since the descent condition is satisfied for $g_k = 0$. Thus, the algorithm chooses $c_{k+1} = c_k$. For this choice, it can simply be verified that Lemma 3.2 and (i) above hold. (iii) follows simply from (i) and (ii), since **Step 3** of the algorithm generates, at each iteration, a finite c_{k+1}. To establish (iv), we note that at each iteration k, **Step 3** generates a $c_{k+1} \in [0, \infty)$ such that the descent condition of Lemma 3.2 and (i) above hold. If c_k is increased infinitely often then, by **Step 3**, we have $\{c_{k+1}\} \to \infty$. As there exists a $c_{k+1} \in [0, \infty)$ that the algorithm can choose at every iteration, the algorithm will not increase c_k indefinitely. There exists a $c_* \in [0, \infty)$, sufficiently large but not infinite, such that the test (3.12) of **Step 3** is always satisfied for $c_k = c_*, \forall k \geqslant k_*$. k_* is the iteration count at which the algorithm reaches c_*. □

4 GLOBAL CONVERGENCE

We consider in this section the global convergence of the method. Convergence of the stepsize sequence $\{\tau_k\}$ to unity and the superlinear rates of convergence are discussed in the next section. We first establish that the stepsize strategy (3.10) can always be satisfied. If the satisfaction of (3.10) was assumed then the final inequality in the proof, given by (4.8) below, could be replaced by (3.10) Lemmas 3.2 and 3.4(i) would provide the monotonicity result of Theorem 4.1.

Assumption 4.1

$$f, g \in \mathbb{C}^2(\mathbb{R}^n).$$

Theorem 4.1
Let

(i) Assumption 4.1 be satisfied;

(ii) the approximate Hessian be such that, for each k,

$$m \|y\|_2^2 \leqslant y^T \hat{\mathfrak{Q}}_k y \leqslant \mathcal{M} \|y\|_2^2; \ \mathcal{M} \geqslant m > 0; 0 \neq y \in \mathbb{R}^n, \tag{4.1}$$

(iii) for each k $\exists$ a bounded optimal point† $(d_k, \vartheta_{k+1}, \lambda_{k+1}, \mu_{k+1}, \eta_{k+1}) \in \mathbb{R}^n \times \mathbb{R}^\varsigma \times \mathbb{R}^e \times \mathbb{R}^\varsigma \times \mathbb{R}^1$ *of the quadratic subproblem (3.5a), and similarly $\exists$ a bounded Kuhn–Tucker point* $(d_k, \nu_{k+1}, \vartheta_{k+1}, \lambda_{k+1}) \in \mathbb{R}^n \times \mathbb{R}^1 \times \mathbb{R}^\varsigma \times \mathbb{R}^e$ *for subproblem (3.5b).*

Then, there exist an integer k_* and a scalar $c_* \geqslant 0$ such that (3.11) is satisfied, $\forall k \geqslant k_*$ with $c_{k+1} = c_k = c_*$; the stepsize computed in **Step 4** *of the algorithm is such that $\tau_k \in (0, 1]$ and the sequence $\{x_k, \vartheta_k\}$, computed by the algorithm, satisfies the stepsize strategy (3.10) and generates a corresponding sequence*

$$\left\{\psi(x_k) + \frac{1}{2}c_*\langle g_k, g_k\rangle\right\},$$

which decreases monotonically for $k \geqslant k_*$.

Remark

As shown in Lemmas 3.2 and 3.4, **Step 3** of the algorithm is always satisfied for finite c_k, and there exists a c_*, sufficiently large but finite for which the test (3.11) is satisfied $\forall k \geqslant k_*$. The integer k_* is the iteration at which c_* is attained by the algorithm.

Proof

By definition of $\psi(x)$, we have, for $x_{k+1} = x_k + \tau_k d_k$,

$$\begin{aligned} \psi(x_{k+1}) &+ c_*\langle g(x_{k+1}), g(x_{k+1})\rangle \\ &= \max_{\vartheta \in \mathbb{E}_+^\varsigma}\{\langle \vartheta, f(x_k + \tau_k d_k)\rangle\} + c_*\langle g_{k+1}, g_{k+1}\rangle \qquad (4.2a) \\ &\leqslant \max_{\vartheta \in \mathbb{E}_+^\varsigma}\{\langle \vartheta, f(x_k) + \tau_k \nabla f_k^{\mathrm{T}} d_k\rangle\} + \frac{1}{2}\tau_k^2\langle d_k, \hat{\mathfrak{Q}}_k d_k\rangle \\ &\quad + \frac{1}{2}c_*[\langle g_k, g_k\rangle + \tau_k\langle g_k, \mathscr{G}_k^{\mathrm{T}} d_k\rangle] + \xi_k \tau_k^2 \|d_k\|^2, \qquad (4.2b) \end{aligned}$$

where, with $x(t) = x_k + t\tau_k d_k$,

$$\begin{aligned} \xi_k = \int_0^1 (1-t)\Bigg\| &\max_{\vartheta \in \mathbb{E}_+^\varsigma}\left\{\sum_{i=1}^{\varsigma} \vartheta^i \nabla^2 f^i(x(t))\right\} + c_*\mathscr{G}(x(t))\mathscr{G}(x(t))^{\mathrm{T}} \\ &+ c_*\sum_{j=1}^{e} g^i(x(t))\nabla^2 g^i(x(t)) - \hat{\mathfrak{Q}}_k\Bigg\| \mathrm{d}t. \end{aligned} \qquad (4.3)$$

The matrix $\hat{\mathfrak{Q}}_k$ is a symmetric positive-definite approximation to (3.4).

Using (3.2), we can write (4.2) as

† The boundedness conditions can be equivalently associated with the properties of the quadratic subproblem. Along with assumptions (i) and (ii), boundedness is ensured when the active constraint normals at the solution of the quadratic subproblem are linearly independent. Another condition implied by (iii) is that the feasible set of the quadratic subproblem is nonempty.

$$\psi(x_{k+1}) + c_*\langle g(x_{k+1}), g(x_{k+1})\rangle \leqslant \max_{\vartheta\in\mathbb{E}_+^\varsigma} \mathscr{F}_k(x_k + \tau_k d_k, \vartheta) + \frac{1}{2}\tau_k^2\langle d_k, \hat{\mathfrak{Q}}_k d_k\rangle$$
$$+ \frac{1}{2}c_*[\langle g_k, g_k\rangle + \tau_k\langle g_k, \mathscr{G}_k^{\mathrm{T}} d_k\rangle] + \xi_k\tau_k^2\|d_k\|^2. \tag{4.4}$$

Since the function $\max_{\vartheta\in\mathbb{E}_+^\varsigma} \mathscr{F}_k(x, \vartheta)$ is convex with respect to x, we have, for $\tau_k \in [0, 1]$,

$$\max_{\vartheta\in\mathbb{E}_+^\varsigma} \mathscr{F}_k(x_k + \tau_k d_k, \vartheta) \leqslant \max_{\vartheta\in\mathbb{E}_+^\varsigma} \mathscr{F}_k(x_k, \vartheta)$$
$$+ \tau_k\left[\max_{\vartheta\in\mathbb{E}_+^\varsigma} \mathscr{F}_k(x_k + d_k, \vartheta) - \max_{\vartheta\in\mathbb{E}_+^\varsigma} \mathscr{F}_k(x_k, \vartheta)\right]. \tag{4.5}$$

Furthermore, we have

$$\max_{\vartheta\in\mathbb{E}_+^\varsigma} \mathscr{F}_k(x_k, \vartheta) = \psi(x_k), \tag{4.6a}$$

$$\max_{\vartheta\in\mathbb{E}_+^\varsigma} \mathscr{F}_k(x_k + d_k, \vartheta) = \psi_k(x_k + d_k). \tag{4.6b}$$

Substituting (4.6) in (4.5) and substituting (4.5) in turn in (4.4) yields

$$\begin{aligned}
\psi(x_{k+1}) &+ \frac{1}{2}c_*\langle g(x_{k+1}), g(x_{k+1})\rangle \\
&\leqslant \psi(x_k) + \tau_k[\psi_k(x_k + d_k) - \psi(x_k)] + \frac{1}{2}\tau_k^2\langle d_k, \hat{\mathfrak{Q}}_k d_k\rangle \\
&\quad + \frac{1}{2}c_*(\langle g_k, g_k\rangle + \tau_k\langle g_k, \mathscr{G}_k^{\mathrm{T}} d_k\rangle) + \xi_k\tau_k^2\|d_k\|^2 \\
&\leqslant \psi(x_k) + \frac{1}{2}c_*\langle g_k, g_k\rangle + \tau_k\Bigg[\psi_k(x_k + d_k) - \psi(x_k) \\
&\quad + c_*\langle g_k, \mathscr{G}_k^{\mathrm{T}} d_k\rangle + \frac{1}{2}\langle d_k, \hat{\mathfrak{Q}}_k d_k\rangle + \xi_k\tau_k\|d_k\|^2\Bigg].
\end{aligned} \tag{4.7}$$

Using the descent property established in Lemma 3.2 and Lemma 3.4(i), we can express (4.7) as

$$\begin{aligned}
\psi(x_{k+1}) &+ \frac{1}{2}c_*\langle g(x_{k+1}), g(x_{k+1})\rangle \\
&\leqslant \psi(x_k) + \frac{1}{2}c_*\langle g_k, g_k\rangle + \tau_k\Phi(d_k, c_*)\left(1 - \frac{\tau_k}{\epsilon m}\xi_k\right).
\end{aligned} \tag{4.8}$$

The scalar $\rho \in (0, 1)$ in the stepsize strategy (3.10) determines τ_k such that

$$\rho \leqslant 1 - \frac{\tau_k}{\epsilon m}\xi_k \leqslant 1.$$

Since, by Lemma 3.2 and Lemma 3.4(i), the descent property $\Phi(d_k, c_{k+1}) \leqslant 0$ holds, there exists a $\tau_k \in (0, 1]$ to ensure that (4.8) and thence (3.10) holds. Suppose that τ^0 is the largest $\tau \in (0, 1]$ satisfying inequalities (4.8) and (3.10).

It follows that all $\tau \leqslant \tau^0$ also satisfy these conditions and that the strategy in **Step 4** of the algorithm selects a $\tau_k \in [\hat{\tau}\tau^0, \tau^0]$. Since $\Phi(d_k, c_{k+1}) \leqslant 0$, it follows that $\{\psi(x_k) + \frac{1}{2}c_*\langle g_k, g_k\rangle\}$ is monotonically decreasing. □

Lemma 4.2†

Let the assumptions of Theorem 4.1 be satisfied and let, for some k_0 and $k \geqslant k_0$, the set

$$\mathbb{F} \equiv \left\{x \in \mathbb{R}^n \mid \psi(x) + \frac{1}{2}c_*\langle g(x), g(x)\rangle \leqslant \psi(x_{k_0}) + \frac{1}{2}c_*\langle g_{k_0}, g_{k_0}\rangle\right\}$$

be bounded. Then, we have

$$\lim_{k\to\infty} \Phi(d_k, c_*) = 0. \tag{4.9}$$

Proof

Given $\rho \in (0, 1)$, by (4.8), the choice $\tau^0 = \min\{1, (1-\rho)\epsilon m/\xi_k\}$ always satisfies the stepsize strategy (3.10). Clearly, $\tau_k = \bar{\tau}^{j_k}$ chosen in **Step 4** of the algorithm is in the range $\tau_k \in [\bar{\tau}\tau^0, \tau^0]$, and thereby also satisfies (3.10). Since f and g are twice continuously differentiable and $\mathbb{F}$ is compact, there is a scalar $\bar{M} < \infty$ such that $\xi_k \leqslant \bar{M}$. Thus, since $m > 0$, we have $\tau_k \geqslant \hat{\tau} > 0, \forall k$, for stepsize strategy (3.10) and for some positive $\hat{\tau}$. The boundedness of $\psi(x) + \frac{1}{2}c_*\langle g(x), g(x)\rangle$ on $\mathbb{F}$ and $\Phi(d_k, c_*) \leqslant 0$ imply

$$\begin{aligned} 0 &\leqslant \rho \sum_{\kappa} \tau_k \mid \Phi(d_k, c_*) \mid \\ &\leqslant \sum_{\kappa}\left\{\left[\psi(x_k) + \frac{1}{2}c_*\langle g_k, g_k\rangle\right] - \left[\psi(x_{k+1}) + \frac{1}{2}c_*\langle g(x_{k+1}), g(x_{k+1})\rangle\right]\right\} \\ &< \infty, \end{aligned}$$

which yields (4.9). □

Lemma 4.3

Let the assumptions of Theorem 4.1 be satisfied. Inequality (3.14) and Lemmas 3.4(i) and 4.2 imply $\lim_{k\to\infty} \|d_k\| = 0$.

Proof

Using (4.9) with (3.14) yields the required result. □

Theorem 4.2

Let the assumptions of Theorem 4.1 be satisfied. Then the following hold.

(i) The algorithm either terminates at a Danskin point‡ (i.e. satisfying the first-order necessary conditions) of problem (1.3) or it generates an infinite

† The convergence results being discussed are dependent on the discussion in CN.2 of Chapter 6.

‡ Following Polak, Mayne and Higgins (1988), we call a point satisfying the first-order necessary conditions of the min–max problem a *Danskin point*.

sequence $\{x_k\}$ *in which there exists a subsequence* $\{x_k\}$ *with* $k_* \leqslant k \in K \subset \{0, 1, \ldots\}$ *such that* $\{\|d_k\|\} \to 0$, *and thus every accumulation point* x_* *(and the other sequences generated by either quadratic subproblem) of the infinite sequence* $\{x_k\}$ *is a Danskin point.*

(ii) If, furthermore, strict complimentarity holds at the solution of the quadratic subproblem (3.5a) then, for large k, μ_k *predicts the active inequality constraints at the solution of (1.3). Similarly, for (3.5b)* ϑ_k *predicts the inequality constraints active at the solution of (2.1b).*

Proof

By Lemmas 3.4, 4.2 and 4.3, there exists a subsequence $\{x_k\}$ such that $\{\|d_k\| \to 0\}$. Let there exist $x_* \in \mathbb{R}^n, \vartheta_* \in \mathbb{R}^\varsigma, \lambda_* \in \mathbb{R}^e, \mu_* \in \mathbb{R}^\varsigma_+, \eta_*, v_* \in \mathbb{R}^1$ such that $\{x_k\} \to x_*, \{\vartheta_k\} \to \vartheta_*, \{\lambda_k\} \to \lambda_* \{\mu_k\} \to \mu_*, \{\eta_k\} \to \eta_*, \{v_k\} \to v_*$, $k_* \leqslant \subset K$. The existence of such points is ensured, since, by Theorem 4.1, the algorithm decreases

$$\psi(x_k) + \frac{1}{2} c_* \langle g_k, g_k \rangle$$

sufficiently at each iteration, thereby ensuring that $x_k \in \mathbb{F}$, with $\mathbb{F}$ compact.

In order to show that $x_*, \vartheta_*, \lambda_*, \mu_*, \eta_*, v_*$ constitute a Danskin point, we consider the optimality conditions (3.13) of the quadratic subproblems (3.5). Applying Lemma 4.3, for $k \to \infty, k_* \leqslant k \in K$, we have

$$g(x_*) = 0, \quad \vartheta_*, \in \mathbb{E}^\varsigma_+, \quad \mu_* \geqslant 0 \quad \vartheta^i_* \mu^i_* = 0, \quad \forall i \in \{1, \ldots, \varsigma\}, \tag{4.10a}$$

$$\begin{aligned} 0 &= \nabla f(x_*)\vartheta_* + \mathcal{G}_*[c_* g(x_*) + \lambda_*] \\ &= \nabla f(x_*)\vartheta_* + \mathcal{G}_* \lambda_*, \end{aligned} \tag{4.10b}$$

$$0 = f(x_*) - \mu_* + I\eta_*. \tag{4.10c}$$

$$I v_* \geqslant f(x_*), \quad \vartheta_*[I v_* - f(x_*)] = 0. \tag{4.10, d}$$

The expressions in (4.10) are the first-order conditions for optimality for problem (1.3) and (2.1b). Hence, x_* is a Danskin point.

To show (ii), we note that, in view of (4.10a), with strict complementarity holding, none of the inactive constraints, i.e.

$$\vartheta^i_k > 0, \quad \mu^i_k = 0, \tag{4.11}$$

are predicted to be active at $\vartheta_*, \vartheta^i_* > 0, \mu^i_* = 0$, for k sufficiently large. The same result can be established using similar arguments for the alternative min–max formulation in (2.1b) and (3.5b). □

As a consequence of the above theorem, for sufficiently large k, the inactive constraints at ϑ_* do not affect the computation of z_k. At that stage, it would make no difference if the active constraints were treated as equality constraints. Thus, if $\bar{\vartheta}$ denotes the vector of active constraints and $\bar{I}$ denotes the corresponding matrix of active constraints then

$$\bar{I}^{\mathrm{T}}\bar{\vartheta}_k = 0 \quad \text{or} \quad \bar{\vartheta}_k = 0, \qquad \bar{\mu}_k > 0, \tag{4.12}$$

for sufficiently large k. With strict complementarity, the multiplier corresponding to $\bar{\vartheta}$ is such that $\bar{\mu}_* > 0$ implies $\bar{\mu}_k > 0$. Similarly, for inactive constraints, (4.11) holds. A similar result for subproblem (3.5b) can be established using the same arguments. The latter is discussed further in Section 6.

5 THE ATTAINMENT OF UNIT STEPSIZES

We consider first the attainment of unit stepsizes. The conditions that ensure unit stepsizes turn out to be the closeness of the algorithm to the solution as well as the accuracy of the Hessian approximation used near this solution. The condition that ensures a superlinear convergence rate is shown in the next section to be the increasing accuracy of the Hessian approximation.

We shall first establish a result in the following lemma. This ensures that

$$\langle g_k, \tilde{\lambda}_{k+1}(\tilde{c}_{k+1})\rangle \geqslant 0$$

for a set of specially constructed multipliers $\tilde{\lambda}_{k+1}(\tilde{c}_{k+1})$ and a specially constructed $\tilde{c}_{k+1}$. It is also shown that $\|\tilde{\lambda}_{k+1}(\tilde{c}_{k+1}) - \lambda_{k+1}\| \to 0$ and $\tilde{c}_{k+1} - c_* \to 0$ as the solution is approached. The values $\tilde{\lambda}$ and $\tilde{c}$ are constructed to show that unit stepsizes are attained such that $\tau_k = 1, \forall k \geqslant k_\tau \geqslant k_*$: they are *not* used in any computational procedure in the algorithm. The integer k_τ is the iteration at which $\tau_k = 1$ is achieved and maintained thereafter.

Assuming, as in Section 3, that the columns of $\mathscr{G}_k$ are linearly independent and that $\hat{\mathfrak{Q}}_k$ is positive-definite, and using (3.13b, e), we have

$$\tilde{\lambda}_{k+1}(\tilde{c}_{k+1}) = -(\mathscr{G}_k^{\mathrm{T}}\hat{\mathfrak{Q}}_k^{-1}\mathscr{G}_k)^{-1}(\mathscr{G}_k^{\mathrm{T}}\hat{\mathfrak{Q}}_k^{-1}\nabla f_k\vartheta_{k+1} - g_k) - \tilde{c}_{k+1}g_k, \tag{5.1}$$

$$\lambda_{k+1} = -(\mathscr{G}_k^{\mathrm{T}}\hat{\mathfrak{Q}}_k^{-1}\mathscr{G}_k)^{-1}(\mathscr{G}_k^{\mathrm{T}}\hat{\mathfrak{Q}}_k^{-1}\nabla f_k\vartheta_{k+1} - g_k) - c_*g_k. \tag{5.2}$$

Lemma 5.1

Let the assumptions of Theorem 4.2 be satisfied, let $\{x_k\}$ be convergent and let the columns of $\mathscr{G}_k$ be linearly independent. Then

(i) for $\tilde{\lambda}_{k+1}$ given by (4.6), $\exists\tilde{c}_{k+1} \in (-\infty, +\infty)$ *such that* $\langle g_k, \tilde{\lambda}_{k+1}(\tilde{c}_{k+1})\rangle \geqslant 0$;

(ii)

$$\|\lambda_{k+1} - \tilde{\lambda}_{k+1}(\tilde{c}_{k+1})\| \to 0. \tag{5.3}$$

Proof

To show (i), we premultiply the optimality condition (3.13b) by d_k and use (3.13e) to obtain

$$\langle d_k, \hat{\mathfrak{Q}}_k d_k\rangle + \langle d_k, \nabla f_k\vartheta_{k+1}\rangle - \tilde{c}_k\langle g_k, g_k\rangle = \langle g_k, \tilde{\lambda}_{k+1}(\tilde{c}_k)\rangle.$$

We note that the parameter $\tilde{c}$ does not affect the solution of the quadratic subproblem. Let $\tilde{c}_{k+1}$ be determined by

$$\tilde{c}_{k+1} = \begin{cases} c_* & \text{if } \langle g_k, \tilde{\lambda}_{k+1}(c_*)\rangle \geqslant 0, \\ \dfrac{\langle d_k, \hat{\mathfrak{Q}}_k d_k\rangle + \langle d_k, \nabla f_k \vartheta_{k+1}\rangle}{\langle g_k, g_k\rangle} & \text{if } \langle g_k, \tilde{\lambda}_{k+1}(\tilde{c}_*)\rangle < 0. \end{cases} \tag{5.4}$$

Since $g_k = 0$, we have $\langle g_k, \tilde{\lambda}_{k+1}(c_*)\rangle = 0$, and it can be verified that this choice of $\tilde{c}_{k+1}$ ensures (i). Furthermore, as $\{\|g_k\|\} \to 0$, it follows from (5.4) that $\{\tilde{c}_k\} \to c_*$.

To show (ii), consider (5.1) and (5.2). As $k \to \infty, k_* \leqslant k \in K$, and, according to Lemma 4.2 and Theorem 4.2, we have $\|d_k\| \to 0, \|g_k\| \to 0$ and

$$\|\lambda_{k+1} - \tilde{\lambda}_{k+1}(\tilde{c}_{k+1})\| = |c_* - \tilde{c}_{k+1}|\,\|g_k\| = 0,$$

since c_* and $\tilde{c}_{k+1}$ are finite. □

Theorem 5.1

Let the assumptions of Lemma 5.1 be satisfied and let $\mathfrak{Q}_*$ denote the value of the Hessian (3.4) at the solution. They there is a small number $\sigma > 0$ such that if $\phi_k \leqslant \sigma$, for all large k, with

$$\phi_k = \frac{\|P_k^{\mathrm{T}}(\mathfrak{Q}_* - \hat{\mathfrak{Q}}_k)P_k d_k\|}{\|d_k\|} + \frac{\|g_k^{\mathrm{T}}\mathcal{G}_k^{\mathrm{T}}(\mathfrak{Q}_* - \hat{\mathfrak{Q}}_k)(\bar{\mathcal{G}}_k g_k - 2P_k d_k)\|}{\|d_k\|^2}$$

and $P_k, \tilde{\mathcal{G}}_k$ given either by

$$P_k = I - \hat{\mathfrak{Q}}_k^{-1}\mathcal{G}_k(\mathcal{G}_k^{\mathrm{T}}\hat{\mathfrak{Q}}_k^{-1}\mathcal{G}_k)^{-1}\mathcal{G}_k^{\mathrm{T}}, \qquad \tilde{\mathcal{G}}_k = \hat{\mathfrak{Q}}_k^{-1}(\mathcal{G}_k^{\mathrm{T}}\hat{\mathfrak{Q}}_k^{-1}\mathcal{G}_k)^{-1}$$

or by

$$P_k = I - \mathcal{G}_k(\mathcal{G}_k^{\mathrm{T}}\mathcal{G}_k)^{-1}\mathcal{G}_k^{\mathrm{T}}, \qquad \tilde{\mathcal{G}}_k = \mathcal{G}_k(\mathcal{G}_k^{\mathrm{T}}\mathcal{G}_k)^{-1},$$

then the stepsize strategy (3.10) is satisfied for $\tau_k = 1, \forall k \geqslant k_\tau \geqslant k_*$.

Proof

Consider the second-order expansion of $\langle g(x), \lambda_{k+1}\rangle$:

$$\langle g_{k+1}, \lambda_{k+1}\rangle = \langle g_k, \lambda_{k+1}\rangle + \langle d_k, \mathcal{G}_k\lambda_{k+1}\rangle + \frac{1}{2}\left\langle d_k, \left(\sum_{i=1}^{e} \nabla^2 g_*^i \lambda_*^i\right) d_k\right\rangle + \nu_k^1,$$

$$\nu_k^1 = \int_0^1 (1-t)\left\langle d_k, \left[\left(\sum_{i=1}^{e} \nabla^2 g^i(x_k + t d_k)\lambda_{k+1}^i\right) - \left(\sum_{i=1}^{e} \nabla^2 g_*^i \lambda_*^i\right)\right] d_k\right\rangle \mathrm{d}t$$

where $\sum_{i=1}^{e} \nabla^2 g_*^i \lambda_*^i$ is the constraint second derivative at x_*, to be subsumed in $\mathfrak{Q}_*$. Using the fact that d_k satisfies the linearized constraints, $\mathcal{G}_k^T d_k + g_k = 0$, and using Lemma 5.1, we have

$$\langle g_{k+1}, \lambda_{k+1} \rangle = \frac{1}{2} \left\langle d_k, \left(\sum_{i=1}^{e} \nabla^2 g_*^i \lambda_*^i \right) d_k \right\rangle + \nu_k^1$$

$$\leqslant \langle g_{k+1}, \bar{\lambda}_{k+2}(\bar{c}_{k+2}) \rangle + \frac{1}{2} \left\langle d_k, \left(\sum_{i=1}^{e} \nabla^2 g_*^i \lambda_*^i \right) d_k \right\rangle + \nu_k^1. \qquad (5.5)$$

Now consider the expression

$$\Gamma_k = \langle g_{k+1}, \bar{\lambda}_{k+2}(\bar{c}_{k+2}) \rangle - \langle g_{k+1}, \lambda_{k+2} \rangle + \langle g_{k+1}, \lambda_{k+2} \rangle - \langle g_{k+1}, \lambda_{k+1} \rangle. \qquad (5.6)$$

Using the second-order expansion of $g(x)$ and (3.13e), we can write

$$\|g_{k+1}\| \leqslant \|g_k + \mathscr{G}_k^{\mathrm{T}} d_k\| + e_1 \|d_k\|_2^2 = e_1 \|d_k\|_2^2,$$

for some $e_1 \geqslant 0$. Furthermore, as $\|\lambda_k - \lambda_{k+1}\| \to 0$ and, by Lemma 5.1,

$$\{\|\lambda_{k+1} - \tilde{\lambda}_{k+1}(\tilde{c}_{k+1})\|\} \to 0,$$

$\|\Gamma_k\|$ can be expressed as

$$\|\Gamma_k\| \leqslant \beta_k \|d_k\|_2^2, \qquad (5.7a)$$

$$\lim_{k\to\infty} \beta_k = \lim_{k\to\infty} e_1 [\|\bar{\lambda}_{k+2}(\bar{c}_{k+2}) - \lambda_{k+2}\| + \|\lambda_{k+2} - \lambda_{k+1}\|] = 0. \qquad (5.7b)$$

Consider expression (4.2a). Let the value of ϑ corresponding to the solution of the right-hand side be given by ϑ_k^*. Thus, using Lemma 2.1,

$$\psi(x_{k+1}) + c_* \langle g(x_{k+1}), g(x_{k+1}) \rangle = \langle \vartheta_k^*, f(x_k + d_k) \rangle + c_* \langle g_{k+1}, g_{k+1} \rangle.$$

By Theorem 4.2, $\{x_k\} \to x_*$. Then, clearly, $\{\vartheta_k^*\} \to \vartheta_*$. Thus, (4.2b) can be rewritten as

$$\psi(x_{k+1}) + c_* \langle g(x_{k+1}), g(x_{k+1}) \rangle$$

$$\leqslant \langle \vartheta_k^*, f(x_k) + \nabla f_k^{\mathrm{T}} d_k \rangle + \frac{1}{2} \langle d_k, \left(\hat{\mathfrak{Q}}_k - \hat{\mathfrak{Q}}_k + \mathfrak{Q}_*^g \right) d_k \rangle$$

$$+ \frac{1}{2} c_* \left(\langle g_k, g_k \rangle + \langle g_k, \mathscr{G}_k^{\mathrm{T}} d_k \rangle \right) + \nu_k^2$$

$$\leqslant \max_{\vartheta \in \mathbb{E}_+^\varsigma} \{ \langle \vartheta, f(x_k) + \nabla f_k^{\mathrm{T}} d_k \rangle \} + \frac{1}{2} \langle d_k, (\hat{\mathfrak{Q}}_k - \hat{\mathfrak{Q}}_k + \mathfrak{Q}_*^g) d_k \rangle$$

$$+ \frac{1}{2} c_* \left(\langle g_k, g_k \rangle + \langle g_k, \mathscr{G}_k^{\mathrm{T}} d_k \rangle \right) + \nu_k^2, \qquad (5.8)$$

where, with $x(t) = x_k + t d_k$,

$$\nu_k^2 = \int_0^1 (1 - t) \left\langle d_k, \left[\sum_{i=1}^{\varsigma} \bar{\vartheta}^i \nabla^2 f^i(x(t)) + c_* \mathscr{G}(x(t)) \mathscr{G}(x(t))^{\mathrm{T}} \right. \right.$$

$$\left. \left. + c_* \sum_{j=1}^{e} g^i(x(t)) \nabla^2 g(x(t)) - \mathfrak{Q}_*^g \right] d_k \right\rangle \mathrm{d}t.$$

Here $\mathfrak{Q}_*^g$ is the the second-derivative matrix evaluated at x_*, ϑ_*:

$$\mathfrak{Q}_*^g = \sum_{i=1}^{\varsigma} \vartheta_*^i \nabla^2 f^i(x_*) + c\mathcal{G}_*\mathcal{G}_*^{\mathrm{T}}.$$

Using (5.5), (5.6) and (5.8), we can rewrite (4.7) as

$$\begin{aligned}\psi(x_{k+1}) + \frac{1}{2}c_*\langle g(x_{k+1}), g(x_{k+1})\rangle \\ &\leqslant \psi(x_k) + \frac{1}{2}c_*\langle g_k, g_k\rangle \\ &\quad + \psi_k(x_k + d_k) - \psi(x_k) + c_*\langle g_k, \mathcal{G}_k^{\mathrm{T}} d_k\rangle \\ &\quad + \frac{1}{2}\langle d_k, (\hat{\mathfrak{Q}}_k - \hat{\mathfrak{Q}}_k + \mathfrak{Q}_*)d_k\rangle + \Gamma_k + \nu_k^1 + \nu_k^2. \end{aligned} \tag{5.9}$$

We note that

$$\begin{aligned} \|\nu_k^1 + \nu_k^2\| &= \|d_k\|_2^2 \xi_k^* \\ \xi_k^* &= \Bigg\| \int_0^1 (1-t)\Bigg[\sum_{i=1}^{\varsigma} \bar{\vartheta}^i \nabla^2 f^i(x(t)\,) \\ &\quad + c_*\mathcal{G}(x(t)\,)\mathcal{G}(x(t))^{\mathrm{T}} + \sum_{i=1}^{e} \nabla^2 g^i(x_k + td_k)\lambda_{k+1}^i \\ &\quad + c_*\sum_{j=1}^{e} g^i(x(t))\nabla^2 g(x(t)) - \mathfrak{Q}_*^g - \sum_{i=1}^{e}\nabla^2 g_*^i \lambda_*^i\Bigg]\Bigg\| \,\mathrm{d}t \end{aligned}$$

and

$$\mathfrak{Q}_* = \mathfrak{Q}_*^g + \sum_{i=1}^{e} \nabla^2 g_*^i \lambda_*^i.$$

Using (5.7), the descent property (3.14) in Lemma 3.2 and also Lemma 3.4(i), we can write (5.9) as

$$\begin{aligned}\psi(x_{k+1}) &+ \frac{1}{2}c_*\langle g(x_{k+1}),\ g(x_{k+1})\rangle \\ &\leqslant \psi(x_k) + \frac{1}{2}c_*\langle g_k, g_k\rangle \\ &\quad + \psi_k(x_k + d_k) - \psi(x_k) + c_*\langle g_k, \mathcal{G}_k^T d_k\rangle \\ &\quad + \frac{1}{2}\langle d_k, (\hat{\mathfrak{Q}}_k - \hat{\mathfrak{Q}}_k + \mathfrak{Q}_*)d_k\rangle + \beta_k\|d_k\|_2^2 + \|d_k\|_2^2\xi_k^* \\ &\leqslant \psi(x_k) + \frac{1}{2}c_*\langle g_k, g_k\rangle + \Phi(d_k, c_*)\left[1 - \frac{1}{m\epsilon}\left(\frac{1}{2}\phi_k + \beta_k + \xi_k^*\right)\right], \end{aligned} \tag{5.10}$$

where the last expression is obtained by invoking the equality $d_k = P_k - \bar{\mathscr{G}}_k g_k$. Since, by Theorem 4.2, $\{x_k\} \to x_*, \{\lambda_k\} \to \lambda_*, \{\mu_k\} \to \mu_*, \{\eta_k\} \to \eta_*, \{v_k\} \to v_*$, we have $\{\xi_k^*\} \to 0$ and $\{\beta_k\} \to 0$. The scalar

$$\rho \in (0,\ 1)$$

in (3.10) is bounded by

$$\rho \leq 1 - \frac{1}{m\varepsilon}\left(\frac{1}{2}\phi_k + \beta_k + \xi_k^*\right). \qquad (5.11).$$

If $\sigma > 0$ is such that $(m\epsilon)^{-1}(\frac{1}{2}\sigma + \beta_k + \xi_k^*) \leqslant 1 - \rho$ (in view of $\xi_k \to 0, \beta_k \to 0$, this defines the number ρ) then (5.11) holds, and therefore, because $\Phi(d_k, c_*) \leqslant 0$, (3.10) is satisfied with $\tau_k = 1$.

6 SUPERLINEAR CONVERGENCE RATES OF THE ALGORITHM

The superlinear convergence of the algorithm is discussed only for the quadratic programming subproblem (3.5b). An interesting aspect of the subproblem and the convergence rate is that the Hessian of the quadratic objective function is singular. The results below discuss the superlinear convergence of $\{x_k\}$, $\{x_k, v_k\}$, $\{x_k, \vartheta_k, v_k\}$, $\{x_k, \vartheta_k, \lambda_k, v_k\}$ and relate these to the accuracy of the projected Hessian. The reason for discussing the convergence rate of $\{x_k\}$, as well as the rates of $\{x_k, v_k\}$, $\{x_k, \vartheta_k, v_k\}$, $\{x_k, \vartheta_k, \lambda_k, v_k\}$ is that, as mentioned by Boggs, Tolle and Wang (1982), the Q-superlinear convergence of the latter three sequences only imply the R-superlinear convergence of the former. The situation is further complicated by the fact that the superlinear convergence results for the latter two sequences are not necessarily the same as the one for the former two sequences when projected Hessians are being considered.

Lemma 6.1
Let $\{x_k\} \to x_*$ and $\|d_k\| \leqslant \kappa_k \|d_{k-1}\|, \lim_{k\to\infty} \kappa_k = 0$. Then the sequence $\{x_k\}$ is Q-superlinearly convergent, i.e.

$$\lim_{k\to\infty} \frac{\|x_* - x_{k+1}\|}{\|x_* - x_k\|} = 0.$$

Proof
We have

$$\begin{aligned}\|x_* - x_k\| &\leqslant \lim_{t\to\infty} \sum_{j=k}^{t-1} \|x_{j+1} - x_j\| \\ &\leqslant \kappa_k \|d_{k-1}\|(1 + \omega + \omega^2 + \omega^3 + \ldots) \\ &\leqslant \frac{\kappa_k}{1-\omega}(\|x_k - x_*\| + \|x_* - x_{k-1}\|),\end{aligned}$$

for some $\omega \in [0, 1)$. As $\{\kappa_k\} \to 0, \omega$ is chosen such that $\kappa_k \leqslant \omega < 1, \forall k \geqslant K_0$. K_0 is an integer and is such that $\kappa_k < 1, \forall k \geqslant K_0$. Rearranging the above expression yields

$$\frac{\|x_k - x_*\|}{\|x_k - x_*\| + \|x_{k-1} - x_*\|} \leqslant \frac{\kappa_k}{1 - \omega},$$

which yields the required result. □

Assumptions 6.1

(i) $\exists x_*, \vartheta_*, \lambda_*, \mu_*, \eta_*$ satisfying the first-order conditions of (1.3), or, equivalently, $\exists x_*, \vartheta_*, \lambda_*, v_*$ satisfying the first-order conditions of (2.1b).

(ii) Strict complementarity holds for the inequality constraints of the quadratic programming subproblem (3.5b).

(iii) $v^{\mathrm{T}} \mathfrak{Q}_* v > 0, \forall v \in \mathbb{R}^n,\ \mathscr{G}_*^{\mathrm{T}} v = 0.$

(iv) The coefficient matrix

$$\nabla \hat{\Gamma}_k = \begin{bmatrix} \hat{\mathfrak{Q}}_k & \nabla \bar{f}_k & \mathscr{G}_k & 0 \\ \nabla \bar{f}_k^{\mathrm{T}} & 0 & 0 & \bar{1} \\ \mathscr{G}_k^{\mathrm{T}} & 0 & 0 & 0 \\ 0 & \bar{1}^{\mathrm{T}} & 0 & 0 \end{bmatrix}, \tag{6.1}$$

arising from the first-order necessary conditions of the quadratic subproblem (3.5b), is nonsingular. The matrix $\nabla \bar{f}_k$ corresponds to the columns of ∇f_k of the inequality constraints in (3.5b) satisfied as equalities at the solution x_* of the min–max problem. Similarly, the vector $\bar{1}$ corresponds to the right-hand sides of the inequality constraints in (3.5b) satisfied as equalities at x_*.

(v) The columns of $\mathscr{G}_k, [\nabla \bar{f}_k^{\mathrm{T}}, \bar{1}]^{\mathrm{T}}$ are linearly independent at the solution x_*.

As a result of the convergence of Algorithm 3.1, established in Theorem 4.2, there is a stage, near the solution of the min–max problem, when the inequality constraints satisfied as equalities at the solution are also satisfied as equalities for the quadratic program (3.5b). At this stage, with strict complementarily holding, the multipliers $\vartheta_k^i > 0$ if the constraint i is satisfied as an equality at the solution x_*. Also, $\langle \nabla f_k^i, d_k \rangle + f_k^i < v_{k+1}$ and $\vartheta_k^i = 0$ if the constraint is strictly satisfied at the solution. Thus, at this stage, the multipliers can predict the constraints active at the solution. These active constraints are written as $\langle \nabla \bar{f}_k^i, d_k \rangle + \bar{f}_k^i = v_{k+1}$. Since the other constraints do not affect the solution x_*, they can be ignored for the convergence rate analysis. Thus, the quadratic subproblem (3.5b) can be rewritten as

$$\min_{d,v} \left\{ v + c_k \langle \mathscr{G}_k g_k, d \rangle + \frac{1}{2} \langle d, \hat{\mathfrak{Q}}_k d \rangle \,\middle|\, \mathscr{G}_k^{\mathrm{T}} d + g_k = 0, \nabla \bar{f}_k^{\mathrm{T}} d + \bar{f}_k = \bar{1} v \right\}. \tag{6.2}$$

The first-order optimality conditions for (6.2) can be written as

$$\nabla\hat{\Gamma}_k \begin{bmatrix} d_k \\ \bar{\vartheta}_{k+1} - \bar{\vartheta}_k \\ \lambda_{k+1} - \lambda_k \\ -(v_{k+1} - v_k) \end{bmatrix} = -\Gamma_k, \tag{6.3}$$

where

$$\Gamma_k = \begin{bmatrix} \nabla\bar{f}_k\bar{\vartheta}_k + \mathscr{G}_k(\lambda_k + c_* g_k) \\ g_k \\ \bar{f}_k - \bar{1}v_k \\ 0 \end{bmatrix}, \tag{6.4}$$

and $\bar{\vartheta}_k$ relate to the strictly positive elements of ϑ_k, which correspond to the inequality constraints satisfied as equalities.

The quadratic subproblem computes d_k such that $\mathscr{G}_{k-1}^{\mathrm{T}} d_{k-1} + g_{k-1} = 0$. Thus, using the second-order expansion of $g(x)$ at x_k, we have, for some $e_1 \geqslant 0$,

$$\|g_k\| = e_1\|d_{k-1}\|^2. \tag{6.5}$$

Since the inequality constraints are satisfied as equalities, we have

$$\bar{1}v_k = \nabla\bar{f}_{k-1}^{\mathrm{T}} d_{k-1} + \bar{f}_{k-1}. \tag{6.6}$$

The second-order expansion of $\bar{f}(x)$ about x_{k-1} yields, for some $e_2 \geqslant 0$,

$$\bar{f}_k - \bar{f}_{k-1} - \nabla\bar{f}_{k-1}^{\mathrm{T}} d_{k-1} = e_2\|d_{k-1}\|^2.$$

Thus, $\bar{f}_k - \bar{1}v_k$ in (6.4) can be written as

$$\bar{f}_k - \bar{1}v_k = e_2\|d_{k-1}\|^2. \tag{6.7}$$

The first-order expansion of the gradient, with respect to x, of the Lagrangian of either min–max formulation, $\nabla\bar{f}(x)\bar{\vartheta} + \mathscr{G}(x)[\lambda + c_* g(x)]$, yields

$$\begin{aligned} \nabla\bar{f}_k\bar{\vartheta}_k + \mathscr{G}_k(\lambda_k + c_* g_k) =& \nabla\bar{f}_{k-1}\bar{\vartheta}_{k-1} + \mathscr{G}_{k-1}(\lambda_{k-1} + c_* g_{k-1}) \\ &+ \hat{\mathfrak{Q}}_{k-1} d_{k-1} + \Delta\mathfrak{Q}(t) d_{k-1}, \end{aligned} \tag{6.8a}$$

$$\Delta\mathfrak{Q}(t) = \int_0^1 [\mathfrak{Q}(x_{k-1} + t d_{k-1}) - \mathfrak{Q}_* + \mathfrak{Q}_* - \hat{\mathfrak{Q}}_{k-1}]\, \mathrm{d}t \tag{6.8b}$$

Clearly, as $\{x_k\} \rightarrow x_*$,

$$\left\{\int_0^1 [\mathfrak{Q}(x_{k-1} + t d_{k-1}) - \mathfrak{Q}_*]\, \mathrm{d}t\right\} \rightarrow 0$$

and, as shown below, the condition for superlinear convergence ensures that the rest of (6.8b) also vanishes at the solution.

It should be noted that the first-order optimality condition (3.13b), for either quadratic subproblem, at x_{k-1}, is

$$\nabla\bar{f}_{k-1}\bar{\vartheta}_{k-1} + \mathcal{G}_{k-1}(\lambda_{k-1} + c_* g_{k-1}) = 0,$$

and thus (6.8a) can be written as

$$\nabla\bar{f}_k\bar{\vartheta}_k + \mathcal{G}_k(\lambda_k + c_* g_k) = \Delta\mathfrak{Q}(t)d_{k-1}. \tag{6.9}$$

We can now state the superlinear convergence theorem. In this result, conditions are discussed that are similar to the characterization of Boggs, Tolle and Wang (1982) and Powell (1983) for nonlinear programming. The results below are established without invoking a prior linear convergence rate argument. As opposed to the necessary and sufficient result of Boggs, Tolle and Wang (1982), the result below only establishes the necessary part in the case of min–max problems. The analogous necessary and sufficient condition is discussed in Theorem 6.2. The following lemma is used in the latter part of Theorem 6.1.

Lemma 6.2
Let Assumptions (2.1), (4.1) and (6.1) hold. Then, for some $e_0 \in [0, \infty)$,

$$\|d_k\| \leqslant e_0\|x_* - x_k\|.$$

Proof
The first-order optimality condition of the quadratic programming subproblem can be written as

$$[\nabla\hat{\Gamma}_k]\begin{bmatrix} d_k \\ \bar{\vartheta}_{k+1} - \bar{\vartheta}_* \\ \lambda_{k+1} - \lambda_* + c_* g_k \\ -(v_{k+1} - v_*) \end{bmatrix} = -\begin{bmatrix} \nabla\bar{f}_k\bar{\vartheta}_* + \mathcal{G}_k\lambda_* \\ g_k \\ \bar{f}_k - \bar{1}v_* \\ 0 \end{bmatrix}.$$

Using the first-order expansion of the right side of the above expression about $(x_*, \bar{\vartheta}_*, \lambda_*, v_*)$, by the nonsingularity of the coefficient matrix on the left and by the first-order optimality condition of (2.1b), i.e.

$$\nabla\bar{f}(x_*)\bar{\vartheta}_* + \mathcal{G}_*\lambda_* = 0, \quad g(x_*) = 0, \quad \bar{f}(x_*) = \bar{1}v_*,$$

we have

$$\|d_k\| \leqslant \left\| \begin{matrix} d_k \\ \bar{\vartheta}_{k+1} - \bar{\vartheta}_* \\ \lambda_{k+1} - \lambda_* + c_* g_k \\ -(v_{k+1} - v_*) \end{matrix} \right\| \leqslant \hat{e}[\nabla\hat{\Gamma}_k]^{-1}\|x_* - x_k\|,$$

for some $\hat{e} \in [0, \infty)$. Hence, the result follows from the above expression. □

Theorem 6.1
Let Assumptions 2.1, 4.1 and 6.1 be satisfied. Then the following hold.

(i) The sequence $\{x_k\}$ *satisfies* $\|d_k\| \leqslant \kappa_k \|d_{k-1}\|$ *with* $\lim_{k\to\infty} \kappa_k = 0$, *and thence converges Q-superlinearly if, for* $P_k = I - \mathcal{G}_k(\mathcal{G}_k^T \mathcal{G}_k)^{-1} \mathcal{G}_k^T$,

$$\lim_{k\to\infty} \frac{\|P_k(\mathfrak{Q}_* - \hat{\mathfrak{Q}}_k)d_k\|}{\|d_k\|} = 0. \tag{6.10}$$

(ii) Let $\Delta\vartheta_k = \vartheta_{k+1} - \vartheta_k$ *and* $\Delta v_k = v_{k+1} - v_k$. *Then the sequence* $\{x_k, \vartheta_k, v_k\}$ *satisfies*

$$\left\| \begin{matrix} d_k \\ \Delta\vartheta_k \\ \Delta v_k \end{matrix} \right\| \leqslant \kappa_k \|d_{k-1}\| \leqslant \kappa_k \left\| \begin{matrix} d_{k-1} \\ \Delta\vartheta_{k-1} \\ \Delta v_{k-1} \end{matrix} \right\|, \tag{6.11}$$

and thence converges Q-superlinearly if (6.10) is satisfied.

(iii) The sequence $\{x_k, \vartheta_k, \lambda_k, v_k\}$ *converges at a two-step Q-superlinear rate if (6.10) holds.*

(iv) The sequence $\{x_k, \vartheta_k, \lambda_k, v_k\}$ *converges at a Q-superlinear rate if*

$$\lim_{k\to\infty} \frac{\|P_k(\mathfrak{Q}_* - \hat{\mathfrak{Q}}_k)d_k\|}{\|d_k\|} = 0 \quad \text{and} \quad \lim_{k\to\infty} \frac{\|(F_k^{\mathrm{T}} F_k)^{-1} F_k^{\mathrm{T}}(\mathfrak{Q}_* - \hat{\mathfrak{Q}}_k)d_k\|}{\|d_k\|} = 0, \tag{6.12}$$

where $F_k = [\mathcal{G}_k \quad \nabla \bar{f}_k]$.

Proof
Using (6.9), expression (6.3) can be written as

$$[\nabla\hat{\Gamma}_k] \begin{bmatrix} d_k \\ \Delta\vartheta_k \\ \Delta\lambda_k + (\mathcal{G}_{k-1}^{\mathrm{T}} \mathcal{G}_{k-1})^{-1} \mathcal{G}_{k-1}^{\mathrm{T}} \Delta\mathfrak{Q}(t) d_{k-1} \\ -\Delta v_k \end{bmatrix} = \begin{bmatrix} P_{k-1}\Delta\mathfrak{Q}(t)d_{k-1} \\ g_k \\ \bar{f}_k - \bar{1} v_k \\ 0 \end{bmatrix}. \tag{6.13}$$

Since the matrix $\nabla\hat{\Gamma}_k$ is nonsingular, (6.13), (6.10), (6.5), (6.7) and $\{\|d_k\|\} \to 0$ yield

$$\|d_k\| \leqslant \left\| \begin{matrix} d_k \\ \Delta\vartheta_k \\ \Delta v_k \end{matrix} \right\| \leqslant \kappa_k \|d_{k-1}\| \leqslant \kappa_k \left\| \begin{matrix} d_{k-1} \\ \Delta\vartheta_{k-1} \\ \Delta v_{k-1} \end{matrix} \right\|, \quad \text{with} \quad \lim_{k\to\infty} \kappa_k = 0. \tag{6.14}$$

Thence, Lemma 6.1 establishes the Q-superlinear convergence of $\{x_k\}$, $\{x_k, \vartheta_k, v_k\}$. This establishes (i) and (ii).

In order to show (iii), we first show that

$$\left\| \begin{matrix} \vartheta_* - \vartheta_k \\ \lambda_* - \lambda_k \\ v_* - v_k \end{matrix} \right\| \leqslant \kappa_k^1 \|x_* - x_{k-2},\|, \quad \text{with} \quad \lim_{k\to\infty} \kappa_k^1 = 0. \tag{6.15}$$

From (6.3), (6.4), (6.5), (6.7) and (6.9), we have

$$\left\| \begin{matrix} \Delta\vartheta_k \\ \Delta\lambda_k \\ \Delta v_k \end{matrix} \right\| \leqslant e_3\|d_k\| + e_4\|d_{k-1}\| + e_4\|d_{k-1}\|^2, \tag{6.16}$$

for some $e_3, e_4, e_5 \geqslant 0$. We can use $\|d_k\| \leqslant \kappa_k\|d_{k-1}\|$ in (6.16) to obtain

$$\left\| \begin{matrix} \Delta\vartheta_k \\ \Delta\lambda_k \\ \Delta v_k \end{matrix} \right\| \leqslant \kappa_k^2\|d_k\|, \qquad \lim_{k\to\infty} \kappa_k^2 = 0. \tag{6.17}$$

This leads to

$$\begin{aligned} \left\| \begin{matrix} \vartheta_* - \vartheta_k \\ \lambda_* - \lambda_k \\ v_* - v_k \end{matrix} \right\| &\leqslant \lim_{t\to\infty} \sum_{j=k}^{t-1} \left\| \begin{matrix} \Delta\vartheta_j \\ \Delta\lambda_j \\ \Delta v_j \end{matrix} \right\| \\ &\leqslant \lim_{t\to\infty} \sum_{j=k}^{t-1} \kappa_j^2\|d_{j-2}\| \\ &\leqslant \kappa_k^2\|d_{k-2}\|(1 + \omega + \omega^2 + \omega^3 \ldots) \end{aligned},$$

for some $\omega \in [0, 1)$. Combining this with Lemma 6.2 yields (6.15). Using (6.15) with the superlinear convergence of $\{x_k\}$, we have, for $\lim_{k\to\infty} \kappa_k^3 = 0$,

$$\left\| \begin{matrix} x_* - x_k \\ \vartheta_* - \vartheta_k \\ \lambda_* - \lambda_k \\ v_* - v_k \end{matrix} \right\| \leqslant \kappa_k\|x_* - x_{k-1}\| + \kappa_k^1\|x_* - x_{k-2}\| \leqslant \kappa_k^3 \left\| \begin{matrix} x_* - x_{k-2} \\ \vartheta_* - \vartheta_{k-2} \\ \lambda_* - \lambda_{k-2} \\ v_* - v_{k-2} \end{matrix} \right\| \tag{6.18}$$

To establish (iv), we consider the first equation of (6.3) with (6.5), (6.7) and (6.9). We have

$$\begin{aligned} F_k \begin{bmatrix} \Delta\vartheta_k \\ \Delta\lambda_k \end{bmatrix} &= -[\hat{\mathfrak{Q}}_k d_k + \Delta\mathfrak{Q}(t) d_{k-1}], \\ \begin{bmatrix} \Delta\vartheta_k \\ \Delta\lambda_k \end{bmatrix} &= -(F_k^{\mathrm{T}} F_k)^{-1} F_k^{\mathrm{T}} [\hat{\mathfrak{Q}}_k d_k + \Delta\mathfrak{Q}(t) d_{k-1}]. \end{aligned}$$

Using $\|d_k\| \leqslant \kappa_k\|d_{k-1}\|$ with $\lim_{k\to\infty} \kappa_k = 0$ established above, we have

$$\left\| \begin{matrix} \Delta\vartheta_k \\ \Delta\lambda_k \end{matrix} \right\| \leqslant \left[\kappa_k\|(F_k^{\mathrm{T}} F_k)^{-1} F_k^{\mathrm{T}} \hat{\mathfrak{Q}}_k\| + \frac{\|(F_k^{\mathrm{T}} F_k)^{-1} F_k^{\mathrm{T}} \Delta\mathfrak{Q}(t) d_{k-1}\|}{\|d_{k-1}\|} \right] \|d_{k-1}\|,$$

which, with (6.11) and Lemma 6.1, yields the Q-superlinear convergence of $\{x_k, \vartheta_k, \lambda_k, v_k\}$. □

We now consider an alternative Q-superlinear convergence theorem that establishes the *necessary and sufficient* condition for this rate. We first define the following matrices that are used below:

$$\hat{\mathfrak{Q}}_k = \begin{bmatrix} \hat{\mathfrak{Q}}_k & 0^{n\times 1} \\ 0^{1\times n} & 0^{1\times 1} \end{bmatrix}, \qquad \Delta\mathfrak{Q}(t) = \begin{bmatrix} \Delta\mathfrak{Q}(t) & 0^{n\times 1} \\ 0^{1\times n} & 0^{1\times 1} \end{bmatrix}, \; \mathfrak{F}_k = \begin{bmatrix} \nabla\bar{f}_k & \mathcal{G}_k \\ \bar{1}^{\mathrm{T}} & 0^{1\times e} \end{bmatrix},$$

and the corresponding projection operator is given by $\mathcal{P}_k = I - \mathfrak{F}_k(\mathfrak{F}_k^{\mathrm{T}}\mathfrak{F}_k)^{-1}\mathfrak{F}_k^{\mathrm{T}}$. The superscripts on the 0s indicate the dimensions.

Theorem 6.2

Under the same assumptions as in Theorem 6.1, the sequences $\{x_k\}$ and $\{x_k, v_k\}$ satisfy $\|d_k\| \leqslant \kappa_k^4 \|d_{k-1}\|$ and

$$\left\| \begin{matrix} d_k \\ \Delta v_k \end{matrix} \right\| \leqslant \kappa_k^5 \left\| \begin{matrix} d_{k-1} \\ \Delta v_{k-1} \end{matrix} \right\|$$

respectively, and thereby converge Q-superlinearly if and only if

$$\lim_{k\to\infty} \frac{\left\| \mathcal{P}_k \begin{bmatrix} \mathfrak{Q}_* - \hat{\mathfrak{Q}}_k \\ 0^{1\times n} \end{bmatrix} d_k \right\|}{\|d_k\|} = 0. \tag{6.19}$$

Proof

The optimality condition given by (6.3) can be rewritten using the above matrices and (6.5), (6.7), (6.9) as

$$\begin{bmatrix} \hat{\mathfrak{Q}}_k & \mathfrak{F}_k \\ \mathfrak{F}_k^{\mathrm{T}} & 0 \end{bmatrix} \begin{bmatrix} d_k \\ v_k \\ \begin{bmatrix} \Delta\vartheta_k \\ \Delta\lambda_k \end{bmatrix} + (\mathfrak{F}_k^{\mathrm{T}}\mathfrak{F}_k)^{-1}\mathfrak{F}_k^{\mathrm{T}} \begin{bmatrix} \Delta\mathfrak{Q}(t) \\ 0^{1\times n} \end{bmatrix} d_{k-1} \end{bmatrix} = - \begin{bmatrix} \mathcal{P}_{k-1} \begin{bmatrix} \Delta\mathfrak{Q}(t) d_{k-1} \\ 0^{1\times n} \end{bmatrix} \\ \bar{f}_k - \bar{1} v_k \\ g_k \end{bmatrix}. \tag{6.20}$$

Using the same arguments as in Theorem 6.1 for (6.13), the Q-superlinear convergence of $\{x_k\}$ and $\{x_k, v_k\}$ can be demonstrated if (6.19) is satisfied.

Suppose, conversely, that $\|d_k\| \leqslant \kappa_k^4 \|d_{k-1}\|$ and

$$\left\| \begin{matrix} d_k \\ \Delta v_k \end{matrix} \right\| \leqslant \kappa_k^5 \left\| \begin{matrix} d_{k-1} \\ \Delta v_{k-1} \end{matrix} \right\|$$

respectively for the two sequences. Premultiplying the first equation of (6.20) and using the identity

$$\mathcal{P}_k \begin{bmatrix} d_k \\ \Delta v_k \end{bmatrix} = \begin{bmatrix} d_k \\ \Delta v_k \end{bmatrix} - \mathfrak{F}_k(\mathfrak{F}_k^{\mathrm{T}}\mathfrak{F}_k)^{-1}\mathfrak{F}_k^{\mathrm{T}} \begin{bmatrix} d_k \\ \Delta v_k \end{bmatrix},$$

we have

$$\mathscr{P}_k \hat{\mathfrak{Q}}_k \mathscr{P}_k \begin{bmatrix} d_k \\ \Delta v_k \end{bmatrix} + \mathfrak{F}_k (\mathfrak{F}_k^{\mathrm{T}} \mathfrak{F}_k)^{-1} \mathfrak{F}_k^{\mathrm{T}} \begin{bmatrix} d_k \\ \Delta v_k \end{bmatrix} = -(\mathscr{P}_k - \mathscr{P}_{k-1}) \Delta \mathfrak{Q}(t) \begin{bmatrix} d_{k-1} \\ \Delta v_{k-1} \end{bmatrix} - \mathscr{P}_{k-1} \Delta \mathfrak{Q}(t) \begin{bmatrix} d_{k-1} \\ \Delta v_{k-1} \end{bmatrix}.$$

Assumptions 6.1(iv, v) ensure that $\|\mathscr{P}_k \hat{\mathfrak{Q}}_k \mathscr{P}_k\| \in (0, \infty)$. Also as $\{\|d_k\|\} \to 0$, $\{\|\mathscr{P}_k - \mathscr{P}_{k-1}\|\} \to 0$.
Using the facts that

$$\|d_k\| \leqslant \left\| \begin{bmatrix} d_k \\ \Delta v_k \end{bmatrix} \right\| \quad \text{and} \quad \Delta \mathfrak{Q}(t) \begin{bmatrix} d_k \\ \Delta v_k \end{bmatrix} = \begin{bmatrix} \Delta \mathfrak{Q}(t) d_{k-1} \\ 0 \end{bmatrix},$$

and dividing the above expression by $\|d_{k-1}\|$ and $\left\| \begin{matrix} d_k \\ \Delta v_k \end{matrix} \right\|$ respectively yields (6.19) for both sequences as required.

REFERENCES

Arrow, K.J., F.J. Gould and S.M. Howe (1973). General saddle point result for constrained optimization. *Math. Programming* **5**, 225–234.

Arrow, K.J. and L. Hurwicz (1956). Reduction of constrained maxima to saddle-point problems. In *Proceedings of 3rd Berkeley Symposium of Mathematical Statistics and Probability* (J. Neyman, ed.).

Arrow, K.J. and R.M. Solow (1958). Gradient methods for constrained maxima, with weakened assumptions. In *Studies in Linear and Nonlinear Programming*. (K. Arrow, L. Hurwicz and H. Uzawa, eds. Stanford University Press.

Becker, R.G., B. Dwolatzky, E. Karakitsos and B. Rustem (1986). The simultaneous use of rival models in policy optimization. *Econ. J.* **96**, 425–448.

Biggs, M.C.B. (1974). The development of a class of constrained minimization algorithms and their application to the problem of power scheduling. PhD Thesis, University of London.

Boggs, P.T., J.W. Tolle and P. Wang (1982). On the local convergence of quasi-Newton methods for constrained optimization. *SIAM J. Control Optim.* **20**, 161–171.

Bowman Jr, V.J. (1976). On the relationship of the Tchebycheff norm and the efficient frontier of multi-criteria objectives. In *Multiple Criteria Decision Making* (H. Thiriez and S. Zionts, eds). Springer-Verlag, Berlin.

Chow, G.C. (1979). Effective use of econometric models in macroeconomic policy formulation. In *Optimal Control for Econometric Models* (S. Holly, B. Rustem and M. Zarrop, eds). Macmillan, London.

Charalambous, C. and J.W. Bandler (1976). Nonlinear minimax optimization as a sequence of least pth optimization with finite values of *p*. *Int. J. Syst. Sci.* **7**, 377–391.

Charalambous, C. and A.R. Conn (1978). An efficient algorithm to solve the min–max problem directly. *SIAM J. Numer. Anal.* **15**, 162–187.

Cohen, G. (1981). An algorithm for convex constrained minmax optimization based on duality. *Appl. Math. Optim.* **7**, 347–372.

Coleman, T.F. (1978). A note on 'New algorithms for constrained minimax optimization'. *Math. Programming* **15**, 239–242.

Conn, A.R. (1979). An efficient second order method to solve the constrained minmax problem. Report, Department of Combinatorics and Optimization, University of Waterloo, January.

Conn, A.R. and Y. Li (1988). The computational structure and characterization of nonlinear discrete Chebychev problems. Technical Report CS-88-39, Department of Computer Science, University of Waterloo.

Demyanov, V.F. and V.N. Malozemov (1974). *Introduction to Minmax*. Wiley, New York.

Demyanov, V.F. and A.B. Pevnyi (1972). Some estimates in minmax problems. *Kibernetika* **1**, 107–112.

Dutta, S.R.K. and M. Vidyasagar (1977). New algorithms for constrained minmax optimization. *Math. Programming* **13**, 140–155.

Fuhrer, J. and J. Haltmaier (1986). Minimum variance pooling of forecasts at different levels of aggregation. Special Studies Paper 208, Federal Reserve Board, Washington, D.C.

Gill, P.E., W. Murray and M.H. Wright (1981). *Practical Optimization*. Academic Press, New York.

Granger, C. and P. Newbold (1977). *Forecasting Economic Time Series*. Academic Press New York.

Hald, J.H. and K. Madsen (1981). Combined LP and quasi-Newton methods for minimax optimization. *Math. Programming* **20**, 49–62.

Han, S.P. (1978). Superlinear convergence of a minimax method. Technical Report 78–336, Department of Computer Science, Cornell University.

Han, S.P. (1981). Variable metric methods for minimizing a class of nondifferentiable functions. *Math. Programming* **20**, 1–13.

Lawrence, M.J., R.H. Edmunson and M.J. O'Connor (1986). The accuracy of combining judgemental and statistical forecasts. *Management Sci.* **32**, 1521–1532.

Makridakis, S. and R. Winkler (1983). Averages of forecasts: some empirical results. *Management Sci.* **29**, 987–996.

Medanic, J. and M. Andjelic (1971). On a class of differential games without saddle-point solutions. *J. Optim. Theory Applic.* **8**, 413–430.

Medanic, J. and M. Andjelic (1972). Minmax solution of the multiple target problem. *IEEE Trans. Autom. Control* **AC-17**, 597–604.

Murray, W. and M.L. Overton (1980). A projected Lagrangian algorithm for nonlinear minmax optimization. *SIAM J. Sci. Statist. Comput.* **1**, 345–370.

Pinto, R.L.V. (1995). A logic-based modelling language and integer-programming framework for multicriteria optimization. PhD Thesis, Department of Computing, Imperial College, London.

Pinto, R.L.V. and B. Rustem (1996a) Building and solving multi-criteria models involving logical conditions. In *Advances in Computational Economics* (M. Gilli, ed.). Kluwer, Dordrecht.

Pinto, R.L.V. and B. Rustem (1996b). Solving a mixed-integer multiobjective bond portfolio model involving logical conditions. Department of Computing, Imperial College.

Polak, E. and D.Q. Mayne (1981). A robust secant method for optimization problems with inequality constraints. *J. Optim. Theory Applic.* **33**, 463–477.

Polak, E. and A.L. Tits (1981). A globally convergent, implementable multiplier method with automatic penalty limitation. *Appl. Math. Optim.* **6**, 335–360.

Polak, E. and D.Q. Mayne and J.E. Higgins (1988). A superlinearly convergent minmax algorithm for min-max problems. Memorandum UCB/ERL M86/103, Berkeley, California.

Powell, M.J.D. (1978). A fast algorithm for nonlinearly constrained optimization problems. In *Numerical Analysis* (G.A. Watson, ed.), Lecture Notes in Mathematics, Vol. 630, Springer-Verlag, Berlin.

Powell, M.J.D. (1983) Variable metric methods for constrained optimization. In *Mathematical Programming: The State of The Art* (A. Bachem, M. Grotschel and B. Korte, eds), Springer-Verlag, Berlin.

Rockafeller, R.T. (1973). A dual approach to solving nonlinear programming problems by unconstrained minimization. *Math. Programming* **5**, 354–373.

Rustem, B. (1981). *Projection Methods in Constrained Optimization and Application to Optimal Policy Decisions*. Springer-Verlag, Berlin.

Rustem, B. (1986). Convergent stepsizes for constrained optimization algorithms. *J. Optim. Theory Applic.* **49**, 136–160.

Rustem, B. (1987). Methods for the simultaneous use of multiple models in optimal policy design. In *Developments in Control Theory for Economic Analysis* (C. Carraro and D. Sartore, eds). Maritnus Nijhoff/Kluwer, Dordrecht.

Rustem, B. (1992). A constrained min–max algorithm for rival models of the same economic system. *Math. Programming* **53**, 279–295.

Rustem, B. (1993). Equality and inequality constrained optimization algorithms with convergent stepsizes. *J. Optim. Theory Applic.* **76**, 429–453.

Rustem, B. (1994). Robust min–max decisions with rival models. In *Computational Techniques for Econometrics and Economic Analysis* (D. Belsley, ed.), Kluwer, Dordrecht.

Tapia, R.A. (1986). On secant updates for use in general constrained optimization. Technical Report 84–3, Rice University.

Womersley, R.S. and R. Fletcher (1986). An algorithm for composite nonsmooth optimization problems. *J. Optim. Theory Applic.* **48**, 493–523.

Index

Index compiled by Geoffrey C. Jones